GROUP THEORY

Finite Discrete Groups and Applications

GROUP THEORY
Finite Discrete Groups and Applications

John Demetrius Vergados

University of Ioannina, Greece

Vasiliki-Ioanna Vergadou-Remediaki

World Scientific

NEW JERSEY · LONDON · SINGAPORE · BEIJING · SHANGHAI · HONG KONG · TAIPEI · CHENNAI · TOKYO

Published by

World Scientific Publishing Co. Pte. Ltd.

5 Toh Tuck Link, Singapore 596224

USA office: 27 Warren Street, Suite 401-402, Hackensack, NJ 07601

UK office: 57 Shelton Street, Covent Garden, London WC2H 9HE

Library of Congress Control Number: 2023940436

British Library Cataloguing-in-Publication Data
A catalogue record for this book is available from the British Library.

GROUP THEORY
Finite Discrete Groups and Applications

ISBN 978-981-127-475-6 (hardcover)
ISBN 978-981-127-476-3 (ebook for institutions)
ISBN 978-981-127-477-0 (ebook for individuals)

For any available supplementary material, please visit
https://www.worldscientific.com/worldscibooks/10.1142/13366#t=suppl

Desk Editor: Nur Syarfeena Binte Mohd Fauzi

Typeset by Stallion Press
Email: enquiries@stallionpress.com

In memory of my teacher

Professor K. T. Hecht

Preface

The role played by symmetry in the understanding of the physical world is well known. Initially, mainly in the era of ancient Greek philosophers and mathematicians, it involved the study and geometric properties of:

(a) plane shapes.

Not only the common ones, such as regular triangles, squares and hexagons, but even more involved ones, such as the dodecagon and icosagon; and

(b) three-dimensional objects.

Regular polyhedra, including the common ones, such as the cube and tetrahedron, as well as those that are more involved, such as the dodecahedron and icosahedron. The recognition of the existence of symmetries led to the notion of transformations, which lead from one state of the system to another. It was then realized that such transformations, under the operation of multiplication, constitute a set called a **group** by mathematicians, a system possessing very interesting mathematical properties. Thus, group theory was developed. This theory became much more interesting and led to some additional applications with the emergence of quantum mechanics. Soon, the internal degrees of freedom were recognized and put into the realm of symmetries, and thus, group theory is no longer purely geometric.

For practical as well as pedagogical reasons, group theory is split into two different parts. The first deals with **discrete groups** and the second with **continuous groups**. In the first case, the elements of the group are countable and usually finite in number. This part deals more with geometric symmetries. The second is characterized by group elements, which depend on continuous parameters. We will not consider the case of continuous groups in the current volume since there exist many books on this subject, including, in particular, a text covering such topics by the first author, recently published by WSPC, which can be considered a companion volume to this book.

The current book deals with discrete groups. It is a translation from a book in Greek with the same title by the first author, suitably updated and extended. It was intended to cover the material given in the first of a two-semester course given at the University of Ioannina. Parts of this material have also been delivered to senior-level physics students at Nanjing University, Nanjing, China, in 2012.

In the long past, this subject was taught in a course on algebra, with the applications considered a part of mathematical methods in physics, with a focus on applications in a specific research area of physics. We have come a long way from that, and this subject is now taught everywhere, as outlined above, i.e. in a *bona fide* group theory course. In many cases, however, after a brief presentation of the basic theoretical ideas, many instructors devise special rules and rush to use them in specific applications. We believe, however, that group theory should be presented from a unified perspective involving both the beauty of symmetries, resulting in elegant mathematical expressions, and using it as a tool for applications, either currently popular or expected to emerge in the future. This approach is not recommended for the Epicurean type who would like to look up a formula or use the results of some table without understanding their derivation. We expect, however, that in the long run, our approach will be more beneficial for the student who will have the patience to go through the material. To this end, we made every effort to include illustrative examples, which are the simplest possible to clarify the basic concepts introduced. Some simple applications, especially those that may contribute to a better understanding of some relevant concepts of the theory, have been included as illustrative examples. A set of useful tables has been included, with the dedicated student expected to construct some segments of them as part of homework. More detailed applications form the material of a whole chapter, which may be skipped at first reading. This approach resulted in a much larger volume for the book. I hope this will not have an adverse effect on the student's decision to register for a one-semester course.

On this occasion, I would like to thank former student of physics, Miltos Christoulakis, currently the president of Happy–Box, for preparing the cover as well as most of the figures in the book.

J. D. Vergados, V-I Vergadou-Remediaki

Ioannina, October 2022,

Contents

Chapter 1

The Role of Symmetries in Physics — A Prelude

The notion of symmetry is not, of course, a new one. It has been known since the very ancient times that the human body possesses right/left symmetry, a "mirror symmetry", i.e. a reflection with respect to a plane; that a cube has a high degree of symmetry, both with respect to rotations around some axes as well as reflections with respect to certain planes; that the sphere is the most symmetric of all bodies. The idea of geometric symmetry affected deeply the thought of many ancient Greek philosophers, including Pythagoras, Plato and others. Indeed, Plato attempted to describe the motion of the heavenly bodies in terms of circles or circles over circles (circles on epicenters).

In spite of its beauty as a theory, however, group theory (GT) did not become a truly useful tool in physics until after the foundation of quantum mechanics in the 1920s. To this end, a crucial role was played by Wigner, Weyl, Gelfand, Racah and others. In other words, GT became very useful to the physical sciences when it was realized that the set of linear transformations, which lead from one state of a system to another, constitute a group.

As outlined above, the ancient notion of symmetry became very useful to the physical sciences. What is a symmetry? Roughly, a symmetry is a property of some object which remains invariant under some operations. Note that the notion of symmetry requires both an object and the operation which acts or operates on the object. Invariance means that the relevant property of the object remains the same before and after the operation has been acted upon it.

1.1 Geometric symmetries

As we have already mentioned, these symmetries were the first to be recognized by the ancient Greeks. They relate to transformations which maintain the distance between two points of a body and map it onto itself. Such transformations are essentially of three types:

(i) Rotations about an axis of symmetry by a given angle.

As a concrete example, let us consider a square. We label the four corners of the square A, B, C, D. Intuitively, if we rotate the square by $90°$ clockwise, we get the picture shown in Fig. 1.1(a). Let us call this operation S. Then, the square remains the same with the original orientation. What do we mean by "the same"? It means that, if we drop the labels A, B, C, D, you do not see any difference. Now, a rotation of $180°$ also leaves the square invariant. Let us call this operation T. Clearly, we can achieve the same result by applying the operator S twice. Thus, we write $T = S \cdot S$. Similarly, a rotation of $270°$, indicated as U, leaves the square invariant, and $U = (S \cdot S)S$. Clearly, a rotation of $360°$ does not change anything, and we identify this with the identity operator and designate it as E.

We call the operation S a rotation of the system by an angle $\pi/4$ around an axis perpendicular to the plane of the square, passing through its center. In the standard notation, we indicate the operation associated with S as C_1. If we apply this operation four times successively, the system returns to its original position. We say that this axis is of fourth order, $\frac{2\pi}{\pi/2} = \frac{360}{90} = 4$. We indicate the successive operations as C_1, C_1^2, C_1^3, C_1^4. Obviously, $C_1^4 = E$, where E is the identity operator

We can easily see that a rotation by $\pi/2$ in the opposite sense, i.e. counter clockwise, which is indicated by $-\pi/2$, gives a similar effect as a clockwise rotation by $3\pi/2$. Indicating this as C_{-1}, we verify that $C_{-1} =$

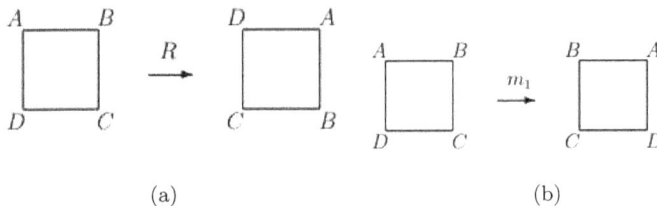

(a)   (b)

Fig. 1.1. Symmetry operations on a square: (a) rotation around a four-fold axis, (b) reflection through a plane which is perpendicular to the square and passes through the middles of AB and CD.

C_1^3. Thus, $C_{-1} \cdot C_1 = E$, i.e. C_{-1} is the inverse of C_1. Similarly, C_{-2} is the inverse of C_2. Furthermore, these operators commute, i.e. $C_1^n C_1^k = C_1^k C_1^n$, and the system $\{C_1, C_1^2, C_1^3, C_1^4 = E\}$ is closed under multiplication.

(ii) Reflections.
Another operation is **reflection**, i.e. a mirror image of the square with respect to a plane that contains the above axis and is perpendicular to the plane of the square passing through the middle of its two opposite sides, e.g. through the middle of AB and CD indicated as m_1, see Fig. 1.1(b). The relevant operator will be indicated by σ_x. There is another reflection corresponding to a similar reflection plane passing through the middle of AC and BD, with the relevant operator designated as σ_y. There exist two additional similar reflection planes passing through the opposite corners of the square, one through A and C and the other through B and D; these will be designated as σ_1 and σ_2, respectively. Obviously, $\sigma_i^2 = E$, $i = x, y, 1, 2$, i.e. each of these elements is its own inverse.

(iii) Inversion (with respect to a center).
An **inversion** causes a transformation of the coordinates

$$(x, y, z) \rightarrow (-x, -y, -z). \tag{1.1}$$

One can see that in our example, inversion is not a new operation, but it coincides with a rotation by π. Both of these operations have the same effect: they interchange $A \leftrightarrow C$ and $B \leftrightarrow D$. Furthermore, one can show that the set

$$\{C_1, C_1^2, C_1^3, C_1^4 = E, \sigma_x, \sigma_y, \sigma_1, \sigma_2\}$$

is a closed set under multiplication. Note that two elements, X and Y, do not always commute, i.e. in some cases, $X \cdot Y \neq Y \cdot X$. Anyway, the above elements constitute a group associated with the symmetry known as C_{4V}.

It was, of course, natural, in view of the ideas of the ancient philosophers, that the geometric symmetries would be recognized as the first applications of group theory in physics, in particular in crystallography and solid-state structures.

Next comes the notion of a space group. By this, we understand the symmetry group which is characteristic of a given periodic system, as for example an ideal crystal. This consists of the set of transformations which carry one point of the system to another. Thus, the space groups should contain point groups, such as the ones we have considered above as well as translation transformations. This imposes conditions on how the elements of the crystal repeat themselves so that they generate the whole crystal. As a result, only a fraction of the point group symmetries are compatible with the required space symmetry. This is because the structure elements of the point group symmetries, when repeated, must cover the whole crystal. We all know that the whole floor can be covered by placing side by side tiles of certain shapes, e.g. triangles, parallelograms (squares, diamonds) or regular hexagons, see Fig. 1.2. This cannot be done by placing circles or

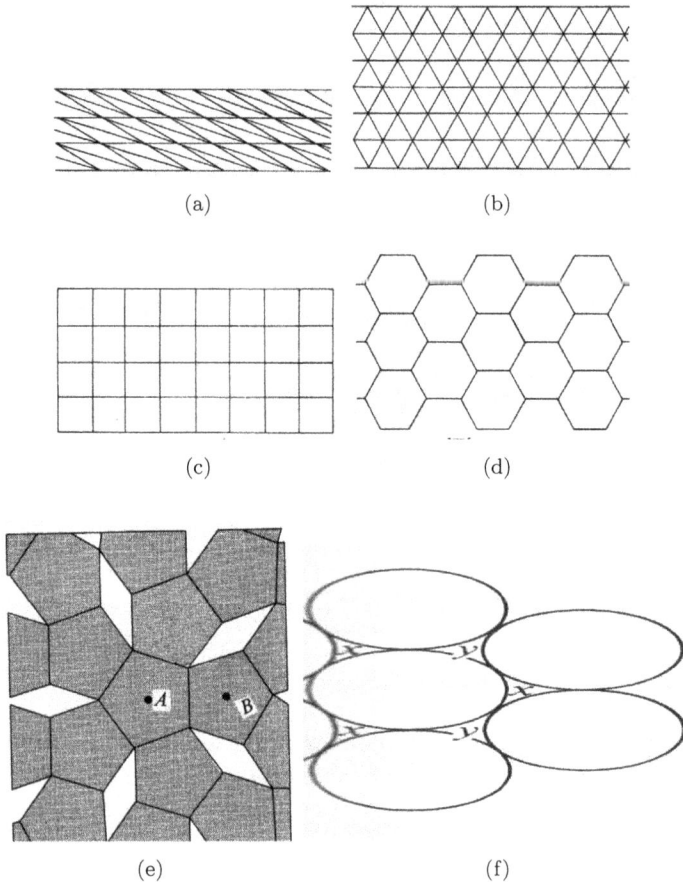

Fig. 1.2. Various shapes that can cover the whole plane: (a) scalene triangles combined to form a parallelogram (second-order axis), (b) equilateral triangles (third-order axis), (c) squares (fourth-order axis) and (d) hexagons (sixth-order axis). This cannot be accomplished by using: (e) regular pentagons and (f) circles or ellipses. Note that in each shape, we have two geometric quantities: two size parameters a and b as well as the angle ϕ between them. Thus, (a) we have $a \neq b$ and ϕ arbitrary, and in the other cases, we have $a = b$, while the angles are $2\pi/3$, $\pi/2$ and $\pi/3$ for (b), (c) and (d), respectively.

regular pentagons. The whole three-dimensional (3D) space can be covered by properly placing prisms with bases of the above shapes.

1.2 Symmetry in the dynamics of classical mechanics

The role of the dynamic symmetries, i.e. those that leave invariant the equations of motion, was recognized later. A first and very interesting example was the motion of a planet around the sun, Kepler's problem. The solution to this problem was, of course, first obtained by solving Newton's equations of motion, using the world gravitational attraction, also invented by Newton. The predicted orbits were in very good agreement with observation. It is worthwhile, however, to look at Kepler's problem from a different perspective and more carefully (see Fig. 1.3). The main points are as follows:

- The orbit lies on a plane.
- The orbit is an ellipse (or in general a conic section) with a fixed axis length.
- The direction of the vector connecting the two foci of the ellipse remains fixed. In other words, the position of the perihelion (point of the orbit closest to the sun) is fixed.

The planarity of the orbit stems from the fact that Newton's force is central, which leads to the conservation of angular momentum. As the angular momentum cannot change and the orbit must always lie at right angles to it,

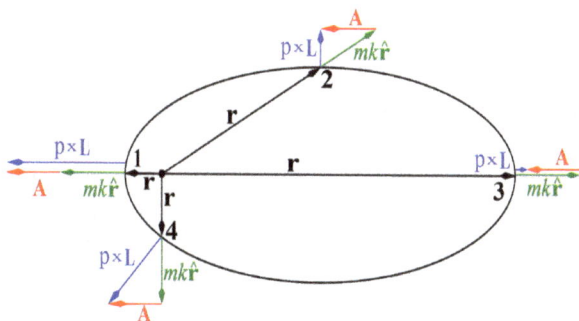

Fig. 1.3. In Kepler's description, the motion of a planet around the sun is exhibited. In particular, the role of the Rünge–Lenz vector, indicted in the picture by **A**, guarantees the stability of perihelion, point 1 in the figure.

the orbit must be planar. Therefore, the symmetry (centrality) of the force leads to the planarity of the orbit. The conservation of angular momentum is related to the fact that the system would remain unchanged after a rotation of space (isotropy).

The length of the longer axis is determined by the energy of the system, which is conserved by the specific form of interaction between two bodies, known as **conservative force**. The eccentricity of the motion is constant because it is determined by the energy and angular momentum of the system, both of which are constant. But why should the perihelion remain constant?

The answer to this question is less obvious. It has to do with the fact that the attractive force is inversely proportional to the square of the distance between the two objects, $F \propto \frac{1}{r^2}$. A small deviation from this, even when $F \propto \frac{1}{r^{2+\epsilon}}$, no matter how small ϵ might be, destroys the stability of the perihelion. This is what actually occurs when we take into account the presence of other planets or generalize Newton's theory using the general theory of relativity developed by Einstein.[1]

The important conclusion to be drawn from this is that by recognizing and invoking symmetry, the problem posed by Kepler can be solved without requiring the solution of a differential equation. We will not discuss this topic here, but it has been treated elsewhere, e.g. in the book Vergados (2017). Even more importantly, the existence of symmetry in a system results in the conservation of some quantity that characterizes the system. Conversely, the conservation of some quantity characterizing a system dictates the existence of a symmetry characterizing the system. This is known as Noether's theorem.

1.3   Symmetry in quantum mechanics

As has already been mentioned, group theory became a more useful tool in solving problems in physics after the development of quantum mechanics. Here, the results were surprising, especially in the case where the group transformations leave the system's Hamilton operator invariant. The analogue in quantum mechanics to Kepler's problem is hydrogen-like atoms. The quantum numbers that describe the states of the system are expected

---

[1] Indeed, the first indication that the general theory of relativity holds true came from studying the movement of the perihelion of Mercury, as the phenomenon could not be explained solely by the influence of the other planets on Mercury's orbit.

to depend on the orbital angular momentum quantum number, ℓ, but they must be independent of m, i.e., the projection of the orbital angular momentum on the quantization axis. This is due to the fact that the Hamilton operator remains unchanged under rotation in 3D space. However, the eigen energies of the system are independent of the azimuthal, ℓ; this is indicative of a greater symmetry, which is rotation in 4D space, (Vergados, 2017). Recognition of this symmetry allows us to calculate the eigen energies of the hydrogen-like atom without having to solve the Schröedinger equation.

The arrival of quantum mechanics was the main reason for the familiarity of physicists with group theory, which had already been developed by mathematicians.

In quantum theory, invariance principles permit even more far-reaching conclusions than in classical mechanics. In quantum mechanics, the state of a physical system is described by a ray in a Hilbert space, $|\Psi\rangle$. A symmetry transformation gives rise to a linear operator, R, that acts on these states and transforms them into new states. Just as in classical physics, the symmetry can be used to generate new allowed states of the system. However, in quantum mechanics, there is a new and powerful twist due to the linearity of the symmetry transformation and the superposition principle. Thus, if $|\Psi\rangle$ is an allowed state, then so is $|R\Psi\rangle$, where R is the operator in the Hilbert space corresponding to the symmetry transformation $\mathcal{R}$. So far, this is similar to classical mechanics. However, we can now superpose these states, i.e. construct a new allowed state: $|\Psi\rangle + |R\Psi\rangle$. (There is no classical analogue for such a superposition, e.g. the superposition of two orbits of the Earth.)

Quantum mechanics also revealed a new kind of symmetry: that of the exchange of identical particles. This led to a classification of all elementary particles as (a) Bosons, characterized by integral spin. Then, the wave function describing a many-identical-particles system must be invariant under the interchange of any two particles; (b) Fermions, with half-integral spin, whose wave function changes sign when any two particles are interchanged. The quantum statistics of a collection of such particles is different, with profound implications for their behavior in aggregate.

At this stage, Eugene Wigner played a very crucial role, connecting group theory with quantum mechanics. In addition, a great contribution toward the familiarity of physicists with group theory can be attributed to Heisenberg and Weyl, mainly for showing the equivalence of Schrödinger's version of quantum mechanics with that of Heisenberg.

One can interpret a symmetry transformation as a change in our point of view when looking at a system that does not alter the results of possible experiments. In quantum mechanics, we know that (pure) states are denoted by wave functions $|\Psi\rangle$ in the Hilbert space. These functions are members of a vector space equipped with a scalar product,[2] i.e. $\langle\Psi_f|\Psi_i\rangle = \langle\Psi_i|\Psi_f\rangle^*$. The measurable quantities are not the wave functions themselves, but the expectation values of any observable quantity. The probability of obtaining an expectation value after measurement depends on the transition probabilities between states, which is given by the square of the modulus of the overlap of two wave functions, $|\langle\Psi_f|\Psi_i\rangle^2|$. Hence, a symmetry transform should keep these probabilities invariant. So, in a quantum scenario, we define as symmetry transformations those which preserve transition probabilities between the states.

So far, the discussion has been about the transition probabilities, but how do the individual states themselves transform under a symmetry transformation? The answer to this particular question was provided by Wigner.

Wigner's theorem: Any symmetry transformation can be represented on the Hilbert space of physical states[3] by an operator which is either linear and unitary or antilinear and antiunitary.

For a symmetry operator U, the theorem can be stated as:

$$\langle U\Psi_A|U\Psi_B\rangle = \langle\Psi_A|\Psi_B\rangle, \, U = \text{unitary} \qquad (1.2)$$

or

$$\langle U\Psi_A|U\Psi_B\rangle = \langle\Psi_A|\Psi_B\rangle^*, \, U = \text{antiunitary}. \qquad (1.3)$$

The theorem was stated and proved for the first time in 1931 by Wigner himself in his book (Wigner, 1959). It has thereafter been proved by many

---

[2]The scalar product is defined through an integral, e.g. in one dimension, as $\langle f|g\rangle = \int_{-\infty}^{+\infty} f^*(x)g(x)dx$. These functions must satisfy the condition $\langle f|f\rangle < \infty$ and similarly for g. Thus, they can be normalized: $\langle f|f\rangle = \langle g|g\rangle = 1$. These relations can easily be generalized in 3D.

[3]Mathematically, however, physical states are denoted by "rays" in the Hilbert space. A set of normalized states whose elements differ only by a complex phase are called rays, i.e., the states ψ and $e^{i\theta}\psi$ are members of the same ray R for some real θ. They correspond to the same physical state. Hence, a symmetry transformation is a ray transformation T such that if $T : R_1 \to TR_1$ and $T : R_2 \to TR_2$, it follows that

$$|\langle\Psi_1'|\Psi_2'\rangle^2| = ||\langle\Psi_1|\Psi_2\rangle^2|$$

for any $\Psi_1 \in R_1$, $\Psi_2 \in R_2$, $\Psi_1' \in TR_1$ and $\Psi_2' \in TR_2$.

people, the most prominent being Bargmann, Ulhorn and, more recently, (Weinberg, 1996).

Unitary transformations are commonly used and discussed extensively in this book. The antiunitary operators are encountered in the case of time reversal, see Section 2.5. One can show that an antiunitary operator can be written as a product of a unitary operator and the complex conjugation operation (Vergados, 2018, Section 12.4.2).

We also owe to Wigner another interesting approach to quantum mechanics:

If we denote as x all the coordinates that describe a system, the time-independent Schrödinger differential equation may take the following form:

$$H(x)\psi(x) = E\psi(x), \tag{1.4}$$

where $H(x)$ is the Hamiltonian describing the system. In this space, we consider a linear transformation, which is an element of a group, $A \in G$:

$$x \to x' = Ax. \tag{1.5}$$

This implies a transformation T_A in functional space:

$$\psi \to \psi' = T_A\psi = \psi(A^{-1}x), \tag{1.6}$$

as was obtained by Wigner. Then, Eq. (1.4) becomes

$$H(A^{-1}x)\psi(A^{-1}x) = E\psi(A^{-1}x). \tag{1.7}$$

Thus, Eq. (1.4(remains unchanged as long as

$$H(A^{-1}x) = H(x). \tag{1.8}$$

This is equivalent to

$$HT_A = T_A H. \tag{1.9}$$

In fact, from Eqs. (1.8) and (1.7), one finds

$$H(x)\psi(A^{-1}x) = E\psi(A^{-1}x) \Rightarrow H(x)T_A\psi(x) = ET_A\psi(x) \Rightarrow \tag{1.10}$$

$$(T_A^{-1}HT_A)\psi(x) = E\psi(x) \Rightarrow T_A^{-1}HT_A = H \Rightarrow HT_A = T_A H. \tag{1.11}$$

Now, considering a representation of T_A, $T_A \to T(A)$, in some basis, we show that

$$T(A)T(H) = T(H)T(A),$$

where $T(H)$ is the representation of the operator H in the same basis.

Let $\psi_i(x)$ be an orthogonal basis in functional space. Then,

$$T(H)_{ij} = \langle \psi_i(x)|H|\psi_j(x)\rangle \equiv H_{ij}, \; T(A)_{ij} = \langle \psi_i(x)|T_A|\psi_j(x)\rangle,$$

$$H'_{ij} = T'(H)_{ij} = \int dx \psi_i'^*(x)|H|\psi_j'(x) = \int dx \psi_i^*(A^{-1}x)|H|\psi_j(A^{-1}x)$$

$$= \int dx' \psi_i^*(x')|H(x')|\psi_j(x') = H_{ij}$$

$$H'_{ij} = T'(H)_{ij} = \int dx T_A \psi_i(x)|H|T_A \psi_j(x)$$

$$= \sum_k \sum_\ell (T(A))^*_{kj}(T(A))_{\ell j} \int \psi_k^*(x)|H|\psi_\ell$$

$$= \sum_k \sum_\ell (T(A))^*_{kj}(T(A))_{\ell j} H_{k\ell} = \left((T(A))^+ T(H) T(A)\right)_{ij} \Rightarrow$$

$$T(A) - (T(\Lambda))^+ T(H) T(\Lambda).$$

Following Wigner, we make the reasonable assumption that the representation is unique, then

$$T(A) \Rightarrow \Gamma(A), \Gamma^+(A) = \Gamma^{-1}(A) = \Gamma(A^{-1}) \Leftrightarrow T(H)$$

$$= \Gamma^{-1}(A) T(H) \Gamma(A) \Rightarrow$$

$$\Gamma(A) T(H) = T(H) \Gamma(A). \tag{1.12}$$

In order to solve Eq. (1.4) the following strategy is used: We expand the function ψ in the basis ψ_i,

$$\psi = \sum_i \alpha_i \psi_i.$$

Equation (1.4) now becomes:

$$\sum_i \alpha_i H \psi_i = \sum_i \alpha_i E \psi_i.$$

Taking the scalar product of both sides of the equation with $\langle \psi_j|$, we find

$$\sum_i \alpha_i \langle \psi_j|H|\psi_i = \sum_i \alpha_i E \langle \psi_j \psi_i \rangle = E \alpha_j \Rightarrow$$

$$\sum_i \alpha_i (H_{ji} - E \delta_{ij}) = 0$$

or, in matrix form,

$$(H)|\alpha\rangle = E|\alpha\rangle. \qquad (1.13)$$

Eq. (1.13) is equivalent to (1.4).

It is sufficient, therefore, to solve Eq. (1.13). If, by utilizing the symmetry, we hit the bull's eye when selecting the basis, we will be able to reduce the matrix (H), i.e.

$$(H) = \begin{pmatrix} (H^{(1)}) & 0 & 0 & \cdots & \cdots & 0 \\ 0 & (H^{(2)}) & 0 & \cdots & \cdots & 0 \\ 0 & 0 & (H^{(3)}) & \cdots & \cdots & 0 \\ 0 & 0 & 0 & (H^{(4)}) & \cdots & 0 \\ 0 & 0 & 0 & \cdots & \cdots & (H^{(k)}) \end{pmatrix}, \qquad (1.14)$$

since the Hamiltonian cannot mix states of different symmetry. This holds true thanks to another famous theorem by Wigner, which will emerge as our group theory evolves, see Section 6.2. Please, be patient!

For the moment, just be satisfied that all we have to do is diagonalize matrices of smaller dimensions.

We may therefore conclude that group theory is recognized as a useful tool for the comprehension and study of various systems. Recognition of symmetry in a system relieves us of the necessity to solve complex equations in order to discover the system's characteristics. This holds true even when the symmetry is not absolute, but approximate. In that case, the symmetry will help us find an approximate solution, which may then be used as a basis for finding a more accurate solution with perturbation theory. The matrices which must be diagonalized in that case are considerably smaller. Moreover, the existence of symmetry allows us to find the form of the Hamilton operator, when that is not already known. In this book, we pay special attention to "geometric symmetries" and their use in quantum mechanics problems (normal modes, etc.).

Some people may argue that physicists need not concern themselves with the abstract group theory that a mathematician might deal with, but the realization in some way of the abstract group elements, i.e their representation, which is nothing but matrices that follow the same multiplication rules with the abstract elements. We consider it necessary, however, to discuss the structure of abstract groups, despite the difficulty posed for those who are not mathematically inclined. It is, after all, necessary to introduce a minimum of mathematical concepts before we go on to their applications. This is what we will do next in Chapter 2.

Chapter 2

Introduction to Discrete Groups

The notion of the group, which is linked with explaining symmetry, became part of mathematics more than two centuries ago. From then on, it developed quite quickly thanks to the creative work of Gauss, Cauchy, Abel, Hamilton, Galois and, especially, Caylay, Cartan, Dynkin and others. In spite of its beauty as a theory, however, group theory (GT) did not become a truly useful tool in physics untill after the foundation of quantum mechanics in the 1920s. To this end, a crucial role was played by Wigner, Weyl, Gelfand, Racah and others. In other words, GT became very useful to the physical sciences when it was realized that the set of linear transformations, which lead from one state of a system to another, constitute a group.

2.1 Definitions and basic concepts

Definition: A group is a set $G \equiv \{E, A, B, C, \ldots\}$, with elements differing from each other, equipped with a binary operation, which is usually a multiplication, denoted by $*$, and having the following properties:

(i) The operation when acting on two elements of G produces a new element of G, i.e.

$$A \in G, B \in G \Rightarrow \exists C' \in G \quad \text{and} \quad C \in G \quad \text{such that} \quad A * B = C,$$

$$B * A = C'. \tag{2.1}$$

It is usual to omit the multiplication symbol, so the above may be written as $AB = C$, etc.

(ii) $A \in G$, $B \in G$, $C \in G$ and $A = B \Rightarrow CA = CB$.

(iii) The operation is associative, i.e.

$$A(BC) = (AB)C, \tag{2.2}$$

which means that the brackets denoting the order in which the operations are to be performed are unnecessary and may be omitted.

(iv) The set G contains an element E such that

$$AE = EA = A, \; A \in G. \tag{2.3}$$

This element E is the identity element and can also be denoted as 1 or 0, depending on the operation, but it should not be confused with the numbers 1 or 0.

(v) For every $A \in G$, there is $B \in G$ such that $AB = E$. B is then referred to as the inverse of A and is denoted by A^{-1}.

The above axioms can be used to prove the following:

- The identity element E is unique. Let us say that $AE = A$ and $AE' = A$, $A \in G$. Then,

$$AE = AE' \Rightarrow A^{-1}AE = A^{-1}AE' \Rightarrow E = E'.$$

- The product of many elements of the group may be defined: $ABCD\ldots$.
- If the group is finite, that is, if the number of elements g of the group is finite, then the product of the operation on an element cannot continue infinitely. At some point, the new product will coincide with one of the elements of the group. Specifically,

If $A \in G$, there exists an integer n, $n > 0$, such that $A^n = E$,

where $A^n = AA, \ldots, A$ (n times product), $n \leq g$.

In general, the operation is not commutative, so that in most cases, $AB \neq BA$. If, however, it happens that

$$A \in G, \; B \in G \Rightarrow AB = BA,$$

the group is then called an Abelian group. We will see that the structure of such a group is very simple. When the elements of the group can be

matched 1:1 to the natural numbers, the group is a **discrete** group.[1]
Otherwise, it is a **continuous** group.

Some simple examples are as follows:

- The set $\{-1, 1\}$ with multiplication as the operation is a discrete group of order $g = 2$.
- The set of complex numbers $\{-1, 1, i, -i\}$ with multiplication compose a discrete group of order $g = 4$.
- The set of real numbers, excluding zero, with 1 as the identity element and multiplication as the operation form a continuous group.
- The set of matrices with elements from the set of real or complex numbers, with a non-zero determinant and with ordinary matrix multiplication as the operation, comprise a continuous group, which, in general, is non-Abelian. This group is too general to be useful; we will, however, discuss more specific cases of such matrices, which are very useful.

In the case of finite discrete groups, it is useful to construct the multiplication table, known as Cayley table.

Example 1: Let us consider the case of the above complex group.

With $E = 1$, $A = -1$, $B = i$, $C = -i$, we find the following Cayley table, (Eq. (2.4)):

$$
\begin{array}{c|cccc}
 & E & A & B & C \\
\hline
E & E & A & B & C \\
A & A & E & C & B \\
B & B & C & A & E \\
C & C & B & E & A
\end{array}
\quad \Longleftrightarrow \quad
\begin{array}{c|cccc}
 & E & A & B & C \\
\hline
E & E & A & B & C \\
A & A & E & C & B \\
C & C & B & E & A \\
B & B & C & A & E
\end{array}
\quad
\begin{array}{c|cccc}
E & B & B^2 & B^3 \\
\hline
E & E & B & B^2 & B^3 \\
B^3 & B^3 & E & B & B^2 \\
B^2 & B^2 & B^3 & E & B \\
B & B & B^2 & B^3 & E
\end{array}
. \qquad (2.4)
$$

We note that it is often more useful to array the multiplications so that the identity element lies along the diagonal.

Additionally, $B = i$ can generate all the elements of the group, and it is called a **generator** of the group:

$$B^2 = -1, B^3 = -i, B^4 = E.$$

---

[1] The binary operation can also be addition. In that case, the identity element is 0, and the inverse of each element is its opposite. The set of integers, with addition as the operation, forms such a group, which is an infinite discrete group.

The group is Abelian, and the Cayley table can also be written as the last table in Eq. (2.4). The set of generators of a group is not always unique. In this example, $B = -i$ is also a generator of the group.

Example 2: Let us consider the matrices

$$A = \begin{pmatrix} -\frac{1}{2} & -\frac{\sqrt{3}}{2} \\ \frac{\sqrt{3}}{2} & -\frac{1}{2} \end{pmatrix}, \ D = \begin{pmatrix} \frac{1}{2} & \frac{\sqrt{3}}{2} \\ \frac{\sqrt{3}}{2} & -\frac{1}{2} \end{pmatrix}. \tag{2.5}$$

Consider the set of matrices that arise beginning with the product of the above matrices and including all the possible products of the matrices that arise subsequently. Examine whether this set constitutes a group.

We see that

$$A^2 = \begin{pmatrix} -\frac{1}{2} & \frac{\sqrt{3}}{2} \\ -\frac{\sqrt{3}}{2} & -\frac{1}{2} \end{pmatrix} = B, \ AB = \begin{pmatrix} 1 & 0 \\ 0 & 1 \end{pmatrix} = E,$$

$$AD = \begin{pmatrix} -1 & 0 \\ 0 & 1 \end{pmatrix} = C, \ DA = \begin{pmatrix} \frac{1}{2} & -\frac{\sqrt{3}}{2} \\ -\frac{\sqrt{3}}{2} & -\frac{1}{2} \end{pmatrix} = F. \tag{2.6}$$

All further multiplications result in one of the above matrices.

It can be seen that

$$A^3 = AB = E, D^2 = E, A^{-1} = B, \text{ etc.}$$

The set of six elements is a closed set. As it satisfies all the criteria, it must be considered a group. One can easily form the Cayley tables:

	E	A	B	C	D	F
E	E	A	B	C	D	F
A	A	B	E	F	C	D
B	B	E	A	D	F	C
C	C	D	F	E	A	B
D	D	F	C	B	E	A
F	F	C	D	A	B	E

$\Longleftrightarrow$

	E	A	B	C	D	F
E	E	A	B	C	D	F
B	B	E	A	D	F	C
A	A	B	E	F	C	D
C	C	D	F	E	A	B
D	D	F	C	B	E	A
F	F	C	D	A	B	E

$\tag{2.7}$

With the Cayley tables, especially the one on the right where the identity element lies on the diagonal, it is easy to find the inverse of an element (i.e.

a matrix in the above example) as it will be in that line and column where the identity element E is. In the right-hand table, the inverse elements lie symmetrically on either side of the diagonal. In any case, if $AB = C$, C^{-1} is the inverse of AB, $ABC^{-1} = CC^{-1} = E$.

The Cayley table may seem, on the surface, to change when the elements are arranged in a different order. However, it remains essentially the same and describes the group fully. The minimal set of elements of the group, which under group operations can generate all elements of the group, comprises the **generators** of the group. It is often preferred to write the elements of a group as expressions of the generators; in our example, in terms of the generators A, D,

$$E = A^3, A, A^2, D, AD, DA.$$

Definition: Two discrete groups

$$G = \{E, A, B, \ldots,\} \quad \text{and} \quad G' = \{E', A', C', \ldots,\}$$

are *isomorphic* if

- they each contain the same number of elements $g = g'$,
- there is a one-to-one correspondence $E \Longleftrightarrow E'$, $A \Longleftrightarrow A'$, $B \Longleftrightarrow B'$, etc., such that

$$AB = C \quad \text{for} \quad G \Longleftrightarrow A'B' = C' \quad \text{for} \quad G'.$$

When the two groups G and G' coincide, this correspondence is called *automorphism*. Two groups sharing the same Cayley table are isomorphic. Mathematically, one cannot distinguish between two isomorphic groups. As we shall see, however, they may be entirely different from a physicist's point of view.

Definition: A set of discrete elements $G' = \{E', A', C', \ldots,\}$ is a subgroup of the discrete group $G = \{E, A, B, \ldots,\}$, if it is a subset of the elements of that group and fulfills all the criteria of a group. We are usually interested in proper subgroups, differing from $\{E\}$ and the set itself, G. The set $G' = \{1, -1\}$ is a subgroup of $G = \{1, -1, i, -i\}$, while the set $\{i, -i\}$ is not even a group.

2.2   The symmetric group S_n

The symmetric group S_n defined for every $n = 1, 2, 3, \ldots$ is perhaps the most important discrete group both mathematically and from the perspective of physics. Mathematically, this is because of Cayley's theorem, further discussed as follows, which states that any discrete group with a finite number of elements must correspond with (be isomorphic to) a subgroup of S_n. As far as physics is concerned, the importance of S_n lies in its usefulness in constructing wave functions for many identical (indistinguishable) particles. The wave functions must have a given symmetry in the event of an exchange of any two particles. Specifically, the function must be anti-symmetric, i.e. change sign, if the particles have a half-odd integer spin (e.g. $1/2$, $3/2$, $\ldots$,), while it must be symmetric, i.e. remain unchanged, for particles with integer spin.

2.2.1   *Brief overview of permutations*

A permutation is a rearrangement of the countable elements of a set. Let us suppose that a set contains three elements, denoted as (1,2,3). These elements may be rearranged in six different ways, (1,2,3), (1,3,2), (2,1,3), (2,3,1), (3,1,2) and (3,2,1). The different rearrangements may be denoted by the following permutations:

$$P_1 = \begin{pmatrix} 1 & 2 & 3 \\ 1 & 2 & 3 \end{pmatrix}, P_2 = \begin{pmatrix} 1 & 2 & 3 \\ 1 & 3 & 2 \end{pmatrix}, P_3 = \begin{pmatrix} 1 & 2 & 3 \\ 2 & 1 & 3 \end{pmatrix},$$

$$P_4 = \begin{pmatrix} 1 & 2 & 3 \\ 2 & 3 & 1 \end{pmatrix}, P_5 = \begin{pmatrix} 1 & 2 & 3 \\ 3 & 1 & 2 \end{pmatrix}, P_6 = \begin{pmatrix} 1 & 2 & 3 \\ 3 & 2 & 1 \end{pmatrix}. \tag{2.8}$$

The first permutation is the identity element of the permutations. In the above notation, the starting point is written on the first line. The arrangement of the elements in the first line could be changed; the horizontal arrangement has no significance. The significance lies in the vertical arrangement. Thus, $P_2 = \begin{pmatrix} 1 & 2 & 3 \\ 1 & 3 & 2 \end{pmatrix} = \begin{pmatrix} 1 & 3 & 2 \\ 1 & 2 & 3 \end{pmatrix}$, etc. The transposition of two elements is a specific form of permutation. For example, the transposition of the elements 2 and 3, which can be symbolized as (23), is none other than P_2.

The above can be generalized for any number of elements. Any permutation of n objects can be symbolized as

$$P = \begin{pmatrix} 1 & 2 & \cdots & n \\ m_1 & m_2 & \cdots & m_n \end{pmatrix} \quad \text{or} \quad Q = \begin{pmatrix} 1 & 2 & \cdots & n \\ \ell_1 & \ell_2 & \cdots & \ell_n \end{pmatrix}.$$

Permutations are algebraic objects and can be combined. The product of two permutations can be found as follows:

$$PQ = \begin{pmatrix} 1 & 2 & \cdots & n \\ m_1 & m_2 & \cdots & m_n \end{pmatrix} \begin{pmatrix} \ell_1 & \ell_2 & \cdots & \ell_n \\ 1 & 2 & \cdots & n \end{pmatrix} = \begin{pmatrix} \ell_1 & \ell_2 & \cdots & \ell_n \\ m_1 & m_2 & \cdots & m_n \end{pmatrix}.$$

By arranging the elements of the first line, i.e. $\ell_1, \ell_2, \ldots, \ell_n$ in the natural (increasing) order, we find

$$PQ = \begin{pmatrix} 1 & 2 & \cdots & n \\ m'_1 & m'_2 & \cdots & m'_n \end{pmatrix}.$$

Generally, $PQ \neq QP$. The inverse of a permutation is easily found:

$$P = \begin{pmatrix} 1 & 2 & \cdots & n \\ m_1 & m_2 & \cdots & m_n \end{pmatrix} \Rightarrow P^{-1} = \begin{pmatrix} m_1 & m_2 & \cdots & m_n \\ 1 & 2 & \cdots & n \end{pmatrix}.$$

As the reader can easily confirm based on the above definition of the product of two permutations,

$$PP^{-1} = \begin{pmatrix} m_1 & m_2 & \cdots & m_n \\ m_1 & m_2 & \cdots & m_n \end{pmatrix} = E.$$

The last symbol is obviously the identity permutation (nothing changes),
    It is clear that in any permutation, it is the changes that are important. We can therefore write

$$P = \begin{pmatrix} 1 & 2 & 3 & 4 & 5 & 6 \\ 5 & 4 & 3 & 2 & 6 & 1 \end{pmatrix} = \begin{pmatrix} 1 & 5 & 6 \\ 5 & 6 & 1 \end{pmatrix} \begin{pmatrix} 2 & 4 \\ 4 & 2 \end{pmatrix} \begin{pmatrix} 3 \\ 3 \end{pmatrix}.$$

The order in which the above three permutations are written is unimportant, since, having no elements in common, they commute. A permutation containing only the elements that change is called a **cycle**, and we can write

$$\begin{pmatrix} 1 & 5 & 6 \\ 5 & 6 & 1 \end{pmatrix} = (156), \quad \begin{pmatrix} 2 & 4 \\ 4 & 2 \end{pmatrix} = (24), \quad \begin{pmatrix} 3 \\ 3 \end{pmatrix} = (3).$$

It is called a cycle because it is a closed set that remains unchanged under a cyclic interchange of its elements. It is clear that

$$(156) = (615) = (561), \quad (156) \neq (516) = (651) = (165).$$

In other words,

$$P = \begin{pmatrix} 1 & 2 & 3 & 4 & 5 & 6 \\ 5 & 4 & 3 & 2 & 6 & 1 \end{pmatrix} = (156)(24)(3).$$

The decomposition of a permutation into cycles is unique.

It is intuitively obvious that any permutation can be achieved given all possible transpositions. This can be proven by considering the possible cycles.

Theorem 1: *Any cycle can be written as a product of transpositions:*

$$(1234,\ldots,n) = (1n)(1,n-1)(1,n-2),\ldots,(13)(12).$$

To prove this, we write the cycles in the form of permutations and find their products as above. For example,

$$(16)(15) = \begin{pmatrix} 1 & 6 & 5 \\ 6 & 1 & 5 \end{pmatrix}\begin{pmatrix} 1 & 5 & 6 \\ 5 & 1 & 6 \end{pmatrix} = \begin{pmatrix} 1 & 5 & 6 \\ 6 & 5 & 1 \end{pmatrix}\begin{pmatrix} 5 & 1 & 6 \\ 1 & 5 & 6 \end{pmatrix}$$

$$= \begin{pmatrix} 5 & 1 & 6 \\ 6 & 5 & 1 \end{pmatrix} = \begin{pmatrix} 1 & 5 & 6 \\ 5 & 6 & 1 \end{pmatrix}.$$

This last permutation is the cycle (156).

The decomposition of a cycle into a product of transpositions is not uniquely defined, as it depends on the original arrangement of the elements in the cycle. For example,

$$(156) = (615) = (65)(61).$$

We will, however, mention the following theorem without proving it.

Theorem 2: *The number of transpositions ℓ into which a particular permutation can be decomposed will either always be an odd number or always be an even number.*

In other words, the number $(-1)^\ell$ characterizes a permutation and is referred to as its **parity**.

2.3   A first examination of S_n

Theorem 3: *The set of permutations of n elements constitutes a group containing $n!$ elements. This group is known as symmetric group, and it will be indicated as S_n.*

The proof of this can be seen above. The product of two elements (permutations) has been defined and the inverse constructed. The identity element was also shown to exist. The decomposition into cycles proves that the set is a closed set. Writing a permutation as the product of cycles and the cycles as transpositions facilitates finding the inverse and expressing it in cycles. For example,

$$P = (m_1, m_1')(m_2, m_2') \cdots (m_{k-1} m_{k-1}')(m_k m_k') \Rightarrow$$

$$P^{-1} = (m_k m_k')^{-1}(m_{k-1} m_{k-1}')^{-1} \cdots (m_2, m_2')^{-1}(m_1, m_1')^{-1} \Rightarrow$$

$$P^{-1} = (m_k m_k')(m_{k-1} m_{k-1}') \cdots (m_2, m_2')(m_1, m_1') \Rightarrow PP^{-1} = E.$$

Example 3: Let us consider S_2.

$$E = \begin{pmatrix} 1 & 2 \\ 1 & 2 \end{pmatrix}, P = \begin{pmatrix} 1 & 2 \\ 2 & 1 \end{pmatrix} \Rightarrow EP = PE = P, P^2 = E, P^{-1} = P.$$

Example 4: Let us consider S_3.

The six elements it contains can be seen in Eq. (2.8). If we write

$$E = P_1 = (11)(22)(33), A = P_4 = (123) = (13)(12),$$

$$B = P_5 = (132) = (12)(13),$$

$$C = P_2 = (1)(23), \ D = P_6 = (2)(13), F = P_3 = (23)(3),$$

we find the same Cayley table as that shown in Table 2.7. The two groups are isomorphic. We will later see that Tables 2.5 and 2.6 are, among other things, a representation of the elements of S_3. The reader may confirm that

$$AB = \begin{pmatrix} 1 & 2 & 3 \\ 2 & 3 & 1 \end{pmatrix} \begin{pmatrix} 1 & 2 & 3 \\ 3 & 1 & 2 \end{pmatrix} = \begin{pmatrix} 3 & 1 & 2 \\ 1 & 2 & 3 \end{pmatrix} \begin{pmatrix} 1 & 2 & 3 \\ 3 & 1 & 2 \end{pmatrix} = \begin{pmatrix} 1 & 2 & 3 \\ 1 & 2 & 3 \end{pmatrix}.$$

This last permutation is $E = (11)(22)(33)$. Of course, this can be more easily derived by observing that: $AB = (13)(12)(12)(13) = (11)(22)(33)$. Similarly,

$$DC = (2)(13)(1)(23) = (2)(13)(23)(1) = (2)(31)(32)(1)$$

$$= (2)(321)(1) = (321) = (132) = B,$$

etc.

2.3.1   The A_4 group as a subgroup of S_4

The S_4 group, the group containing the permutations of four elements, is a special case of the group S_n, and as it is of particular interest in physics, it will be studied further in Chapter 11. It consists of 24 elements, which can be arranged into classes as follows:

(1) The identity element:

$$C_1 \Leftrightarrow E = \begin{pmatrix} 1 & 2 & 3 & 4 \\ 1 & 2 & 3 & 4 \end{pmatrix} = (1\,1),\ (2\,2),\ (3\,3),\ (4\,4).$$

(2) Six transpositions:

$$C_2 \Leftrightarrow (1\,2),\ (1\,3),\ (1\,4),\ (2\,3),\ (2\,4),\ (3\,4).$$

(3) Eight cyclic permutations with a length of 3:

$$C_3 \Leftrightarrow (1\,2\,3),\ (3\,2\,1),\ (1\,2\,4),\ (4\,2\,1),\ (1\,3\,4),\ (4\,3\,1),\ (2\,3\,4),\ (4\,3\,2).$$

(4) Six cyclic permutations with a length of 4:

$$C_4 \Leftrightarrow (1\,2\,3\,4),\ (1\,2\,4\,3),\ (1\,3\,2\,4),\ (1\,3\,4\,2),\ (1\,4\,2\,3),\ (1\,4\,3\,2).$$

(5) Three products of commuting transpositions:

$$C_5 \Leftrightarrow (1\,2)(3\,4),\ (1\,3)(2\,4),\ (1\,4)(2\,3).$$

The 12 elements of the classes C_1, C_3, C_5, whose parity is positive, compose a group, $A_4 = C_1 + C_3 + C_5$, and the elements of the classes C_1, C_5 compose another group, an Abelian group of four elements known as A_2 or group 4, as the reader can confirm. The group A_4 is isomorphic to tetrahedral symmetry.[2] The group A_4 has proven very useful lately in particle physics and will be studied elsewhere; see Chapter 13. Also, this group, as well as many other discrete groups, derive from the elegant theory proposed by Heisenberg, Wigner and Weyl, which we will elaborate on in Chapter 14.

2.3.2   Application to the construction of (fully) symmetric and antisymmetric functions

One application of the symmetric group S_n is in the construction of functions involving many particles in quantum mechanics, where the function

---

[2]See point symmetry groups in Section 2.4.2.

must have a definite symmetry under the exchange of any two particles. More specifically, it is necessary for the function to remain the same if the particles are of integer spin, or change sign when their spin is a half-odd integer. Usually, a basis is given in function space, which is the single-particle functions (orbits)

$$\psi_\alpha, \psi_\beta, \ldots, \psi_\omega.$$

The question then is to find the form of the wave function n, where $1, 2, \ldots, n$ is the number of particles. We examine the following cases:

(i) $n = 2$. The symmetry is S_2.

We need only consider any two of the states, e.g. ψ_α, ψ_β. The following cases are possible:

- The two particles have integer spin (they are Bosons). The two particles may then occupy the same state, i.e.

$$\Psi_B(12) = \psi_\alpha(1)\psi_\alpha(2) \quad \text{or} \quad \Psi_B(12) = \psi_\beta(1)\psi_\beta(2),$$

  which is symmetric.

- The particles have integer spin but may be in two states, $\psi_\alpha \neq \psi_\beta$. We now consider the operator $\mathcal{S} = E + (12)$. Clearly, $(12)\mathcal{S} = \mathcal{S}$. Consequently,

$$\Psi_B(12) = \tfrac{1}{\sqrt{2}}\mathcal{S}\left(\psi_\alpha(1)\psi_\beta(2)\right) = \tfrac{1}{\sqrt{2}}\left(\psi_\alpha(1)\psi_\beta(2) + \psi_\alpha(2)\psi_\beta(1)\right) \Rightarrow$$
$$(12)\Psi_B(12) = \Psi_B(12),$$

  which is symmetric.

- The particles have half-odd integer spin (and are therefore Fermions), in which case they must necessarily occupy two states $\psi_\alpha \neq \psi_\beta$. We now consider the operator $\mathcal{A} = E - (12)$. Clearly, $(12)\mathcal{A} = -\mathcal{A}$. As a result,

$$\Psi_F(12) = \frac{1}{\sqrt{2}}\mathcal{A}\left(\psi_\alpha(1)\psi_\beta(2)\right) = \frac{1}{\sqrt{2}}\left(\psi_\alpha(1)\psi_\beta(2) - \psi_\alpha(2)\psi_\beta(1)\right) \Rightarrow$$
$$(12)\Psi_F(12) = -\Psi_F(12),$$

  which is antisymmetric.

(ii) $n = 3$. The symmetry is S_3. The problem is now less trivial.

We need only consider three states, e.g. $\psi_\alpha, \psi_\beta, \psi_\gamma$. The following cases are possible:

- The particles have integer spin, and all occupy the same state; there are then three states:

$$\Psi_B(123) = \psi_\alpha(1)\psi_\alpha(2)\psi_\alpha(3) \quad \text{or} \quad \Psi_B(123) = \psi_\beta(1)\psi_\beta(2), \psi_\beta(3) \quad \text{or}$$

$$\Psi_B(123) = \psi_\gamma(1)\psi_\gamma(2)\psi_\gamma(3),$$

which, obviously, are symmetric.
- The particles have integer spin but occupy three different states. We then start with the state $\psi_\alpha(1)\psi_\beta(2)\psi_\gamma(3)$ and consider the operator

$$S = E + (12) + (13) + (23) + (123) + (213).$$

By looking at the Cayley table of the S_3 group, we see that all the elements are contained in each of its lines. Consequently,

$$(ij)S = S,$$

$$\Psi_B(123) = \frac{1}{\sqrt{6}} S \left(\psi_\alpha(1)\psi_\beta(2)\psi_\gamma(3) \right) \Rightarrow (ij)\Psi_B(123) - \Psi_B(123),$$

where (ij) is any transposition and $\frac{1}{\sqrt{6}}$ is a normalization factor. More directly,

$$\Psi_B(123) = \frac{1}{\sqrt{6}}(\psi_\alpha(1)\psi_\beta(2)\psi_\gamma(3) + \psi_\alpha(2)\psi_\beta(1)\psi_\gamma(3)$$

$$+ \psi_\alpha(3)\psi_\beta(2)\psi_\gamma(1) + \psi_\alpha(1)\psi_\beta(3)\psi_\gamma(2) + \psi_\alpha(2)\psi_\beta(3)\psi_\gamma(1)$$

$$+ \psi_\alpha(3)\psi_\beta(2)\psi_\gamma(1)).$$

In the case that two particles occupy the same state, e.g. $\psi_\alpha = \psi_\beta$, the six addends of the above expression become three, so that

$$\Psi_B(123) = \frac{1}{\sqrt{3}}(\psi_\alpha(1)\psi_\alpha(2)\psi_\gamma(3) + \psi_\alpha(1)\psi_\alpha(3)\psi_\gamma(2)$$

$$+ \psi_\alpha(2)\psi_\alpha(3)\psi_\gamma(1)),$$

and there is a similar expression for $\psi_\alpha \Rightarrow \psi_\gamma$.
- The particles have half-odd integer spin and therefore must occupy three different states. We now consider the operator

$$A = E - (12) - (13) - (23) + (123) + (213).$$

We observe that as long as the decomposing of cycles into transpositions has a defined parity $(ij)A = -A$ (for proof, see problem (2.4.2)).

For example,

$$(12)E = (12), (12)(12) = E, (12)(13) = (132) = (213), (12)(23)$$
$$= (21)(23) = (231) = (123), (12)(123) = (12)(231)$$
$$= (21)(21)(23) = (23),$$
$$(12)(213) = (12)(132) = (12)(12)(13) = (13) \Rightarrow (12)\mathcal{A} = -\mathcal{A}.$$

Consequently,

$$\Psi_F(123) = \frac{1}{\sqrt{6}} \mathcal{A} \left(\psi_\alpha(1)\psi_\beta(2)\psi_\gamma(3) \right) \Rightarrow (ij)\Psi_F(123) = -\Psi_F(123),$$

which is to say

$$\Psi_F(123) = \frac{1}{\sqrt{6}} (\psi_\alpha(1)\psi_\beta(2)\psi_\gamma(3) - \psi_\alpha(2)\psi_\beta(1)\psi_\gamma(3)$$
$$- \psi_\alpha(3)\psi_\beta(2)\psi_\gamma(1) - \psi_\alpha(1)\psi_\beta(3)\psi_\gamma(2) + \psi_\alpha(2)\psi_\beta(3)\psi_\gamma(1)$$
$$+ \psi_\alpha(3)\psi_\beta(2)\psi_\gamma(1)).$$

A detailed analysis of such topics for $n > 3$ makes use of the Young tableaux, see Chapter 11, and, for a more detailed exposition, see Vergados (2017) and Hammermesh (1964).

2.4 Point symmetry groups

These symmetries relate to transformations which maintain the distance between two points of a body and map it onto itself. Intuitively one expects that these transformations will comprise a group. If we perform two such transformations consecutively, we find a result that could be produced by a single transformation. The identity element exists; it is a transformation that leaves the body in its original state. Every such transformation being 1–1 (one-to-one or injective) has an inverse. Obviously, the transformations are also associative.

This geometric concept of transformations may concern not only macroscopic systems but also the building blocks from which such a system can be assembled, such as its molecules.

Symmetry transformations, mathematically, can be separated into two categories:

(i) Rotations about an axis by an angle ϕ.

An example of rotation can be seen in a cube; one axis of rotation is the line connecting the centers of two opposite sides of the cube.

The cube may be rotated around this axis, say clockwise, by an angle $\phi = \frac{\pi}{2}k$, $k = 1, 2, 3, 4$.

In general, in the case of a rotation, there is a maximum integer n such that $\phi = (2\pi)/n$ is the least (minimum) angle of rotation. Consecutive rotations correspond to $\phi_k = \frac{2\pi}{n}k$, $k = 1, 2, \ldots, n$, and are generally written as

$$C_n^k \equiv C_n C_n, \ldots, C_n \text{ (ktimes)} = (C_n)^k \Leftrightarrow \phi_k = \frac{2\pi}{n}k, \ k = 1, 2, \ldots, n.$$

The integer n is termed order of the axis. Of course, $C_n^n = E$. The inverse of a transformation C_n^k corresponds to a rotation angle of $\phi_k' = 2\pi(1 - \frac{n}{k})$. In the case of the aforementioned rotation of a cube, as the order of the rotation was 4, the axis is four-fold. Of course, the cube has additional axes of rotation (see Fig. 2.1).

(ii) Reflection in a plane of symmetry (mirror plane).

An example of this is the left/right symmetry seen in the human body (a look in the mirror can verify its existence). Reflection symmetry is very common in nature; see, for example, Fig. 2.3. Each symmetry plane corresponds to one transformation symbolized as σ. Of course, $\sigma^2 = E$.

One could consider inversion through a center of symmetry as a separate symmetry transformation. In many cases, however, this transformation can be expressed as a special case of rotation by π combined with a reflection in a plane perpendicular to the rotational axis, σ_h, see Fig. 2.2. In the presence of a rotation axis of order n, characterized by a minimal rotation, $C_n = \frac{2\pi}{n}$. From these two, one generates a group of order $g = 2n$:

$$C_n, (C_n)^2, \ldots, (C_n)^{n-1}, (C_n)^n = E, \sigma_h C_n, \sigma_h (C_n)^2, \ldots, \sigma_h (C_n)^{n-1}, \sigma_h.$$

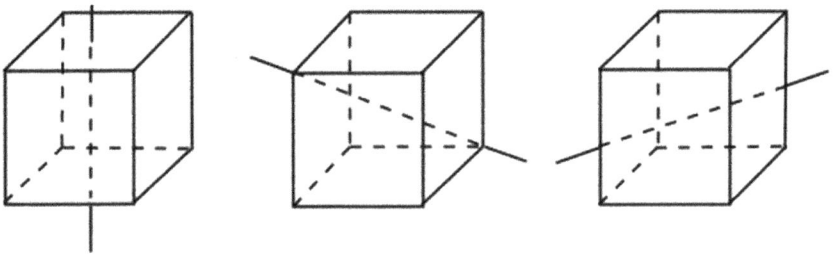

Fig. 2.1. The symmetry axes of a cube: four-fold (left), three-fold (center) and two-fold (right).

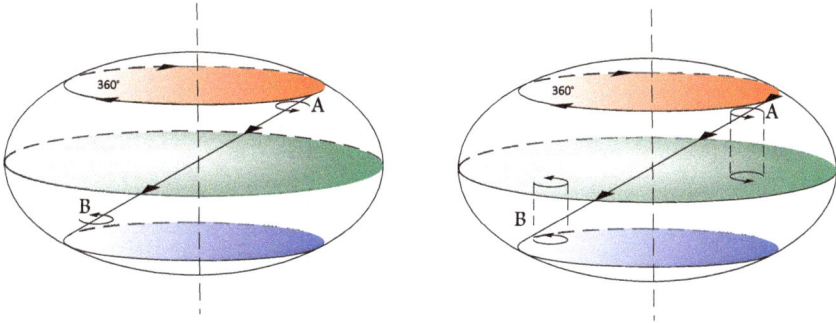

Fig. 2.2. Inversion through a center of symmetry (left) and the equivalent action of a rotation by π and a reflection (right).

As the rotation axis and the reflection plane are perpendicular, rotations and reflections are commutative, $\sigma_h C_n = \sigma_h C_n$, which means that these groups are Abelian. Then, suppose that there exists a rotation $(C_n)^k$ equivalent to an angle π around the axes, i.e.

$$(2\pi)\frac{k}{n} = \pi \Rightarrow \frac{k}{n} = \frac{1}{2}, k, n \text{ integers.} \qquad (2.9)$$

This can occur only if n is even. In such a case, $k = \frac{n}{2}$. Then, $\sigma_h(C_n)^k$ corresponds to a rotation π followed by a reflection, and one can see from Fig. 2.2 that this corresponds to a central inversion, i.e.

$$\sigma_h(C_n)^k = I, k = \frac{n}{2}, n = \text{even.} \qquad (2.10)$$

This symmetry, for obvious reasons, is referred to as rotation–reflection, and the corresponding axis is called a rotation–reflection axis. The inversion symmetry I, associated with rotation–reflection ($n = 2, k = 1$), is possessed by the left and middle shapes in Fig. 2.3. The right shape in this figure does not possess such a symmetry.

Rotation–reflection is very useful in the study of optically active materials, which rotate plane-polarized light (Bishop, 2017).

Translational symmetry (symmetry as to movement in a particular direction), which characterizes a periodic system such as a crystal, will be discussed elsewhere, see Chapter 8.

2.4.1 *The symmetry of ammonia*

This symmetry is known as C_{3v}. It is characterized by rotations $(C_3)^k$, $k = 1, 2, 3; (C_k)^3 = E$, where C_3 is a rotation by $2\pi/3$ around an axis

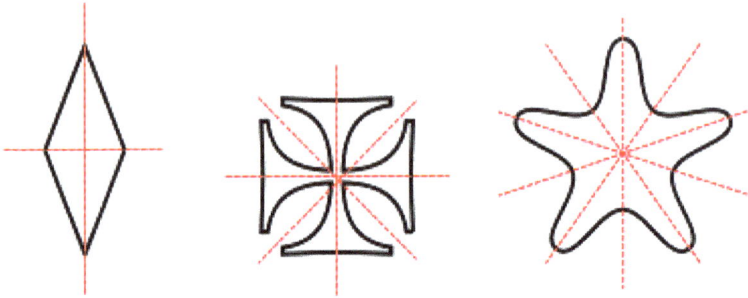

Fig. 2.3. Reflection planes (perpendicular to the plane of the figure) which can be recognized by the line segments denoting the intersections of the mirror planes and the plane of the figure. Two reflection planes in the case of a rhombus (left), four reflection planes in a cross (center) and five in a pentagonoid (right).

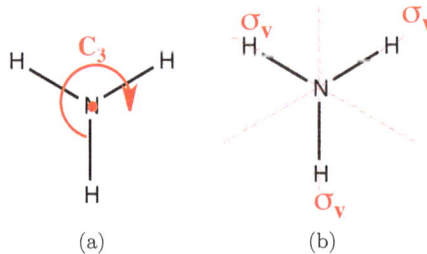

Fig. 2.4. Rotation about an axis perpendicular to the plane of the figure by an angle of $2\pi/3$ (left) and three reflections on the mirror planes perpendicular to the plane of the figure and intersecting with it at NH (right).

perpendicular to the plane defined by the three hydrogen atoms, i.e. perpendicular to the plane of Fig. 2.4(a), and three reflections $\sigma_{1v}, \sigma_{2v}, \sigma_{3v}$. The three reflection planes lie perpendicular to the plane of Fig. 2.4(a), and intersect with it at OH_1, OH_2 and OH_3, respectively, O being the point of intersection between the rotational axis and the plane the figure lies on. In Fig. 2.4(a), O coincides with the center of the nitrogen atom N. The N atom does not, however, actually lie on the same plane as $H_1H_2H_3$ but at a different point of the rotational axis. For this reason, the plane containing the hydrogen atoms is not a reflection plane.

Thus, the C_{3v} symmetry characterizes a right pyramid whose base is an equilateral triangle. In any case, the elements of C_{3v} are

$$C_{3v} : E, C_3, (C_3)^2, \sigma_{1v}, \sigma_{2v}, \sigma_{3v}.$$

Of course, $(\sigma_{1v})^2 = (\sigma_{2v})^2 = (\sigma_{3v})^2 = E$. It is also relatively easy to see that

$$\sigma_{1v}\sigma_{2v} = (C_3)^2, \sigma_{2v}\sigma_{1v} = C_3, \sigma_{1v}\sigma_{3v} = C_3, \sigma_{3v}\sigma_{1v}$$
$$= (C_3)^2, \sigma_{2v}\sigma_{3v} = (C_3)^2,$$
$$\sigma_{3v}\sigma_{2v} = C_3.$$

With a little more difficulty, we can also show that

$$\sigma_{1v}C_3 = \sigma_{2v}, \; C_3\sigma_{1v} = \sigma_{3v}, \; \sigma_{2v}C_3 = \sigma_{3v}, \; C_3\sigma_{2v} = \sigma_{1v},$$

etc. (see Figure 2.5). Figure 2.5(a) shows clearly that the final result when combining one reflection, σ_1, and one rotation, C_3, is that H_1 and H_2 exchange positions—the same result that one sees if the reflection σ_3 is applied to the original state.

We can similarly show that

$$\sigma_{1v}(C_3)^2 = \sigma_{3v}, (C_3)^2\sigma_{1v} = \sigma_{2v}, \sigma_{2v}(C_3)^2 = \sigma_{1v}, (C_3)^2\sigma_{2v} = \sigma_{3v},$$
$$\sigma_{3v}(C_3)^2 = \sigma_{2v}, (C_3)^2\sigma_{3v} = \sigma_{2v}, \sigma_{3v}C_3 = \sigma_{2v}, C_3\sigma_{3v} = \sigma_{2v}.$$

The associative property holds true: $\sigma_{2v}(C_3\sigma_{1v}) = \sigma_{2v}\sigma_{3v} = (C_3)^2$, $(\sigma_{2v}C_3)\sigma_{1v} = \sigma_{3v}\sigma_{1v} = (C_3)^2$, etc. Thus, we may construct the Cayley table:

	E	C_3	$(C_3)^2$	σ_{1v}	σ_{2v}	σ_{3v}
E	E	C_3	$(C_3)^2$	σ_{1v}	σ_{2v}	σ_{3v}
$(C_3)^2$	$(C_3)^2$	E	C_3	σ_{2v}	σ_{3v}	σ_{1v}
C_3	C_3	$(C_3)^2$	E	σ_{3v}	σ_{1v}	σ_{2v}
σ_{1v}	σ_{1v}	σ_{2v}	σ_{3v}	E	$(C_3)^2$	C_3
σ_{2v}	σ_{2v}	σ_{3v}	σ_{1v}	C_3	E	$(C_3)^2$
σ_{3v}	σ_{3v}	σ_{1v}	σ_{2v}	$(C_3)^2$	C_3	E

By writing

$$A \Leftrightarrow C_3, B \Leftrightarrow (C_3)^2, C \Leftrightarrow \sigma_{1v}, D \Leftrightarrow \sigma_{2v}, F \Leftrightarrow \sigma_{3v},$$

we can see that the Cayley table coincides with that in Table 2.7. Therefore, the two groups, C_{3v} and the group containing the matrices, are isomorphic. We will see later in Chapter 3 that those matrices are what one calls a representation of the group C_{3v}.

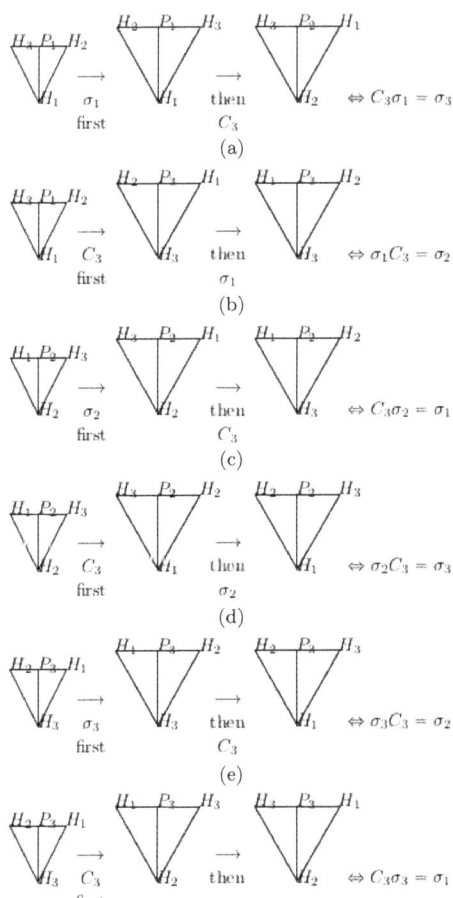

Fig. 2.5. Geometric determination of the multiplication table of the symmetry C_{3v}.

Note: the reflection plane does not change position (rotate) when a rotation is applied to the molecule!

2.4.2 Some categories of point groups

Point groups have important applications in molecular physics and crystal structure. The symmetry transformations found in them are rotations about axes and reflections in planes. The position of a reflection plane is correlated in some way to the axes; the plane may be perpendicular to a rotation axis or may contain such an axis, see Fig. 2.6.

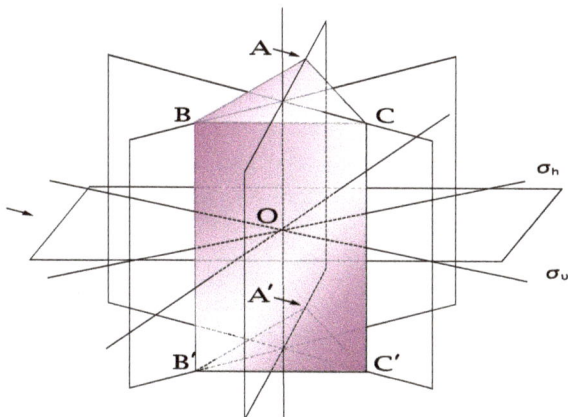

Fig. 2.6. Various symmetry operations: Rotation axes of different orders and reflections in planes both horizontal and vertical, relative to the main (highest order) axis, σ_h and σ_v, respectively.

The point groups are usually denoted using the Schoenflies notation system, and can be divided into the following 12 categories:

(1) C_n groups. These consist of the elements $C_n^k = (C_n)^k$, which correspond to rotations around a given axis by an angle $\phi_k = (2\pi/n)k$, $k = 1, 2, \ldots, n, n \geq 2$:

$$C_n : C_n, (C_n)^2, \ldots, (C_n)^{n-1}, (C_n)^n = E.$$

Each such group contains $g = n$ elements. Crystallographically, these groups are of interest when $n = 1, 2, 4, 6$ (see Table 8.1).

(2) C_{nh} groups. These groups contain all the elements of C_n as well as a reflection σ_h in a plane perpendicular to C_n. As the rotation axis and reflection plane are perpendicular, rotations and reflections are commutative, $\sigma_h C_n = C_n \sigma_h$, which means that these groups are Abelian. The elements of the group C_{nh} are:

$$C_n, (C_n)^2, \ldots, (C_n)^{n-1}, (C_n)^n = E, \sigma_h C_n, \sigma_h (C_n)^2, \ldots, \sigma_h (C_n)^{n-1}, \sigma_h.$$

Thus, the order of the group is $g = 2n$. We have seen from Eq. (2.10), that if n is an even number, there is inversion symmetry I (central inversion).

A molecule with such a symmetry may be in the shape of square, regular hexagon, equilateral triangle, etc., but may also be in the shape of, for example, a rhomboid or a star. The molecule needn't be flat; however, for any atom lying a certain distance above the plane of reflection, there must be another atom of the same type lying at the same distance below that plane.

This is not a simple group, as it can be written as a direct product (in the sense described in Section 3.7) as follows:

$$C_{nh} = C_n \otimes \sigma_h. \tag{2.11}$$

Of interest in crystallography are those groups with $n = 2, 4, 6$ (see Table 8.1). For a set of their generators, see Tables 17.1–17.3.

(3) C_{nv} groups. Here, besides the n-order rotation axis, there are also reflection planes that contain that axis. This is possible when the planes intersect at an angle of π/n. These groups are not Abelian; in fact, $(C_n)^k \sigma_v = \sigma_v (C_n)^{n-k} \neq \sigma_v (C_n)^k$. One such point group is C_{3v}, which we studied above. We encourage the reader to construct the Cayley table for C_{4v}, the symmetry of a right pyramid with a square base (see Bishop (2017)).

If a molecule has such symmetry, its elements will generally be arranged symmetrically in, for example, a square or a regular hexagon, but any part of the molecule lying above the plane containing this shape need not be symmetric to what lies below it.

In crystallography, the groups of interest are those with $n = 2, 3, 4, 6$ (see Table 8.1). A system of generators is given in Tables 17.1–17.3.

(4) S_{2n} groups. Not to be confused with the symmetric group for an even number of objects seen in Section 2.2, S_{2n} contains rotation–reflection symmetry (see Fig. 2.2) with an axis of even order. The elements of such a group are:

$$S_{2n}, (S_{2n})^2, \ldots, (S_{2n})^{2n-1}, (S_{2n})^{2n} = E.$$

Of interest in crystal structure are the point groups with $n = 2, 3$ (see Table 8.1 and a system of generators in Tables 17.1–17.3). The case $n = 3$ is discussed in Section 4.7.3.

(5) D_n groups. This group contains all the elements of C_n of an n order axis and, in addition, n rotational axes of order 2 perpendicular to it. This group is isomorphic to C_{nv}.

Of interest in crystallography are those groups where $n = 2, 3, 4, 6$ (see Table 8.1. For a system of their generators, see Tables 17.1–17.3).

(6) D_{nh} groups. This characterizes the symmetry of a regular n-sided prism. In addition to the rotational elements found in D_n, D_{nh}

includes reflection planes containing the axes of rotation, not only the n-order rotational axis but also the second-order axes. This group is not a simple group; it can be written as a direct product (in the sense of Section 3.7), i.e.

$$D_{nh} = \begin{cases} D_n \otimes \sigma_h, & n = \text{even} \\ D_n \otimes I, & n = \text{odd} \end{cases}, \quad I = \text{central inversion}. \qquad (2.12)$$

Of interest in crystal structure are the groups with $n = 2, 4, 6$ (see Table 8.1 and a system of their generators in Tables 17.1–17.3).

(7) D_{nd} groups. Here, we find the symmetry of a body composed of two regular n-gonal prisms rotated at a relative angle of π/n. When n is an odd number, $D_{nd} = D_n \otimes I$. Of interest crystallographically are those groups with $n = 2, 3$ (see Table 8.1 and a system of their generators in Tables 17.1–17.3).

(8) T (tetrahedral) group. Here, we find all those rotational symmetries which characterize a regular tetrahedron (Fig. 2.7):

- Three-fold axes through one vertex and the center of the opposite face. There are four such axes ($4 \times 2 = 8$ elements).
- Two-fold axes, each going through the centers of two non-intersecting edges. There are three such elements.
- The identity element.

(See Fig. 2.7.)

The order of this group is therefore 12 ($g = 12$), and it is isomorphic to A_4. This symmetry is a subsymmetry of the cubic one (Fig. 2.11).

(9) T_h group. This group contains all the elements of T as well as the central inversion I, i.e.

$$\{X_k, IX_k, X_k \in T, I = \text{central inversion}, g = 24\}.$$

This group is of interest in crystal structure (see Table 8.1 and a system of its generators in Table 17.3).

(10) T_d group. This group contains all the elements of T. In addition, it contains the symmetry elements:

- Planes, each defined by two of the three-fold axes. There are $6 \times 1 = 6$ such planes.
- Rotation–reflection second-order axes. There are six such elements. The group is therefore of order $g = 24$.

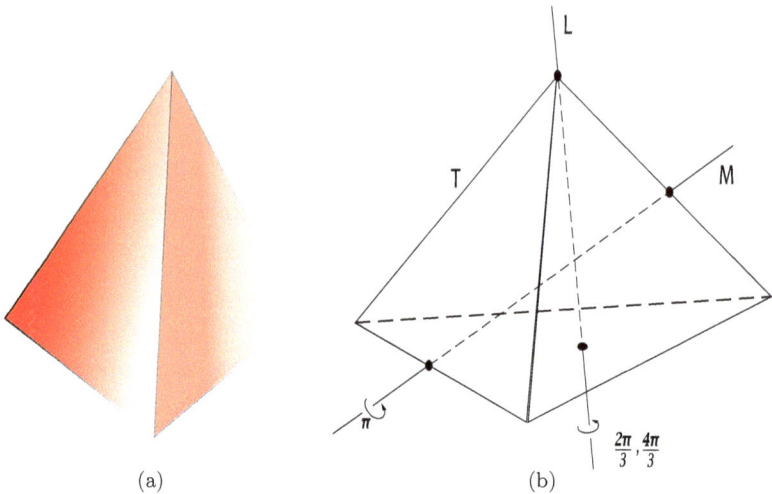

(a)   (b)

Fig. 2.7. A regular tetrahedron (a) and its symmetry axes (b). There are four 3-fold axes, L, each going through one vertex and the center of the opposite face, and three 2-fold axes, M, each going through the centers of two non-intersecting edges.

This group is of interest in crystallography (cubic crystal system) (see Table 8.1). A system of its generators can be found in Table 17.3.

(11) O (octahedral) group. The symmetry axes of the octahedral group are exhibited in Fig. 2.8. This group is of interest in crystallography (cubic crystal system) (see Table 8.1 and a system of its generators in Table 17.3).

It contains all the rotational symmetries:

- Three 4-fold axes through the opposite vertices (corners) of the octahedron (see Fig. 2.8(a)). There are six such elements.
- Four 3-fold axes through the centers of the opposite triangular faces (see Fig. 2.8(b)). There are eight such elements.
- Three 2-fold axes where the C_4 axes run (see Fig. 2.8(a)). There are three such elements.
- Six 2-fold axes each of which goes through two opposite edges of the octahedron (see Fig. 2.8(c)). There are six such elements.
- The identity element.

Therefore, this group is of order $g = 24$.

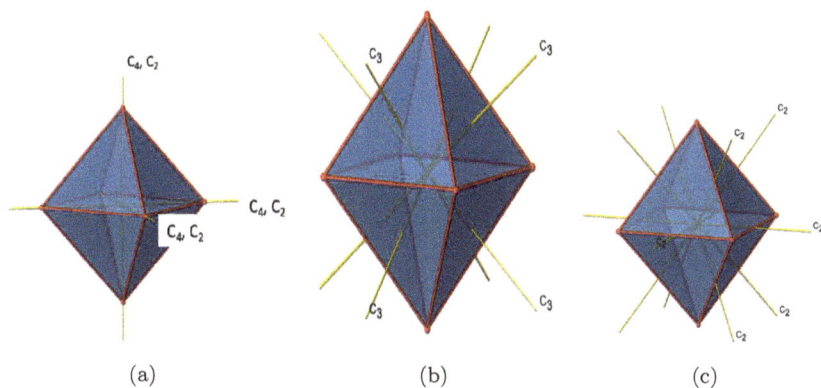

Fig. 2.8. The symmetry axes of the octahedral group O.

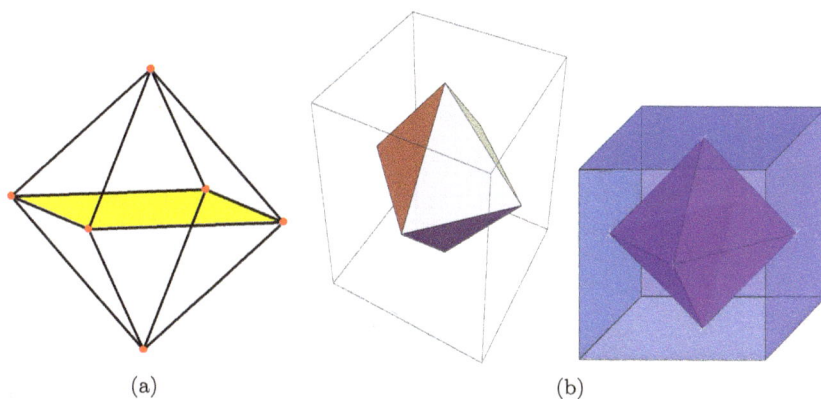

Fig. 2.9. (a) A regular octahedron and (b) the octahedral symmetry as a subsymmetry of cubic symmetry.

The octahedral symmetry is of interest in crystallography (cubic crystal system) (see Table 8.1 and a system of its generators can be seen in Tables 17.1–17.3).

The tetrahedral and octahedral groups are subgroups of the cubic symmetry, see Fig. 2.9. This is more explicitly exhibited in Figs. 2.10 and 2.11.

(12) O_h. This contains all the symmetry operations of a cube. That is, besides all the elements of O, it also contains the element of inversion

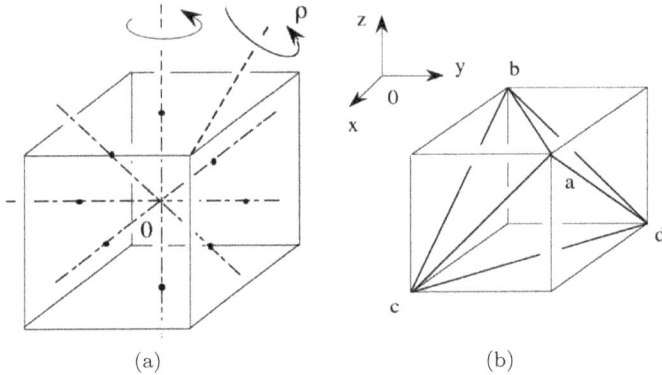

Fig. 2.10. (a) The symmetry axes of a cube. The tetrahedron inscribed in a cube. The faces of the tetrahedron are (abc), (abd), (acd) and (bcd).

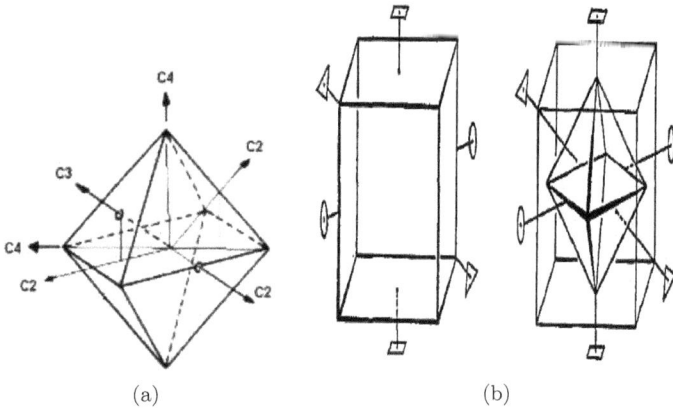

Fig. 2.11. (a) The octahedral symmetry axes C_4, C_3 and C_2. (b) The octahedral as a sub-symmetry of cubic symmetry.

through the center of the cube. Thus,

$$\{X_k, \, IX_k, X_k \in O, I = \text{central inversion}, g = 48\}.$$

This group is of crystallographic interest (cubic crystal system), see Table 8.1. A system of its generators can be found in Table 17.3.

Other symmetries of theoretical interest are those of the Platonic solids (regular polyhedra first studied in ancient Greek times) not discussed above, i.e. the regular dodecahedron and the regular icosahedron, see Fig. 2.12.

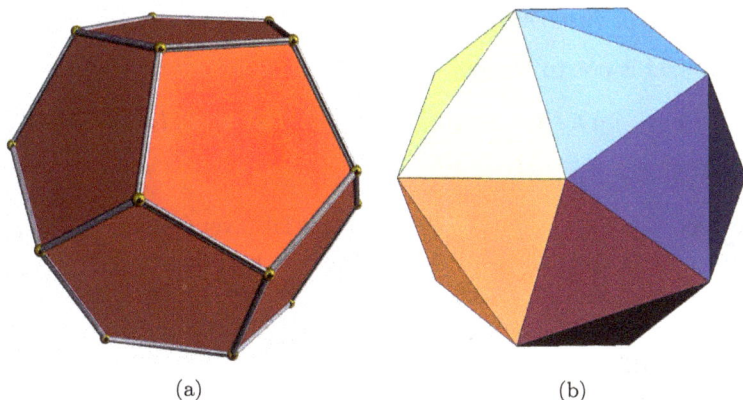

(a)   (b)

Fig. 2.12. Two further regular polyhedra studied by ancient Greek philosophers: the dodecahedron and the icosahedron.

2.5 Extended point symmetry groups

2.5.1 *Time reversal*

If we wish to be thorough, we must mention that in the study of magnetic materials, it is useful to enrich certain of the above symmetries (Nowick, 1995) with **time reversal**, i.e. the transformation[3] $t \to -t$. The classical equations of motion remain unchanged as the force is time- invariant. In quantum mechanics, this is described by an operator containing a factor Θ, which acting upon a wave function produces its complex conjugate (Vergados, 2018). These new transformations are described by an **antiunitary** transformation $R = U\Theta$, where U is a unitary transformation and $\Theta = \Theta^{-1}$. U and Θ refer to two separate actions and are commutative $U\Theta = \Theta U$. Some interesting cases of transformations are

$$R|\mathbf{r}\rangle = |\mathbf{r}\rangle,\ R|\mathbf{p}\rangle = -|\mathbf{p}\rangle,\ R|\mathbf{L}\rangle = -|\mathbf{L}\rangle,\ R|\mathbf{s}\rangle = -|\mathbf{s}\rangle,\ R|\mathbf{J}\rangle = -|\mathbf{J}\rangle,$$

$$R|\mathbf{E}\rangle = |\mathbf{E}\rangle,\ R|\mathbf{B}\rangle = -|\mathbf{B}\rangle \tag{2.13}$$

for the operators of position, linear momentum, orbital angular momentum, spin, total angular momentum, electric field and magnetic field, respectively. Of particular interest in the study of magnetic materials is the above behavior of spin.

----

[3]This section may be omitted when first perusing this book.

2.5.2 *Central inversion*

Time reversal is not to be confused with central inversion (i or I):

$$I|\mathbf{r}\rangle = -|\mathbf{r}\rangle, I|\mathbf{p}\rangle = -|\mathbf{p}\rangle, I|\mathbf{L}\rangle = |\mathbf{L}\rangle, I|\mathbf{s}\rangle = |\mathbf{s}\rangle, I|\mathbf{J}\rangle = |\mathbf{J}\rangle,$$

$$I|\mathbf{E}\rangle = -|\mathbf{E}\rangle, I|\mathbf{B}\rangle = |\mathbf{B}\rangle. \tag{2.14}$$

Those vectors which change sign under spatial inversion are called **polar vectors** or true vectors, while those that do not change sign are called **pseudovectors** or **axial vectors**. In the extension under discussion here, both transformations play a part. For an instructive approach to this topic, see (Graef, 2012).

Let us consider as an example the symmetry group C_{2h} with the elements $\{E, C_2, \sigma_h, I\}$. The action of the required operators on an axial vector can be seen in Fig. 2.13. The group must therefore be expanded to include the elements $\{R, C_2', \sigma_h', I'\}$. The extended group is

$$\{E, C_2, \sigma_h, I, R, C_2', \sigma_h', I'\}$$

The reader may confirm that the above elements comprise a group, known as C_{2h}' or gray C_{2h} or $2/m, 1'$ ($1' = R$) in the Hermann–Mauguin (international notation, see Section 8.4.1) system.

Of the 32 crystallographic groups, only

$$E, C_2, C_{2h}, C_3, C_4, S_4, C_{4h} C_6, S_6, C_{6h}$$

can be thus extended. There are other extensions of this type (Nowick, 1995) leading to 90 point groups, but the discussion of that topic is beyond the scope of this book.

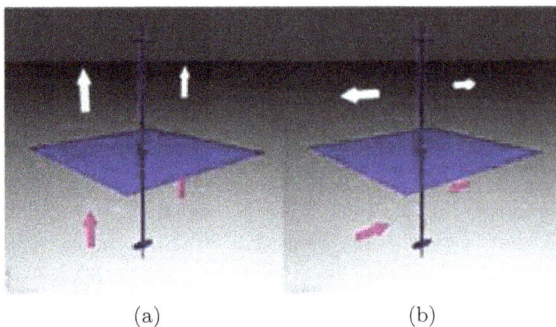

(a)   (b)

Fig. 2.13. Representation of the action of an axial vector lying parallel to the symmetry axis (left) and perpendicular to it (right).

2.6 Problems

2.2.1 Show that the group $G = \{1, -1, i, -i\}$ is isomorphic to the group in the following table:

	E	A	B	C
E	E	A	B	C
A	A	E	C	B
C	C	B	E	A
B	B	C	A	E

2.2.2 Examine whether the elements $Z^m, m =$ integer, $Z \neq 0$ comprise a group with usual multiplication as the operation. If so, of what order is the group?

2.2.3 Examine whether a group is formed by the elements

$$e^{2\pi i(m/n)}, \; m = 1, 2, \ldots, n-1, n \text{ integer} > 1.$$

2.2.4 Examine whether the integers, equipped with addition, form a group.

2.2.5 Show that the set of real numbers comprise a group as to addition. Is this set a group as to multiplication?

2.2.6 Examine whether, equipped with matrix multiplication, the following matrices comprise a group:

$$\pm \begin{pmatrix} 1 & 0 \\ 0 & 1 \end{pmatrix}, \pm \begin{pmatrix} 1 & 0 \\ 0 & -1 \end{pmatrix}, \pm \begin{pmatrix} 0 & 1 \\ -1 & 0 \end{pmatrix}, \pm \begin{pmatrix} 0 & 1 \\ 1 & 0 \end{pmatrix}.$$

If so, construct the Cayley table of that group.

2.2.7 Examine whether the matrices

$$A = \begin{pmatrix} 1 & 0 & 0 & 0 \\ 0 & 1 & 0 & 0 \\ 0 & 0 & 1 & 0 \\ 0 & 0 & 0 & 1 \end{pmatrix}, B = \begin{pmatrix} 0 & 0 & 0 & 1 \\ 1 & 0 & 0 & 0 \\ 0 & 1 & 0 & 0 \\ 0 & 0 & 1 & 0 \end{pmatrix}, C = \begin{pmatrix} 0 & 0 & 1 & 0 \\ 0 & 0 & 0 & 1 \\ 1 & 0 & 0 & 0 \\ 0 & 1 & 0 & 0 \end{pmatrix}.$$

form a group. If so, construct the Cayley table. If not, complete the set with the minimum number of matrices so that the new set comprises a group and construct its Cayley table. Is it Abelian? Of what order is it?

2.2.8 Examine whether the matrix that follows is an element of the previous group:

$$\gamma_5 = \begin{pmatrix} 0 & 0 & 0 & 1 \\ 0 & 0 & 1 & 0 \\ 0 & 1 & 0 & 0 \\ 1 & 0 & 0 & 0 \end{pmatrix}.$$

If not, does a new group arise by including it?

2.2.9 If the previous set comprises a group, is that group isomorphic to the group in problem 2.2.6?

2.2.10 Consider the functions

$$f_1(x) = x,\ f_2(x) = 1 - x,\ f_3(x) = \frac{x}{x-1},\ f_4(x) = \frac{1}{x},$$
$$f_5(x) = \frac{1}{x-1},\ f_6(x) = \frac{x-1}{x},$$

with the usual composition of functions as an operation:

$$(f_5 f_4)(x) = f_5(f_4(x)) = f_5(1/x) = \frac{1}{1 - 1/x}$$
$$= \frac{x}{x-1} = f_3(x) \Leftrightarrow f_5 f_4 = f_3,$$

etc. Show that

- $f_3^{-1} = f_6 \Rightarrow f_6^{-1} = f_3,\ f_4^{-1} = f_1,\ f_2^{-1} = f_5, \text{etc.},$
- the above set comprises a group.

Is this group isomorphic to the group in Example 2?

2.3.1 Examine whether the permutations

$$P_E = \begin{pmatrix} 1 & 2 & 3 \\ 1 & 2 & 3 \end{pmatrix},\ P_A = \begin{pmatrix} 1 & 2 & 3 \\ 2 & 3 & 1 \end{pmatrix},\ P_B = \begin{pmatrix} 1 & 2 & 3 \\ 3 & 1 & 2 \end{pmatrix}$$

comprise a group.

2.3.2 Examine whether the cyclic permutations of four elements with the identity permutation compose a group. If so, what is its order? Is it Abelian?

2.4.1 Show that

$$S = E + (12) + (13) + (23) + (123) + (132) \Rightarrow (ij)S = S,$$

$$i \neq j, \ i, j = (1, 2, 3).$$

2.4.2 Show that

$$A = E - (12) - (13) - (23) + (123) + (132) \Rightarrow (ij)A = -A,$$

$$i \neq j, \ i, j = (1, 2, 3).$$

2.4.3 Find the subgroup of S_4 which leaves the following expression unchanged:

$$x(1)x(2) + x(3) + x(4).$$

2.4.4 Do the same for the expression

$$x(1)x(2) + x(3)x(4).$$

Does it contain that found in problem 2.4.3?

2.5.1 Find all the symmetry elements of a square.

2.5.2 Find all the symmetry elements of a rectangle.

2.5.3 Find all the symmetry elements of a rhombus.

2.5.4 Find the symmetry elements of the group C_{4v} and construct its Cayley table. The use of a diagram similar to Fig. 2.5 for a square may be helpful.

2.5.5 Construct the Cayley table for D_3. Is D_3 isomorphic to C_{3v}?

2.5.6 Construct the Cayley table for D_4. Is D_4 isomorphic to C_{4v}?

2.5.7 The planar molecule $C_2H_2Cl_2$ can be found in various forms:

(a) In one form, the two Cl atoms lie on the same long edge of the approximate rectangle formed by them and the two H atoms (cis-$C_2H_2Cl_2$). The two carbon atoms lie near the center of the rectangle on a line parallel to that formed by the Cl atoms. Show that the symmetry of this molecule is C_{2v} with the elements $\{E, C_2, \sigma_{1v}, \sigma_{2v}\}$

and the following Cayley table:

	E	C_2	σ_{1v}	σ_{2v}
E	E	C_2	σ_{1v}	σ_{2v}
C_2	C_2	E	σ_{2v}	σ_{1v}.
σ_{1v}	σ_{1v}	σ_{2v}	E	C_2
σ_{2v}	σ_{2v}	σ_{1v}	C_2	E

(b) Another form is when the molecule is otherwise the same, but the two Cl atoms lie on the same diagonal (on opposite corners) of the rectangle, with the remaining two corners taken up by the hydrogen atoms (trans-$C_2H_2Cl_2$). The symmetry elements are now $\{E, C_2, I, \sigma_h\}$. Show that the Cayley table for C_{2h} is

	E	C_2	I	σ_h
E	E	C_2	I	σ_h
C_2	C_2	E	σ_h	I.
I	I	σ_h	E	C_2
σ_h	σ_h	I	C_2	E

Is the group Abelian?

Is C_{2v} isomorphic to C_{2h}?

(c) As in (a), but the two Cl atoms lie at the corners on the shorter edge of the rectangle, that is, the edge that is perpendicular to the line through the two carbon atoms (1,1-dichloroethylene). Identify the symmetry elements, and construct the Cayley table for this molecule.

2.5.8 Construct the Cayley table for C_{3h}.

2.5.9 Construct the Cayley table for the tetrahedral group T.

2.5.10 Identify all the symmetry elements of a cube.

2.5.11 Construct the Cayley table for D_6. Is D_6 isomorphic to C_{6v}? Is it isomorphic to C_{3h}? To T?

2.5.12 A rotation axis AA$'$ is called two-sided if there is a rotation that transforms it to its opposite, i.e. the axis A$'$A. Show that if there is a symmetry plane lying perpendicular to an axis, that axis is not two-sided. Show also that if, on the other hand, there is a symmetry plane containing the axis, the axis may still be two-sided.

2.5.13 Examine which of the axes in Exercises 2.5.1–2.5.11 are two-sided.

Chapter 3

Discrete Groups — Basic Theorems

3.1 Cayley's reordering theorem

From the multiplication tables considered in Chapter 2, one may conclude the following:

- Every row and column contains a given element only once.
- Each row or column contains all the elements of the group in some order.

The above can be stated more accurately as follows.

Theorem 1: *Given a group G of order $g < \infty$ with elements $E, A_1, A_2, \ldots, A_{g-1}$, the sequences*

$$EA_i, A_1 A_i, \ldots, A_{g-1} A_i \quad \text{and} \quad A_i E, A_i A_1, \ldots, A_i A_{g-1},$$

$$i = 1, 2, \ldots, g - 1,$$

are simply a reordering of the elements of the group G.

Proof: Suppose that in one row, e.g. A, which corresponds to the element A, the element D is found twice, e.g. in columns B and C associated with the elements B, C such that $B \neq C$. Then,

$$AB = D, \quad AC = D,$$

and hence,

$$B = A^{-1}D, \quad C = A^{-1}D \Rightarrow B = C,$$

which is absurd.

Since all the elements are g and each can be placed in one of the g positions, each can appear only once. Furthermore. $A \neq E \Rightarrow AB \neq B$ and $B \neq E \Rightarrow AB \neq A$.

Application 1: We will find all possible groups of order $g = 3$.

Let the elements of the group be E, A, B. The first row is trivial, containing the elements E, A, B. Suppose that the second row begins with the element A. Then, the second element $A^2 = E$ or $A^2 = B$. The first choice leaves no room for the element AB, since excluded are the cases $AB = E$, $AB = B$, $AB = A$. The second choice $A^2 = B$ leads to the choice $AB = E$, i.e. to $A^3 = E$, which is the acceptable solution. Thus, a discrete group of three elements is unique with elements $A, A^2, A^3 = E$, and thus, it is Abelian.

Application 2: Find all discrete groups of order $g = 4$.

One of the elements is the identity E. Suppose the other elements are A, B, C. The first row and the first column are obvious. To proceed further we consider the cases:

(i) $A^2 = E$:

Since the elements of row A are all different, we should have $AB \neq A$, and since $A \neq E$, it follows that $AB \neq B$. As a result, since the elements of the group are 4, $AB = C$ and $AC = B$. This way the second row is complete.

From the second column, in a similar fashion, we find $BA = C$ and $CA = B$. Proceeding to the third row, we have the following options:

(1) $B^2 = A$. Then,

$$B^2 = A \Rightarrow BC = E \Rightarrow CB = E \Rightarrow$$

$$C^2 = A \Rightarrow C = B^3 \Rightarrow B^4 = E.$$

In other words, the group is Abelian with elements E, A, A^2, A^3, $A^4 = E$. The multiplication table is shown in Table 3.1(b).

(2) $B^2 = E$.

Then $BC = A$, hence $CB = A$ and $C^2 = E$. The matrix is complete as shown in Table 3.1(a). Since those two tables are different these two groups are not isomorphic.

(ii) Start with $A^2 \neq E$:

This does not give anything new. Since it coincides with the group 3.1(b) discussed above.

We therefore conclude that all the elements of the group can come from a subset of the elements of the group; in the Abelian case from one element, as shown in Table 3.1(b), while in the case of Table 3.1(a), from the elements

Table 3.1.

	E A B C			E	A	A^2	A^3
E	E A B C			E	E	A	A^2	A^3
A	A E C B			A	A	A^2	A^3	E
B	B C E A			A^2	A^2	A^3	E	A
C	C B A E			A^3	A^3	E	A	A^2

(a)   (b)

A and B since $AB = C$ and $A^2 = E$. The last group is known as the Klein 4 group. Both groups are Abelian.

The set that contains the minimum number of elements that can generate all the elements of the group is called **set of generators** of G and the elements of this set are called **generators** of G. In the case of the group C_{3v} considered in Chapter 2, the generators are the rotation C_3 and one reflection, e.g. σ_1. In the case of the group S_n the generators are the transpositions $(12), (1,3), \ldots, (1,n)$.

The minimum integer k such that $A^k = E$ is called order of the element A of G. In case the group has a generator with order g, i.e. $A^g = E$, the group is Abelian.

Theorem 2 (Cayley's theorem): *Every discrete group G of order g is isomorphic to some subgroup of the symmetric group S_g,* see Section 2.2.

This theorem is very important from a mathematical perspective since the symmetric group has been extensively studied. From a practical point of view, it is of limited importance since the study of the symmetric group is by no means simple.

Proof: Suppose that g elements of the group G are numbered and ordered as follows:

$$Q_1 : \{A_1 = E, A_2, \ldots, A_g\}.$$

Let A_i be a specific element of G. Then, the set

$$Q_2 : \{A_i A_1, A_i A_2, \ldots, A_i A_g\}$$

is just a mere reordering of the elements of Q_1. Consider the 1–1 correspondence

$$A_i \Leftrightarrow P_{A_i} = \begin{pmatrix} A_1 & A_2 & \cdots & A_g \\ A_i A_1 & A_i A_2 & \cdots & A_i A_g \end{pmatrix}, \quad i = 1, 2, \ldots, g. \qquad (3.1)$$

Then, by definition,

$$A_i A_j \Leftrightarrow P_{A_i A_j} = \begin{pmatrix} A_1 & A_2 & \cdots & A_g \\ A_i A_j A_1 & A_i A_j A_2 & \cdots & A_i A_j A_g \end{pmatrix}, \quad i, j = 1, 2, \ldots, g.$$

(3.2)

It remains to be shown that

$$P_{A_i A_j} = P_{A_i} P_{A_j}.$$

Recall that according to the discussion in Section 2.2, the horizontal translation of the pairs of a permutation is without any consequence. Thus,

$$P_{A_i} = \begin{pmatrix} A_1 & A_2 & \cdots & A_g \\ A_i A_1 & A_i A_2 & \cdots & A_i A_g \end{pmatrix}$$

$$P_{A_i} = \begin{pmatrix} A_j A_1 & A_j A_2 & \cdots & A_j A_g \\ A_i(A_j A_1) & A_i(A_j A_2) & \cdots & A_i(A_j A_g) \end{pmatrix}.$$

But

$$P_{A_j} = \begin{pmatrix} A_1 & A_2 & \cdots & A_g \\ A_j A_1 & A_j A_2 & \cdots & A_j A_g \end{pmatrix} \Rightarrow$$

$$P_{A_i} P_{A_j} = \begin{pmatrix} A_j A_1 & A_j A_2 & \cdots & A_j A_g \\ A_i A_j A_1 & A_i A_j A_2 & \cdots & A_i A_j A_g \end{pmatrix}$$

$$\times \begin{pmatrix} A_1 & A_2 & \cdots & A_g \\ A_j A_1 & A_j A_2 & \cdots & A_j A_g \end{pmatrix}$$

$$= \begin{pmatrix} A_1 & A_2 & \cdots & A_g \\ A_i A_j A_1 & A_i A_j A_2 & \cdots & A_i A_j A_g \end{pmatrix} \Rightarrow P_{A_i} P_{A_j} = P_{A_i A_j}.$$

Thus, the above 1–1 correspondence preserves the product, and the theorem is proved.

As a simple application, let us consider the case of G with three elements, which is Abelian. This is isomorphic to a subgroup of S_3 with the elements $\{E, P_A = (123), P_B = (132)\}$. Indeed, it can easily be seen that $P_A P_B = E$ and $P_A^2 = P_B$, which implies $(P_A)^3 = E$, i.e this is Abelian as well.

We have also seen in Chapter 2 that the tetrahedral group is isomorphic to A_4, i.e. the group of even permutations, and clearly a subgroup of S_4.

3.2   Conjugate elements and classes of conjugate elements

Given two elements A and B of a group G, one can define the element $A^{-1}BA$. This belongs to G, i.e. there exists an element $C \in G$ such that $C = A^{-1}BA$. Furthermore,

$$C = A^{-1}BA \Rightarrow B = ACA^{-1}.$$

Two elements B and C of a group G connected this way are called **conjugate**. The transformation that connects them is known as similarity transformation. The elements thus connected are also called similar. If the element B is conjugate to C and C is conjugate to D, then B is conjugate of D. In other words, there exist elements $X \in G$ and $Y \in G$ such that

$$C = XBX^{-1} \quad \text{and} \quad C = YDY^{-1}.$$

Then, however,

$$C = XBX^{-1} = YDY^{-1} \Rightarrow Y^{-1}XBX^{-1}Y = D.$$

Then, setting $A = Y^{-1}X$, we obtain:

$$A^{-1} = (Y^{-1}X)^{-1} = (X^{-1})(Y^{-1})^{-1} = X^{-1}Y \Rightarrow$$

$$D = ABA^{-1} \Rightarrow B = A^{-1}DA.$$

This procedure allows us to arrange the elements of a group G into sets consisting of similar elements. These sets constitute the **conjugate classes** or simply **classes** of the group G. Elements belonging to different classes are not similar to each other.

One can easily see that the identity element E constitutes a class by itself $(X^{-1}EX = X^{-1}XE = E)$. Similarly, every element of an Abelian group is a class by itself. Indeed, $XAX^{-1} = AXX^{-1} = A$. Finally, every element is conjugate to itself $(E^{-1}AE = EAE = EA = A)$.

To find all possible classes of a group, one follows an algorithm of the following type:

- We select an element $A \in G$ and form the set Y_A, which contains all elements conjugate to it:

$$\{Y_A\} = XAX^{-1}, \quad X \in G.$$

- From all the resulting elements, we keep only those that are different from each other.

- We select an element $B \in G$, which does not belong to the above set, and we repeat the same procedure.
- We continue this process until all group elements of the group G are exhausted.

Application 3: Find the conjugate classes of the group C_{3v}.

The first class Y_1 consists of only the identity E.

We consider the element $A = C_3$ and form the elements XAX^{-1}, $X \in C_{3v}$. We have

$$X = C_3 \Leftrightarrow XC_3X^{-1} = C_3C_3(C_3)^{-1} = C_3,$$

$$X = (C_3)^2 \Leftrightarrow C_3^2C_3((C_3)^{-2})^{-1} = EC_3 = C_3,$$

$$X = \sigma_1 \Rightarrow \sigma_1C_3(\sigma_1)^{-1} = \sigma_2\sigma_1 = C_3^2,$$

$$X = \sigma_2 \Rightarrow \sigma_2C_3(\sigma_2)^{-1} = \sigma_1\sigma_2 = C_3^2,$$

$$X = \sigma_3 \Rightarrow \sigma_3C_3(\sigma_3)^{-1} = \sigma_1\sigma_3 = (C_3)^2.$$

Thus, we obtain the second class Y_2 consisting of the set $\{C_3, C_3^2\}$.

We consider next the element σ_1. Then, we obtain

$$X = \sigma_1 \Rightarrow \sigma_1\sigma_1(\sigma_1)^{-1} = \sigma_1\sigma_1\sigma_1 = \sigma_1,$$

$$X = \sigma_2 \Rightarrow \sigma_2\sigma_1(\sigma_2)^{-1} = \sigma_2\sigma_1\sigma_2 = \sigma_1C_3 = \sigma_3,$$

$$X = \sigma_3 \Rightarrow \sigma_3\sigma_1(\sigma_3)^{-1} = \sigma_3\sigma_1\sigma_3 = C_3\sigma_3 = \sigma_2.$$

Thus, the class Y_3 contains the elements of the set $\{\sigma_1, \sigma_2, \sigma_3\}$. At this point, we observe that all elements of the group have been put into classes, and we are finished. Thus, C_{3v} consists of three classes. The geometric similarity of the members of each class is remarkable.

3.3   The conjugate classes of S_n

The algorithm for obtaining the classes of the symmetric group S_n is very neat, see, for example, the classic book on the subject (Hammermesh, 1964), and quite useful since every discrete group is isomorphic to a subgroup of S_n.

Let us consider the permutations

$$A = \begin{pmatrix} 1 & 2 & \cdots & n \\ A_1 & A_2 & \cdots & A_n \end{pmatrix}, \quad B = \begin{pmatrix} 1 & 2 & \cdots & n \\ B_1 & B_2 & \cdots & B_n \end{pmatrix},$$

$$B^{-1} = \begin{pmatrix} B_1 & B_2 & \cdots & B_n \\ 1 & 2 & \cdots & n \end{pmatrix}.$$

Permutation B acting on $A_1, A_2, \ldots, A_n$ yields

$$B = \begin{pmatrix} A_1 & A_2 & \cdots & A_n \\ A_{b_1} & A_{b_2} & \cdots & A_{b_n} \end{pmatrix}.$$

Hence,

$$BAB^{-1} = \begin{pmatrix} A_1 & A_2 & \cdots & A_n \\ A_{b_1} & A_{b_2} & \cdots & A_{b_n} \end{pmatrix} \begin{pmatrix} 1 & 2 & \cdots & n \\ A_1 & A_2 & \cdots & A_n \end{pmatrix}$$

$$\times \begin{pmatrix} B_1 & B_2 & \cdots & B_n \\ 1 & 2 & \cdots & n \end{pmatrix}$$

or

$$BAB^{-1} = \begin{pmatrix} B_1 & B_2 & \cdots & B_n \\ A_{b_1} & A_{b_2} & \cdots & A_{b_n} \end{pmatrix}.$$

This shows that the action of BAB^{-1} is equivalent to the permutation of B both in the upper as well as lower lines of A. Thus, for example,

$$A = \begin{pmatrix} 1 & 2 & 3 & 4 & 5 \\ 3 & 5 & 1 & 2 & 4 \end{pmatrix} = (13)(254), \quad B = \begin{pmatrix} 1 & 2 & 3 & 4 & 5 \\ 2 & 1 & 3 & 5 & 4 \end{pmatrix} = (23)(145).$$

Obviously, with the action of B, the upper line of A becomes $2\,1\,3\,5\,4$. The lower line becomes $3\,4\,2\,1\,5$, and hence,

$$\begin{pmatrix} 2 & 1 & 3 & 5 & 4 \\ 3 & 4 & 2 & 1 & 5 \end{pmatrix} = (23)(145).$$

The above can be applied when the permutations are expressed in terms of cycles without common parts, as, for example, when $A = (13)(254)$,

$B = (12)(45)$. Then,

$$BAB^{-1} = B(13)B^{-1}B(254)B^{-1} = (23)(145),$$

making use of the relations

$$B(13)B^{-1} = B \begin{pmatrix} 1 & 3 \\ 3 & 1 \end{pmatrix} B^{-1} = \begin{pmatrix} 2 & 3 \\ 3 & 2 \end{pmatrix} = (23),$$

$$B(254)B^{-1} = B \begin{pmatrix} 2 & 5 & 4 \\ 5 & 4 & 2 \end{pmatrix} B^{-1} = \begin{pmatrix} 1 & 4 & 5 \\ 4 & 5 & 1 \end{pmatrix} = (145).$$

In other words, the action of conjugation does not change the cyclic structure of a permutation. Thus, the elements of different structures, e.g. even and odd, cannot belong to the same class.

Theorem 3: *The symmetric group S_n contains as many conjugate classes as its different cyclic structures.*

In the case of S_3, we can easily verify the following:

- First class: cycles of length 1, $\{(1)(2)(3)\} = \{E\}$, one element.
- Second class: cycles of length 2, $\{(12), (13), (23)\}$, three elements.
- Third class: cycles of length 3, $\{(123), (132)\}$, two elements.

This is in agreement with the results for C_{3v}, which is isomorphic to S_3.

In the case of S_n, there can be cycles of length 1 up to n. We must, therefore, know

the number ν_1 1-cycles (cycles of length 1)
the number ν_2 2-cycles (cycles of length 2)
the number ν_3 3-cycles (cycles of length 3)
....
....
the number ν_n n-cycles (cycles of length n).

The numbers $\nu = \nu_1, \nu_2, \ldots, \nu_n$ characterize the class completely. We usually write

$$\{\nu \equiv \{1^{\nu_1}, 2^{\nu_2}, 3^{\nu_3}, \ldots, n^{\nu_n}\} \Leftrightarrow \text{ conjugate classes of } S_n\}.$$

It is obvious that, since every cycle of length k contains k symbols and the total number of symbols is n, we must have

$$\nu_1 + 2\nu_2 + 3\nu_3 + \cdots + n\nu_n = n, \nu_i, \; i = 1, 2, \ldots, n, \text{ non-negative integers.}$$

$$(3.3)$$

As a result, the number of classes is equal to the number of solutions ν_i of the previous equation. The reader is encouraged to confirm that the number of possible solutions for S_3 is equal to 3: $\{\nu_1 = 3, \nu_2 = 0, \nu_3 = 0\}$, $\{\nu_1 = 1, \nu_2 = 1, \nu_3 = 0\}$ and $\{\nu_1 = 0, \nu_0 = 0, \nu_3 = 1\}$, which correspond to the classes $\{E\}$, $\{(12)(3), (13)(2), (1)(23)\} \Leftrightarrow \{(12), (13), (23)\}$ and $\{(123), (213)\}$

Application 4: Find the conjugate classes of S_4.

It is adequate to find the non-negative integer solutions of the equation

$$\nu_1 + 2\nu_2 + 3\nu_3 + 4\nu_4 = 4.$$

We find:

(1) $[1^4] = \{\nu_1 = 4, \nu_2 = 0, \nu_3 = 0, \nu_4 = 0\} \Leftrightarrow \{(1)(2)(3)(4)\} = \{E\}$.

(2) $[1^2, 2] = \{\nu_1 = 2, \nu_2 = 1, \nu_3 = 0, \nu_4 = 0\} \Leftrightarrow \{(12), (13), (14), (23), (24), (34)\}$.

It is understood that $(12)(3)(4) = (12)$, etc.

(3) $[2^2] = \{\nu_1 = 0, \nu_2 = 2, \nu_3 = 0, \nu_4 = 0\} \Leftrightarrow \{(12)(34), (13)(24), (14)(23)\}$.

(4) $[1, 3] = \{\nu_1 = 1, \nu_2 = 0, \nu_3 = 1, \nu_4 = 0\} \Leftrightarrow$
$\{(123), (132), (124)(142), (143)(134), (234), (243)\}$
It is understood that $(123)(4) = (123)$, etc.

(5) $[4] = \{\nu_1 = 0, \nu_2 = 0, \nu_3 = 0, \nu_4 = 1\} \Leftrightarrow$
$\{(1234), (1243), (1324), (1342), (1423), (1432)\}$.

We verify that $1 + 6 + 3 + 8 + 6 = 24$, i.e. no element of S_4 has been ignored. Thus, S_4 is composed of five classes C_1, C_2, C_3, C_4 and C_5 as above. The classes C_1, C_3 and C_5 constitute a subgroup with 12 elements (the even permutations of S_4), the tetrahedral group A_4.

For $n > 4$, the reader is referred to the book by Hammermesh (1964).

Applications for S_n, $n > 3$, can be simplified with the use of Young tableaux, see Chapter 11 and, in more detail, Vergados (2017) and Hammermesh (1964).

3.4 Subgroups and cosets

We recall that a set H is a subgroup of G if

- the set H forms a group.
- $X \in H \Rightarrow X \in G$.

Clearly, for every group G, there exist two trivial subgroups, the identity element E and the whole group G. It is possible that, beyond those trivial subgroups, there exist some proper subgroups. In the case of C_{3v} we have the following five subgroups:

$$H_1 = \{E\}, \quad H_2 = \{E, C_3, C_3^2\}, \quad H_3 = \{E, \sigma_1\},$$

$$H_4 = \{E, \sigma_2\}, \quad H_5 = \{E, \sigma_3\}.$$

The same is true for S_3, which is isomorphic to it.

Note that two elements are similar with respect to elements of the whole group, but not as members of the subgroup, since the element responsible for the similarity may be absent.

Suppose that the subgroup H of order h includes the elements

$$H = \{A_1 = E, A_2, \ldots, A_h\},$$

and we construct the sets

$$H_X^R = \{A_1 X, A_2 X, \ldots, A_h X\}, \quad X \in G \Leftrightarrow H_X^R \equiv HX.$$

Then, we distinguish the following cases:

- $X \in H$. Then, $H_X^R = H$, i.e. the set H_X^R coincides with H, modulo a reordering of its members.
- $X \notin H$. In this case, none of the elements of H_X^R belong to the set H, i.e. the two sets have no common element. Indeed, let us suppose that an element, e.g. $A_k X, 1 \le k \le h$, is also in H. Then, since A_k belongs to H, there exists an element A_k^{-1} such that the element $A_k^{-1}(A_k X)$ belongs to H. This implies that $X \in H$, contrary to our assumption. Consequently, if $X \notin H$, the sets H_X^R and H have no common element, and we write $H_X^R \cap H = 0$.

In case $HR_X \cap H = 0$, the set H_X^R is called right coset of H with respect to X. Similarly, one can construct the left coset of H with respect to X:

$$H_X^L = \{XA_1, XA_2, \ldots, XA_h\}, \quad X \in G \Leftrightarrow H_X^L \equiv XH.$$

Clearly, since the identity element does not appear in the above cosets, a coset cannot constitute a subgroup.

Remark: For the trivial subgroup $\{E\}$ of any group G of order g, one finds

$$(E)_X^R = (E)_X^L = \{X\}, \quad X \in G.$$

In other words, there exist $g - 1$ cosets each containing only one element. In this case, the right and left cosets coincide.

Example 1: Consider the subgroup $H_2 = \{E, C_3, C_3^2\}$ of C_{3v}. Then,

$$(H_2)_{\sigma_1}^R = \{E\sigma_1, C_3\sigma_1, C_3^2\sigma_1\} = \{\sigma_1, \sigma_3, \sigma_2\},$$
$$(H_2)_{\sigma_2}^R = \{E\sigma_2, C_3\sigma_2, C_3^2\sigma_2\} = \{\sigma_2, \sigma_1, \sigma_3\},$$
$$(H_2)_{\sigma_3}^R = \{E\sigma_3, C_3\sigma_3, C_3^2\sigma_3\} = \{\sigma_3, \sigma_2, \sigma_1\}.$$

In this case, the three cosets coincide, i.e. they become identical. Similarly, one finds the left cosets of H_2, which in this case coincide with the right ones. The subgroup H_2 is characterized by only one coset, which consists of the set of all reflections. In other words,

$$(H_2)_{\sigma_i}^R \cup H = G, \quad (H_2)_{\sigma_i}^L \cup H = G, \quad i = 1, 2, 3.$$

Example 2: Consider the subgroup $H_3 = \{E, \sigma_1\}$ of C_{3v}:

$$(H_3)_{\sigma_2}^R = \{E\sigma_2, \sigma_1\sigma_2\} = \{\sigma_2, C_3\},$$
$$(H_3)_{\sigma_3}^R = \{E\sigma_3, \sigma_1\sigma_3\} = \{\sigma_3, C_3^2\},$$
$$(H_3)_{C_3}^R = \{EC_3, \sigma_1C_3\} = \{C_3, \sigma_2\},$$
$$(H_3)_{C_3^2}^R = \{EC_3^2, \sigma_1C_3^2\} = \{C_3^2, \sigma_3\}.$$

We now notice that there exist two different right cosets, which have no common element.

Example 3: Consider the left cosets of the subgroup $H_3 = \{E, \sigma_1\}$ of C_{3v}:

$$(H_3)_{\sigma_2}^L = \{\sigma_2E, \sigma_2\sigma_1\} = \{\sigma_2, C_3^2\},$$
$$(H_3)_{\sigma_3}^L = \{\sigma_3E, \sigma_3\sigma_1\} = \{\sigma_3, C_3\},$$
$$(H_3)_{C_3}^L = \{C_3E, C_3\sigma_1\} = \{C_3, \sigma_3\},$$
$$(H_3)_{C_3^2}^L = \{C_3^2E, C_3^2\sigma_1\} = \{C_3^2, \sigma_2\}.$$

We observe now that there exist two different left cosets without a common element. We note, however, that the left coset with respect to one element does not coincide with the corresponding right one.

Theorem 4: *Two cosets either both left or both right are identical, or have no common element.*

Proof: We consider the cosets HX and HY. We distinguish two cases:

(1) $Y \notin HX$. Then, $HX \cap HY = 0$. Indeed, if there were A_k and A_m such that $A_k X = A_m Y$, then $Y = A_m^{-1} A_k X \Rightarrow Y \in HX$, which is absurd.

(2) $Y \in HX$. Then, $HX = HY$,

In this case, there exists an element $A_k \in H$ such that $Y = A_k X$ and $X = A_k^{-1} Y$. In this case, however, we find

$$\begin{cases} HY = HA_k X & \Rightarrow & HY \subset HX \\ HX = HA_k^{-1} Y & \Rightarrow & HX \subset HY \end{cases} \Rightarrow HX = HY.$$

Theorem 5 (theorem of Lagrange): *The order h of a subgroup H of G of order $g < \infty$ is a perfect divisor of g, i.e. the number g/h is a positive integer.*

Proof: Let there be a subgroup $H = \{A_1 = E, A_2, \ldots, A_h\}, h < g$. Then, there exists at least one element $X \in G$, $X \notin H$. We then construct the coset

$$\{A_1 X, A_2 X, \ldots, H_h X\} \equiv HX.$$

This has no element in common with H. If $g = 2h$, we are finished. If not, there exists some element Y, such that $Y \in G$, but $Y \notin H$ and $Y \notin HX$. We now construct the coset

$$\{A_1 Y, A_2 Y, \ldots, H_h Y\} \equiv HY.$$

Then, $HX \cap HY = 0$. In this case, if $g = 3h$, we are finished. If not, there exists an element Z, such that $Z \in G$, but $Z \notin H$, $Z \notin HX$ and $Z \notin HY$. With this element, we repeat the above procedure. Obviously, this cannot continue for ever, since g is finite. Thus, there will come a moment such that after k efforts, we will find $hk = g$, $k = $ positive integer.

Remark 1: This theorem corrects a common, at first sight, misconception, that is, a large order g of a group G implies a great number of possible subgroups of G. In fact, if g is a prime number, it does not have any proper divisors, and as a result, G does not have any proper subgroups. Similarly, for example, for $g = 9797$, there exist only two possible divisors, 97 and 101, so G has two proper subgroups with $h = 97$ and $h = 101$.

Remark 2: Let us consider a cyclic group of order 6 with the element A as generator, i.e. $A^6 = E$. Then, $G = Z_6 = \{E, A, A^2, A^3, A^4, A^5\}$. Its two subgroups are $H_1 = Z_2 = \{E, A^3\}$ and $H_2 = Z_3 = \{E, A^2, A^4\}$. This is, of course, expected from Lagrange's theorem since they have orders the integer divisors of 6, which are 2 and 3, respectively. Furthermore, $Z_6 = Z_2 \otimes Z_3$ in the sense of Section 3.7.

3.5 Regular subgroups and simple groups — The factor group

Definition: A subgroup H of G is **regular** if

$$HX = XH \quad \text{for every } X \in G,$$

i.e. when the right and left cosets with respect to every element X coincide. In this case, the sets H and XHX^{-1} with

$$XHX^{-1} \equiv \{XA_1X^{-1}, XA_2X^{-1}, \ldots, XA_hX^{-1}\}$$

are the same, i.e. the set XHX^{-1} is at most a reordering of the elements of H. Whenever $H = XHX^{-1}$ is valid for every $X \in G$, H is called an **invariant subgroup of** G.

We recall that two elements $A_i \in G$ and $A_j \in G$ are called conjugate, if $XA_jX^{-1} = A_i$ for every $X \in G$. As a result, if a regular subgroup contains one element, it must also contain its conjugate. In other words, a **regular** subgroup contains complete classes of the elements of G in which it is contained. The opposite is also true, i.e. if a subgroup H of G, $H \subset G$, contains full classes of G, it must be regular.

Definition: A group G is **simple** if it does not contain any non-trivial invariant subgroups.

We have seen in Example 1 above that the subgroup $H_2 = \{E, C_3, C_3^2\}$ has only one coset, namely $\{\sigma_1, \sigma_2, \sigma_3\}$, which constitutes a full class of C_{3v}. Therefore C_{3v} is not simple.

On the other hand the subgroups $H_3 = \{E, \sigma_1\}$, $H_4 = \{E, \sigma_2\}$ and $H_5 = \{E, \sigma_3\}$ are not regular since they contain only one element of the class $\{\sigma_1, \sigma_2, \sigma_3\}$.

Consider now a regular subgroup H of G. We form the possible cosets:

$$K_i^R = \{A_1X_i, A_2X_i, \ldots, A_hX_i\} = HX_i,$$

$$K_i^L = \{X_iA_1, X_iA_2, \ldots, X_iA_h\} = X_iH,$$

$$A_k \in H, \, k = 1, 2, \ldots, h, \, X_i \in G, X_i \notin H, \, i = 1, 2, \ldots, k, \, k = \frac{g}{h} - 1.$$

Suppose further that the left and right cosets coincide, $K_i^R = K_i^L \equiv K_i$, $i = 1, 2, \ldots, k$. In such a case, one can define **the product of cosets** as follows:

$$K_{ij} = \{B_\alpha B_\beta, \ B_\alpha \in K_i, \ B_\beta \in K_j, \ \alpha, \beta = 1, 2, \ldots, h\}. \tag{3.4}$$

We also define the product:

$$K_i H = \{B_\alpha A_\beta, \ B_\alpha \in K_i, \ A_\beta \in H, \ \alpha, \beta = 1, 2, \ldots, h\}. \tag{3.5}$$

The set of these cosets complete with the element H of the regular subgroup constitutes a group with $k + 1 = q/h$, which is known as the **factor group** or **quotient group** and is represented as

$$K = \frac{G}{H}. \tag{3.6}$$

Example 4: We study the set of cosets in the case of the subgroup $H = E, C_3, C_3^2$ of the group C_{3v}.

We already know (Example 1 above) that only one element suffices to generate all cosets, which confirms the relation $k = g/h - 1 = 6/3 - 1 = 1$. In this case, we encounter only one coset $K_1 = \{\sigma_1, \sigma_2, \sigma_3\}$. Hence,

$$\begin{aligned} K_{11} = K_1 K_1 &= \{\sigma_1^2, \sigma_1\sigma_2, \sigma_1\sigma_3, \sigma_2\sigma_1, \sigma_2^2, \sigma_2\sigma_3, \sigma_3\sigma_1, \sigma_3\sigma_2, \sigma_3^2\} \\ &= \{E, C_3, C_3^2\} = H. \end{aligned}$$

Similarly,

$$K_1 H = \{(\sigma_1, \sigma_2, \sigma_3)(E, C_3, C_3^2)\} = \{\sigma_1, \sigma_2, \sigma_3\} = K_1.$$

In other words,

$$K_1 K_1 = H, K_1 H = K_1,$$

i.e. H plays the role of the identity element in a discrete group of two elements $\{E, K_1\}$, which is isomorphic to the group $\{1, -1\}$. This group is homomorphic to C_{3v} in the following sense.

Homomorphism is the mapping of a group G of order g to another group G' of order $g' < g$ without necessarily having the same number of elements. The simplest such mapping results if we associate to every element $G' = \{A_1' = E', A_2', \ldots, A_{g'}'\}$ at least one element of $G = \{A_1 = E, A_2, \ldots, A_g\}$, partitioning the elements of G into g' sets with $k_{i'}$ elements in each such that

$$\sum_{i'=1}^{g'} k_{i'} = g, \quad C_{i'} = \{A_{i'\alpha}, \alpha = 1, 2, \dots, k_{i'}\}, \quad i' = 1, 2, \dots, g'.$$

We now make the following 1-to-k_i' correspondences:

$$A_{i'}' \Leftrightarrow A_{i'1}, A_{i'}' \Leftrightarrow A_{i'2}, \dots, A_{i'}' \Leftrightarrow A_{i'k_i'}$$

in such a way that

$$A_{i'}' A_{j'}' = A_{m'}' \Rightarrow A_{i'\alpha} A_{j'\beta} = A_{m'\gamma} \quad \text{for every}$$

$$A_{i'\alpha} \in C_{i'}, A_{j'\beta} \in C_{j'}, A_{m'\gamma} \in C_{m'}.$$

As a simple application, we consider the group $G = C_{3v}$:

(i) The mapping

$$E \to \epsilon, C_3 \to \epsilon, (C_3)^2 \to \epsilon, \sigma_1 \to -\epsilon, \sigma_2 \to -\epsilon, \sigma_3 \to -\epsilon,$$

with ϵ the identity and $-\epsilon$ its opposite, constitute a group of two elements, $G' = \{\epsilon, -\epsilon\}$, homomorphic to C_{3v}.

(ii) The group $G' = \{H, K_1\}$ of Example 4 above. In this case, we have

$$C_{1'} = \{E, C_3, c_3^2\}, \quad C_{2'} = \{\sigma_1, \sigma_2, \sigma_3\}.$$

Again, the two groups are homomorphic since

$$HH = H \Leftrightarrow \{EE = E, EC_3 = C_3E = C_3, EC_3^2 = C_3^2E = C_3^2,$$

$$C_3C_3 = C_3^2, C_3C_3^2 = C_3^2C_3 = E, (C_3)^2 = C_3\},$$

i.e. all the products are contained in $C_{1'} \Leftrightarrow H$. Similarly,

$$K_1H = K_1 \Leftrightarrow \{\sigma_1E = \sigma_1, \sigma_1C_3 = \sigma_2, \sigma_1C_3^2 = \sigma_3, \sigma_2E = \sigma_2,$$

$$\sigma_2C_3 = \sigma_3, \sigma_2C_3^2 = \sigma_1, \sigma_3E = \sigma_3, \sigma_3C_3 = \sigma_1, \sigma_3C_3^2 = \sigma_2\},$$

i.e. all the products are contained in $C_{2'} \Leftrightarrow K_1$. The same is true in the case of $HK_1 = K_1$. In a similar fashion,

$$K_1K_1 = H \Leftrightarrow$$

$$\{\sigma_1\sigma_1 = E, \sigma_1\sigma_2 = C_3, \sigma_1\sigma_3 = C_3^2, \sigma_2\sigma_1 = C_3^2, \sigma_2\sigma_2 = E,$$

$$\sigma_2\sigma_3 = C_3, \sigma_3\sigma_1 = C_3, \sigma_3\sigma_2 = C_3^2, \sigma_3\sigma_3 = E\}.$$

Again, all the products are included in $C_{1'} \Leftrightarrow H$.

Example 5: We examine to what extent the group S_4 has invariant subgroups and which ones. Recall that we have found the classes of S_4 in Application 4 above.

In order for a subgroup to be invariant, it must contain full classes. The classes C_2, C_3, C_4 and C_5 cannot constitute a group since they are lacking the identity element. From these, only the class C_5 with the addition of the identity element has the right number of elements, 4, which is a divisor of $g = 24$, and it can be the order h of a subgroup H, i.e.

$$H = \{E, (12)(34), (13)(24), (14)(23)\}.$$

The reader can confirm that

$$\begin{aligned}
H_{12}^L &= \{(12)E, (12)(12)(34), (12)(13)(24), (12)(14)(23)\} \\
&= \{(12), (34), (1234)(1423)\} = K_1,
\end{aligned}$$

$$\begin{aligned}
H_{13}^L &= \{(13)E, (13)(12)(34), (13)(13)(24), (13)(14)(23)\} \\
&= \{(13), (1234), (24)(1423)\} = K_2,
\end{aligned}$$

$$\begin{aligned}
H_{23}^L &= \{(23)E, (23)(12)(34), (23)(13)(24), (23)(14)(23)\} \\
&= \{(23), (1324), (1342)(14)\} = K_3,
\end{aligned}$$

$$H_{12}^L = H_{12}^R, H_{13}^L = H_{13}^R, H_{23}^L = H_{23}^R.$$

Furthermore,

$$H_{34}^L = H_{12}^L, H_{24}^L = H_{13}^L, H_{14}^L = H_{23}^L \Rightarrow H_{34}^R = H_{12}^R, H_{24}^R = H_{12}^R, H_{14}^R = H_{23}^R,$$

that is, from the transpositions, we obtain three cosets. Thus,

$$\begin{aligned}
H_{123}^L &= \{(123)E, (123)(12)(34), (123)(13)(24), (123)(14)(23)\} \\
&= \{(123), (134), (432)(421)\} = K_4,
\end{aligned}$$

$$\begin{aligned}
H_{321}^L &= \{(321)E, (321)(12)(34), (321)(13)(24), (321)(14)(23)\} \\
&= \{(321), (234), (124)(431)\} = K_5.
\end{aligned}$$

The remaining cyclic permutations give nothing new. As a result, the possible cosets of H are K_1, K_2, K_3, K_4, K_5. With the multiplication as defined above, the set $H, K_1, K_2, K_3, K_4, K_5$ constitutes a group isomorphic to S_3:

$$K = \frac{S_4}{H} = \{H, K_1, K_2, K_3, K_4, K_5\} \Leftrightarrow C_{3v} \Leftrightarrow S_3.$$

It is easily seen that the group $K = S_4/H$ is homomorphic to S_4.

3.6 Possible groups of a given order g

We have seen that the isomorphic groups have the same mathematical structure and the same multiplication table, but from a physical point of view, their elements may have different meanings, e.g. matrices, permutations, geometric transformations (rotations, reflections, inversions), symmetries of the wave functions under interchange of particles, etc. Even in applications, of course, an understanding of the mathematical structure is very important. The researcher would be at a complete loss if, every time the importance of a symmetry were recognized, he would have to start from scratch.

In this section, we attempt to determine all the isomorphic groups of a given order g. Clearly, this is simple if g is small or in case g has a small number of integer divisors.

It is clear that for any order $g \geq 2$, there exists at least one element A and at least one group with the elements

$$\{A, A^2, \ldots, A^g = E\} \text{ (Abelian)}.$$

This Abelian group is the only possibility, whenever g is a prime number. We, therefore, restrict our search to the case where g is not a prime number:

(i) $g = 4$. In this case, we have already found that there exist two possibilities. One is, of course, the Abelian case, $\{A, A^2, A^3, A^4 = E\}$ (Abelian). The other is given in Table 3.1(a).

(ii) $g = 6$. The possible order of its elements are the divisors of 6, i.e. 3 and 2.

In this case, we have found a solution such that the order of its elements is different and, in fact, two representatives S_3 and C_{3v}, which are isomorphic. We show now that this is the only possibility. Obviously, at least one element must be of order 3, since if two elements are of second order, the only possibility is E, A, B, AB, BA, which contains only five elements.

Let us now suppose that there exist two elements A and B with $A^3 = E$. Then, we encounter the following products:

$$\{A, A^2, A^3 = E, B, AB, BA, AB^2, B^2A, A^2B, BA^2, B^2\}. \qquad (3.7)$$

Obviously, if the order of B is 2, $B^2 = E$, the different elements are 6, that is, the elements of the group can be chosen to be $\{A, A^2, A^3 =$

$E, B, AB, BA^2\}$ (the products BA and A^2B do not give anything new). The problem is reduced to the previous case. Let us, therefore, suppose that $B^3 = E$. Now,

$B^2 = B \Rightarrow B = E$, unacceptable; $B^2 = BA^2 \Rightarrow B = A^2$, unacceptable;

$$B^2 = A^2B \Rightarrow B = A^2B^2 \Rightarrow AB = E \Rightarrow BA = E, \text{ unacceptable;}$$

$$A^2 = B^2 \Rightarrow BA^2 = AB^2 = E, \text{ unacceptable.}$$

Furthermore,

$$AB = BA \Rightarrow (AB)^2 = ABAB = ABBA = A^2,$$

$$(AB)^3 = A^2(AB) = B, (AB)^4 = (AB)(AB)^3 = AB^2 = A,$$

$$(AB)^5 = BA(AB)^4 = BA^2 = B = (AB)^6 = BA^3B = B^2 = E.$$

In this case, however, the element AB would be of sixth order and the group would be Abelian, something we have already considered. There remains the case: $AB = BA^2$, but then $BA = (BA^2)A^2 = ABA^2 = A(AB) = A^2B$. In this case, the elements of the group are the following six: $\{E, A, A^2, B, BA^2, A^2B\}$. In this case, we find that the elements of row B of the multiplication table are

$$B, BA^2, A^2B, B^2, AB^2, A^2B^2.$$

The only allowed solution for them to be all different is constrained to be $B^2 = E$, that is, the element B cannot be of third order. Thus, the resulting multiplication table is given by Table 3.2:

Table 3.2. (a) The multiplication table of for the case of $g = 6$ discussed in the text. (b) The same as in (a) with the identity element along the diagonal.

	E	A	A^2	B	A^2B	BA^2
E	E	A	A^2	B	A^2B	BA^2
A	A	A^2	E	BA^2	B	A^2B
A^2	A^2	E	A	A^2B	BA^2	A
B	B	A^2B	BA^2	E	A	A^2
A^2B	A^2B	BA^2	B	A^2E	E	A
BA^2	BA^2	B	A^2B	A	A^2	E

(a)

	E	A	A^2	B	A^2B	BA^2
E	E	A	A^2	B	A^2B	BA^2
A^2	A^2	E	A	A^2B	BA^2	A
A	A	A^2	E	BA^2	B	A^2B
B	B	A^2B	BA^2	E	A	A^2
A^2B	A^2B	BA^2	B	A^2	E	A
BA^2	BA^2	B	A^2B	A	A^2	E

(b)

Therefore, there is no other group except those two considered above, since the last one we found is isomorphic to S_3 and C_{3v}.

3.7 Direct group product

Let there be a group $G = \{A_1 = E, A_2, \ldots, A_g\}$ of order g and another one $G' = \{A_1' = E, A_2', \ldots, A_{g'}'\}$ of order g'. In addition, suppose that:

(i) the only common element of the two groups is the identity element E.
(ii) One can define the product $XY = YX$ for every $X \in G, Y \in G'$.

The direct product of two groups is defined as follows:

$$G \otimes G'$$
$$= \{A_1 A_1', A_1 A_2', A_1 A_3', \ldots, A_1 A_{g'}', A_2 A_1', A_2 A_2', A_2 A_3', \ldots,$$
$$A_2 A_{g'}', \ldots, A_g A_1', A_g A_2', A_g A_3', \ldots, A_g A_{g'}'\}. \tag{3.8}$$

This set constitutes a group of order gg'. G and G' are regular subgroups of $G \otimes G'$.

Example 6: $G = \{E, P\}$, $G' = \{E, Q\}$, $P \neq Q, PQ = QP, P^2 = E, Q^2 = E$. Then, $G \otimes G' = \{E, P, Q, PQ\}$. One can easily construct the multiplication table:

$$
\begin{array}{c|cccc}
 & E & P & Q & PQ \\
\hline
E & E & P & Q & PQ \\
P & P & E & PQ & Q \\
Q & Q & PQ & E & P \\
PQ & PQ & Q & P & E \\
\end{array}
\tag{3.9}
$$

The group $G \otimes G'$ is Abelian. The reader can easily verify that a realization of the group is given by the matrices:

$$E = \begin{pmatrix} 1 & 0 \\ 0 & 1 \end{pmatrix}, \quad P = \begin{pmatrix} 0 & 1 \\ 1 & 0 \end{pmatrix}, \quad Q = \begin{pmatrix} -1 & 0 \\ 0 & -1 \end{pmatrix}, \quad PQ = \begin{pmatrix} 0 & -1 \\ -1 & 0 \end{pmatrix}.$$

Example 7: A useful example is the group $S_6 = \{E, C_3, (C_3)^2\} \otimes \{E, I\}$ discussed in problem 3.7.7.

The direct product is one way of extending a given symmetry. Many times in applications, we observe in a system a given symmetry. Is this the maximum symmetry of the system? If not, it is usually desirable to broaden it.[1]

Another possibility is the double covering group of a discrete group, discussed in Section 5.10.

The direct product allows the splitting of one group into others of smaller order. Let us, for example, consider the cyclic group of order 6, $G = \{E, A, A^2, \ldots, A^6 = E\}$. Then, we can take $G = H_1 \otimes H_2$, with $H_1 = \{E, A^2, A^4\}$ and $H_2 = \{E, A, A^3\}$. The reader should verify that all the above demands are satisfied.

3.8 Problems

3.1.1 Consider an Abelian group S_4 of order 4. Find a subgroup of S_4.

3.1.2 A group of order 5 is given. Find a subgroup of S_5.

3.1.3 Find the order of all elements of the group generated by the matrices

$$\begin{pmatrix} 0 & 1 \\ -1 & 0 \end{pmatrix}, \quad \begin{pmatrix} 0 & i \\ i & 0 \end{pmatrix}.$$

3.1.4 Find the order of all elements of the group C_{4v}. Then, find the minimum set of its generators.

3.1.5 Show that the group D_3 can be generated from its elements A and B satisfying

$$A^3 = B^2 = (AB)^2 = E.$$

3.1.6 Show that the group D_n can be generated from its elements A and B satisfying

$$A^n = B^2 = (AB)^2 = E.$$

---

[1]We will see later that, for example, the set of rotations in 3D constitutes a group, known as $R(3)$. When a quantum mechanical system remains invariant under such rotations, it is characterized by the quantum numbers (J, M), characterizing the angular momentum and its projection on the quantization axis, respectively. Let us suppose further that the system is invariant under inversion I with respect to a center. This additional symmetry I is characterized by a group of two elements, $G_I = \{E, I\}$. The enlarged symmetry is $O(3) = R(3) \otimes G_I$ of orthogonal transformations in 3D. Since the system remains invariant under the extended symmetry, it is characterized by an additional quantum number, called parity, indicated by π, which takes the values $\pi = \pm 1$. Thus, the system is characterized by the quantum numbers (π, J, M) due to the symmetry $O(3)$.

3.1.7 Find the minimum number of generators of problem 2.2.2.

3.1.8 Do the same for problem 2.2.3.

3.1.9 Do the same for problem 2.2.7.

3.1.10 Show that the group of four elements E, A, B, C with $C^2 = E$ is isomorphic with one of the groups discussed in Application 2.

3.1.11 You are given an Abelian group of order 6. Find a subgroup of S_6 isomorphic to the given one. Is your answer unique?

3.2.1 Show that the elements C_n^k and C_n^{k-1} are conjugate to each other. These are members of a group that contains only rotations around an axis of order n.

3.2.2 Show that a rotation around a two-sided axis (see problem 2.5.12) belongs to the same class as its inverse.

3.2.3 Equivalent elements of a symmetry, axes or planes, are those that can coincide with the action of a symmetry operation of the group. Are there any such elements in the case of C_{3v}?

3.2.4 Proceeding as in the previous problem, examine whether there exist equivalent symmetry elements in the case of rotation–reflection axes around the same axes. Do the same in the case of a two-sided axis (see problem 2.5.12).

3.2.5 Consider the group D_n. Show that:

(i) the number of its elements is $2n$.
(ii) if $n = 2k$ (even):

(1) the elements E, C_n^k form a class by themselves.
(2) the remaining rotations of order n also belong in pairs in the same class.
(3) the rotations of order 2 form two classes with k elements each.
(4) the total number of classes is $(n + 6)/2$.

(iii) if $n = 2k + 1$ (odd):

(1) the identity element forms a class by itself.
(2) the rotations around the axis of order n form k classes with two elements each.
(3) all the $2k + 1$ axes of order 2 are equivalent, that is they belong to the same class.
(4) the total number of classes is $(n + 3)/2$.

3.2.6 Using the results of 2.2.1–2.2.6, find the conjugate classes of 2.5.1–2.5.7.

3.2.7 Find the classes of S_4 from the multiplication table. Are the results in agreement with those of Section 2.2?

3.2.8 Obtain the classes of S_5.

3.2.9 Do the same for C_{4v}.

3.2.10 Do the same for the group of problem 3.1.3. Show that this has the same number of classes with that of the previous problem but that they are not isomorphic.

Suggestion: It may help if you consider the order of their elements.

3.2.11 Show that all elements of a class have the same order. Give an example showing that the reverse is not true.

3.2.12 Let K_i^{-1} be the set of all elements, which are the inverse of those of a class K_i. Show that the elements K_i^{-1} belong to the same class. This class is known as *inverse*.

3.4.1 Find all the cosets of S_3 with respect to one of its subgroups.

3.4.2 Show that the coset of S_4

$$H_2 = \{E, (12)(34), (13)(24), (23)(14)\}$$

constitutes a group. Show that the only left cosets of H_2 are

$$K_1 = (12)H_2, K_2 = (13)H_2, K_3 = (23)H_2, K_4 = (123)H_2,$$
$$K_5 = (321)H_2.$$

Find its right cosets.

3.4.3 A group is described by the multiplication table

	E	A	B	C	D	F
E	E	A	B	C	D	F
A	A	E	D	F	B	C
B	B	F	E	D	C	A
C	C	D	F	E	A	B
D	D	C	A	B	F	E
F	F	B	C	A	E	D

(i) Find the order of each element.

(ii) Find all of its subgroups.

(iii) Find the left and right cosets of each subgroup.

3.4.4 Do the same for the group C_{4v}.

3.4.5 Do the same for the group of problem 3.1.3.

3.4.6 Let H be a subgroup of G and S any subset of the elements of G. Then:

(i) The set $N(S, H) = \{X \in H, XA = AX$ for every $A \in S\}$ constitutes a group known as *the center of S in H*.

(ii) The set $N'(S, H) = \{X \in S, X^{-1}YX \in S$ for every $Y \in S\}$ constitutes a group known *regularization of S in H*.

3.5.1 Find a homomorphism $G \to G/H$ in the case of an Abelian group G.

3.5.2 Consider the homomorphism $G \to G'$. Show that elements mapped to the identity E belong to the same class.

3.5.3 Show that the elements $\{H, K_1, K_2, K_3, K_4, K_5\}$ of problem 2.4.2 constitute a group isomorphic to S_3.

3.5.4 Show that there exists a homomorphism $S_n \to S_2$.

3.5.5 Show that there exists a homomorphism $C_{4v} \to S_2$. Show that this correspondence can be done in three ways.

3.5.6 Consider a cyclic group of order $g = 12$ with a generator A, and let H be the group with the generator A^3. Find all the cosets of H in G and the multiplication table of the quotient group G/H.

3.5.7 Examine which of the groups of problems 3.4.3–3.4.5 are regular.

3.5.8 Consider the homomorphism $G \to G'$. Show that:

(i) The inverses of G are inverses of G'.

(ii) The elements of G associated with the identity element of G' form an invariant subgroup N of G.

(iii) Show that the group G' is isomorphic to G/N.

3.5.9 Find the elements of S_6, which form an invariant subgroup of S_6. Is it Abelian?

3.6.1 Find all the subgroups of a group of order 7.

3.6.2 Find all the subgroups of a group of order 10.

3.6.3 Find all the subgroups of a group of order 9.

3.6.4 Find all the subgroups of a group of order 8. Is C_{4v} a special case?

3.6.5 Show that the elements A and B such that $A^2 = B^2 = E$ generate two subgroups starting with the elements:

(i) A and AB,

(ii) B^2 and BA.

Are the resulting groups isomorphic?

3.7.1 Consider the group $G = H \otimes K$. Then:

(i) K is isomorphic to G/H.

(ii) H is isomorphic to G/K.

(iii) G is homomorphic to both H and K.

(iv) The number of classes of G is equal to the product of the number of the classes of H and K.

3.7.2 Find the group $G_1 \otimes G_2$ where

$$G_1 = \{E, (12)\}, \; G_2 = \{E, (34)\}.$$

3.7.3 Write down all the elements of the group $\{E, C_3, (C_3)^2\} \otimes \{E, \sigma_1\}$. What is this group known as?

3.7.4 Study the group $D_3 \otimes \{E, I\}$ (multiplication table and classes). Is it related to the group D_{3d}?

3.7.5 Do the same for $\{E, C_4, (C_4)^2, (C_4)^3\} \otimes \{E, I\}$. Does it coincide with D_{4h}?

3.7.6 Do the same for $D_4 \otimes \{E, I\}$. Does it coincide with D_{4h}?

3.7.7 Do the same for $\{E, C_3, (C_3)^2\} \otimes \{E, I\}$. Does it coincide with S_6? Hints: S_6, see Section 2.4.2, should not be confused with the symmetric group of Section 2.2.

$$(S_6) = \begin{pmatrix} -1 & 0 & 0 \\ 0 & -\frac{1}{2} & -\frac{\sqrt{3}}{2} \\ 0 & \frac{\sqrt{3}}{2} & -\frac{1}{2} \end{pmatrix}, \quad (C_3)^{-1}(S_6) = \begin{pmatrix} -1 & 0 & 0 \\ 0 & -1 & 0 \\ 0 & 0 & -1 \end{pmatrix} = I,$$

where (S_6) is a generator of the group and I is the space inversion element. Also, $(S_6)^2 = C_3$, $(S_6)^4 = C_3^2$, $(S_6)^6 = E$.

For an application of this group, see Section 4.7.3.

Chapter 4

Elements of Representation Theory

In the previous chapters, we discussed the structure of some rather abstract discrete groups. We emphasized, however, that in many applications, the elements of a group are transformations in the space describing the system. In fact, these are just operators acting on a Hilbert space, which represents the system. If we select a basis in that space, the operator is described by a matrix, which depends on the operator and the chosen basis. In this sense, we say that the matrix represents the operator in the chosen basis.

4.1 Group representations: Illustrative examples

Definition: Suppose a group G of order g with elements $A_1 = E, A_2, \ldots, A_g$, with E being the identity element. Suppose, in addition, a set of $n \times n$ matrices $T_G = \{T(E) = \epsilon, T(A_1), T(A_2) \cdots T(A_g)\}$, with ϵ the identity. These matrices are characterized by a determinant different from zero, not necessarily different from each other, such that

$$
\begin{cases}
X \to T(X) \\
Y \to T(Y) \\
XY = Z
\end{cases}
\Rightarrow
\begin{cases}
T(X)T(Y) = T(Z) \\
T(E) = \epsilon \\
T(X^{-1}) = (T(X))^{-1}
\end{cases}
,\quad
\begin{array}{l}
X, Y, Z \in G \\
T(X), T(Y), T(Z) \in T_G.
\end{array}
$$

Then, we say that the set of matrices T_G constitutes a representation of the group G.

A trivial representation is one such that to all elements of G one associates the unit $n \times n$ matrix. This representation is not, however, of any use. If, however, it so happens that the matrices of the set T_G are different, $X \neq Y \Rightarrow T(X) \neq T(Y)$ and T_G contains exactly g elements, the representation is called *faithful*. Then, the set T_G constitutes a group G', which is isomorphic to G.

The group described in Example 2 of Section 2.1, Chapter 2, constitutes a faithful representation of S_3, as well as of C_{3v}, Section 2.4.1. On the contrary, the representation of C_{3v},

$$E \to \epsilon, C_3 \to \epsilon, (C_3)^2 \to \epsilon, \sigma_1 \to -\epsilon, \sigma_2 \to -\epsilon, \sigma_3 \to -\epsilon,$$

is not faithful since it contains only two elements, the identity ϵ and its opposite, $-\epsilon$, which constitute a group of two elements $G' = \{\epsilon, -\epsilon\}$ homomorphic to C_{3v}.

4.2   Representations of group operators

Let A be an operator acting on an N-dimensional Hilbert space and $|i\rangle$, $i = 1, 2, \ldots, N$ a basis in this space. Then we define the matrix $T(A) = (\alpha)$ as follows:

$$A|i\rangle - \sum_{j=1}^{N} (\alpha)_{ji}|j\rangle, \ (\alpha)_{ji} \equiv \alpha_{ji},$$

with the usual definition of the matrix elements. In other words, A acting on the vector $|i\rangle$ of the basis yields a vector with projections given by the column i of the matrix.

We have

$$AB|i\rangle = A\sum_{j=1}^{N}(\beta)_{ji}|j\rangle = \sum_{j}^{N}\sum_{k}^{N}(\alpha)_{kj}(\beta)_{ji}|k\rangle = \sum_{k}^{N}((\alpha)(\beta))_{ki}|k\rangle = \sum_{k}^{N}(\gamma)_{ki}|k\rangle,$$

with the usual definition of the matrix multiplication:

$$(\gamma)_{ki} = \sum_{j}(\alpha)_{kj}(\beta)_{ji}$$

(matrix multiplication perhaps resulted from the accommodation of the above product of operators). In other words,

$$A \to T(A) \Leftrightarrow (\alpha), \ B \to T(B) \Leftrightarrow (\beta), \ C \to T(C) \Leftrightarrow (\gamma)$$
$$\Rightarrow AB = C \to T(A)T(B) = T(C) \Leftrightarrow (\alpha)(\beta) = (\gamma).$$

As a result, a simple, but very useful, way of producing a representation of the group results from the multiplication table of the group itself, putting one in the position the resulting element appears and zero every place else. This is possible, of course, if the multiplication table is arranged so that the identity appears along the diagonal. For an Abelian group, as given, for example, by the table given in Eq. (2.4) in the middle, one finds

$$T(E) = \epsilon = \begin{pmatrix} 1 & 0 & 0 & 0 \\ 0 & 1 & 0 & 0 \\ 0 & 0 & 1 & 0 \\ 0 & 0 & 0 & 1 \end{pmatrix}, \quad T(A) = (\alpha) = \begin{pmatrix} 0 & 1 & 0 & 0 \\ 1 & 0 & 0 & 0 \\ 0 & 0 & 0 & 1 \\ 0 & 0 & 1 & 0 \end{pmatrix},$$

$$T(B) = (\beta) = \begin{pmatrix} 0 & 0 & 1 & 0 \\ 0 & 0 & 0 & 1 \\ 0 & 1 & 0 & 0 \\ 1 & 0 & 0 & 0 \end{pmatrix}, \quad T(C) = (\gamma) = \begin{pmatrix} 0 & 0 & 0 & 1 \\ 0 & 0 & 1 & 0 \\ 1 & 0 & 0 & 0 \\ 0 & 1 & 0 & 0 \end{pmatrix}.$$

The selection of $T(E)$ is obvious. To proceed further, we make the correspondence

$$E \Leftrightarrow |1\rangle, \ A \Leftrightarrow |2\rangle, \ B \Leftrightarrow |3\rangle, \ C \Leftrightarrow |4\rangle.$$

We now observe that

	column 1		column 2
$A|1\rangle = AE = A = |2\rangle \Rightarrow$	$\begin{matrix} 0 \\ 1 \\ 0 \\ 0 \end{matrix}$	$A|2\rangle = AA = E = |1\rangle \Rightarrow$	$\begin{matrix} 1 \\ 0 \\ 0 \\ 0 \end{matrix}$

	column 3		column 4
$A|3\rangle = AB = C = |4\rangle \Rightarrow$	$\begin{matrix} 0 \\ 0 \\ 0 \\ 1 \end{matrix}$	$A|4\rangle = AC = B = |3\rangle \Rightarrow$	$\begin{matrix} 0 \\ 0 \\ 1 \\ 0 \end{matrix}$

The matrix with these columns is $T(A) = (\alpha)$. The procedure of constructing the other matrices proceeds in a similar fashion.

This is the *regular* representation, to be discussed in Section 4.3, resulting directly from the group structure. The matrices involved are quite large: $g \times g$ for a group of order g, e.g. 24×24 for a group of order 4!, such as S_4. We examine this representation in some detail in the following section because of its great theoretical interest.

4.3   The regular representation of a group

This representation depends, of course, on the order the elements of the
group are arranged in. Let us, therefore, consider a specific order of the
group elements, say $A_1, A_2, \ldots, A_g$, and suppose for three elements, we
have $A_m A_j = A_k$. In other words, the action (multiplication) of A_m is
nothing but the transformation $\Gamma(A_m) : A_j \to A_k$. This becomes a bit
clearer via the correspondence

$$A_m \Leftrightarrow |m\rangle, \; A_j \Leftrightarrow |j\rangle, \; A_k \Leftrightarrow |k\rangle; \; \Gamma(A_m)||j\rangle = |k\rangle, \quad m, k, j = 1, 2, \ldots, g.$$

Furthermore, the result is quite simple; the column j has only one non-zero
element in position k, which is 1. In other words,

$$\Gamma(A_m)_{kj} = \begin{cases} 1 & A_m A_j = A_k \\ 0 & \text{otherwise} \end{cases} \Rightarrow \Gamma(E)_{kj} = \delta_{kj} = \begin{cases} 1 & k = j \\ 0 & k \neq j. \end{cases} \tag{4.1}$$

Let us now consider the product

$$\Gamma(m)\Gamma(A_n) \Leftrightarrow (\Gamma(m)\Gamma(A_n))_{ij} = \sum_\alpha \Gamma(m)_{i\alpha}\Gamma(n)_{\alpha j}.$$

The sum has only one non-zero term, whenever the conditions $A_n A_j = A_\alpha$
$A_m A_\alpha = A_i$ hold, i.e. $A_m A_n A_j = A_i$. Then, however, $A_m A_n = A_\ell$, $A_\ell = A_i(A_j)^{-1}$. In other words,

$$(\Gamma(m)\Gamma(A_n))_{ij} = \begin{cases} 1 & A_m A_n = A_\ell \\ 0 & \text{otherwise.} \end{cases} \tag{4.2}$$

As a result,

$$A_m A_n = A_\ell \Rightarrow \Gamma(A_m)\Gamma(A_n) = \Gamma(A_\ell), \tag{4.3}$$

which shows that this is indeed a representation. To prevent a risk of con-
fusion, this representation will be indicated as Γ^{reg}.

Application 1: We construct the regular representation of C_{3v}.

We assume the ordering

$$|1\rangle = E, |2\rangle = C_3, |3\rangle = (C_3)^2, |4\rangle = \sigma_1, |5\rangle = \sigma_2, |6\rangle = \sigma_3.$$

From the multiplication table, Eq. (2.7), we easily obtain

$$\Gamma(C_3)|1\rangle = C_3 E = C_3 = |2\rangle, \Gamma(C_3)|2\rangle = C_3 C_3 = (C_3)^2 = |3\rangle,$$

$$\Gamma(C_3)|3\rangle = (C_3)^3 = E = |1\rangle, \Gamma(C_3)|4\rangle = C_3\sigma_1 = \sigma_3 = |6\rangle,$$

$$\Gamma(C_3)|5\rangle = C_3\sigma_2 = \sigma_1 = |4\rangle, \Gamma(C_3)|6\rangle = C_3\sigma_3 = \sigma_2 = |5\rangle,$$

$$\Gamma(\sigma_1)|1\rangle = \sigma_1 E = \sigma_1 = |4\rangle, \Gamma(\sigma_1)|2\rangle = \sigma_1 C_3 = |5\rangle,$$

$$\Gamma(\sigma_1)|3\rangle = \sigma_1 (C_3)^2 = \sigma_3 = |6\rangle, \Gamma(\sigma_1)|4\rangle = \sigma_1\sigma_1 = E = |1\rangle,$$

$$\Gamma(\sigma_1)|5\rangle = \sigma_1\sigma_2 = C_3 = |2\rangle, \Gamma(\sigma_1)|6\rangle = \sigma_1\sigma_3 = (C_3)^2 = |3\rangle.$$

Hence,

$$\Gamma(C_3) = \begin{pmatrix} 0 & 0 & 1 & 0 & 0 & 0 \\ 1 & 0 & 0 & 0 & 0 & 0 \\ 0 & 1 & 0 & 0 & 0 & 0 \\ 0 & 0 & 0 & 0 & 1 & 0 \\ 0 & 0 & 0 & 0 & 0 & 1 \\ 0 & 0 & 0 & 1 & 0 & 0 \end{pmatrix}, \quad \Gamma(\sigma_1) = \begin{pmatrix} 0 & 0 & 0 & 1 & 0 & 0 \\ 0 & 0 & 0 & 0 & 1 & 0 \\ 0 & 0 & 0 & 0 & 0 & 1 \\ 1 & 0 & 0 & 0 & 0 & 0 \\ 0 & 1 & 0 & 0 & 0 & 0 \\ 0 & 0 & 1 & 0 & 0 & 0 \end{pmatrix}.$$

One can similarly proceed with the other elements. One, however, can make use of the fact that that the above two elements are generators of the group, which means that the remaining representations can be obtained by multiplication of these two as well as those obtained from them in all possible ways. Thus,

$$\Gamma\left((C_3)^2\right) = \Gamma(C_3)\Gamma(C_3), \Gamma\left((C_3)^3\right) = \Gamma(E),$$

$$\Gamma(C_3)\Gamma(\sigma_1) = \Gamma(\sigma_3), \Gamma(C_3)\Gamma(\sigma_3) = \Gamma(\sigma_2).$$

This way, with the aid of packages like Mathematica, we obtain

$$\Gamma\left((C_3)^2\right) = \begin{pmatrix} 0 & 1 & 0 & 0 & 0 & 0 \\ 0 & 0 & 1 & 0 & 0 & 0 \\ 1 & 0 & 0 & 0 & 0 & 0 \\ 0 & 0 & 0 & 0 & 0 & 1 \\ 0 & 0 & 0 & 1 & 0 & 0 \\ 0 & 0 & 0 & 0 & 1 & 0 \end{pmatrix}, \quad \Gamma(E) = \begin{pmatrix} 1 & 0 & 0 & 0 & 0 & 0 \\ 0 & 1 & 0 & 0 & 0 & 0 \\ 0 & 0 & 1 & 0 & 0 & 0 \\ 0 & 0 & 0 & 1 & 0 & 0 \\ 0 & 0 & 0 & 0 & 1 & 0 \\ 0 & 0 & 0 & 0 & 0 & 1 \end{pmatrix},$$

$$\Gamma(\sigma_3) = \begin{pmatrix} 0 & 0 & 0 & 0 & 0 & 1 \\ 0 & 0 & 0 & 1 & 0 & 0 \\ 0 & 0 & 0 & 0 & 1 & 0 \\ 0 & 1 & 0 & 0 & 0 & 0 \\ 0 & 0 & 1 & 0 & 0 & 0 \\ 1 & 0 & 0 & 0 & 0 & 0 \end{pmatrix}, \quad \Gamma(\sigma_2) = \begin{pmatrix} 0 & 0 & 0 & 0 & 1 & 0 \\ 0 & 0 & 0 & 0 & 0 & 1 \\ 0 & 0 & 0 & 1 & 0 & 0 \\ 0 & 0 & 1 & 0 & 0 & 0 \\ 1 & 0 & 0 & 0 & 0 & 0 \\ 0 & 1 & 0 & 0 & 0 & 0 \end{pmatrix}.$$

The reader can verify that the resulting multiplication table coincides with that given by Eq. (2.7). The reader can also verify that, by changing the order of the elements of the group, we obtain a different representation. We will see that this representation is equivalent to the previous one.

We can thus verify what we had mentioned earlier, namely that the regular representation can be directly read from the multiplication table, provided, of course, that it is written in such a way that the identity appears along the diagonal. Thus, from Eq. (2.7), one obtains

$$\Gamma'(E) = \begin{pmatrix} 1 & 0 & 0 & 0 & 0 & 0 \\ 0 & 1 & 0 & 0 & 0 & 0 \\ 0 & 0 & 1 & 0 & 0 & 0 \\ 0 & 0 & 0 & 1 & 0 & 0 \\ 0 & 0 & 0 & 0 & 1 & 0 \\ 0 & 0 & 0 & 0 & 0 & 1 \end{pmatrix}, \quad \Gamma'(C_3) = \begin{pmatrix} 0 & 1 & 0 & 0 & 0 & 0 \\ 0 & 0 & 1 & 0 & 0 & 0 \\ 1 & 0 & 0 & 0 & 0 & 0 \\ 0 & 0 & 0 & 0 & 1 & 0 \\ 0 & 0 & 0 & 0 & 0 & 1 \\ 0 & 0 & 0 & 1 & 0 & 0 \end{pmatrix},$$

$$\Gamma'\left((C_3)^2\right) = \begin{pmatrix} 0 & 0 & 1 & 0 & 0 & 0 \\ 1 & 0 & 0 & 0 & 0 & 0 \\ 0 & 1 & 0 & 0 & 0 & 0 \\ 0 & 0 & 0 & 0 & 0 & 1 \\ 0 & 0 & 0 & 1 & 0 & 0 \\ 0 & 0 & 0 & 0 & 1 & 0 \end{pmatrix}, \quad \Gamma'(\sigma_1) = \begin{pmatrix} 0 & 0 & 0 & 1 & 0 & 0 \\ 0 & 0 & 0 & 0 & 0 & 1 \\ 0 & 0 & 0 & 0 & 1 & 0 \\ 1 & 0 & 0 & 0 & 0 & 0 \\ 0 & 0 & 1 & 0 & 0 & 0 \\ 0 & 1 & 0 & 0 & 0 & 0 \end{pmatrix},$$

$$\Gamma'(\sigma_2) = \begin{pmatrix} 0 & 0 & 0 & 0 & 1 & 0 \\ 0 & 0 & 0 & 1 & 0 & 0 \\ 0 & 0 & 0 & 0 & 0 & 1 \\ 0 & 1 & 0 & 0 & 0 & 0 \\ 1 & 0 & 0 & 0 & 0 & 0 \\ 0 & 0 & 1 & 0 & 0 & 0 \end{pmatrix}, \quad \Gamma'(\sigma_3) = \begin{pmatrix} 0 & 0 & 0 & 0 & 0 & 1 \\ 0 & 0 & 0 & 0 & 1 & 0 \\ 0 & 0 & 0 & 1 & 0 & 0 \\ 0 & 0 & 1 & 0 & 0 & 0 \\ 0 & 1 & 0 & 0 & 0 & 0 \\ 1 & 0 & 0 & 0 & 0 & 0 \end{pmatrix}.$$

The resulting matrices are different. They yield, of course, the same multiplication table as before, which we include here, Table 4.1, omitting, of course, the accents.

Table 4.1.

	$\Gamma(E)$	$\Gamma(C_3)$	$((C_3)^2)$	$\Gamma(\sigma_1)$	$\Gamma(\sigma_2)$	$\Gamma(\sigma_3)$
$\Gamma(E)$	$\Gamma(E)$	$\Gamma(C_3)$	$((C_3)^2)$	$\Gamma(\sigma_1)$	$\Gamma(\sigma_2)$	$\Gamma(\sigma_3)$
$((C_3)^2)$	$((C_3)^2)$	$\Gamma(E)$	$\Gamma(C_3)$	$\Gamma(\sigma_2)$	$\Gamma(\sigma_3)$	$\Gamma(\sigma_1)$
$\Gamma(C_3)$	$\Gamma(C_3)$	$((C_3)^2)$	$\Gamma(E)$	$\Gamma(\sigma_3)$	$\Gamma(\sigma_1)$	$\Gamma(\sigma_2)$
$\Gamma(C_3)$	$\Gamma(C_3)$	$((C_3)^2)$	$\Gamma(E)$	$\Gamma(\sigma_3)$	$\Gamma(\sigma_1)$	$\Gamma(\sigma_2)$
$\Gamma(\sigma_1)$	$\Gamma(\sigma_1)$	$\Gamma(\sigma_2)$	$\Gamma(\sigma_3)$	$\Gamma(E)$	$\Gamma(C_3)$	$((C_3)^2)$
$\Gamma(\sigma_2)$	$\Gamma(\sigma_2)$	$\Gamma(\sigma_3)$	$\Gamma(\sigma_1)$	$((C_3)^2)$	$\Gamma(E)$	$\Gamma(C_3)$
$\Gamma(\sigma_3)$	$\Gamma(\sigma_3)$	$\Gamma(\sigma_1)$	$\Gamma(\sigma_2)$	$\Gamma(C_3)$	$((C_3)^2)$	$\Gamma(E)$

4.4 Various ways of constructing representations

If a representation of the group is given, we can construct others in various ways:

(1) Increasing the dimension of a representation.

This can be achieved by increasing the number of rows and putting 1 along the extended diagonal and 0 every place else, e.g.

$$T(E) = \begin{pmatrix} 1 & 0 \\ 0 & 1 \end{pmatrix} \Rightarrow T'(E) \begin{pmatrix} 1 & 0 & 0 \\ 0 & 1 & 0 \\ 0 & 0 & 1 \end{pmatrix},$$

$$T(A) = \begin{pmatrix} -\frac{1}{2} & -\frac{\sqrt{3}}{2} \\ \frac{\sqrt{3}}{2} & -\frac{1}{2} \end{pmatrix} \Rightarrow T'(A) \begin{pmatrix} -\frac{1}{2} & -\frac{\sqrt{3}}{2} & 0 \\ \frac{\sqrt{3}}{2} & -\frac{1}{2} & 0 \\ 0 & 0 & 1 \end{pmatrix}, \text{ etc.}$$

Another possibility arises by putting the 1 in the upper-left corner:

$$T(E) = \begin{pmatrix} 1 & 0 \\ 0 & 1 \end{pmatrix} \Rightarrow T'(E) \begin{pmatrix} 1 & 0 & 0 \\ 0 & 1 & 0 \\ 0 & 0 & 1 \end{pmatrix},$$

$$T(A) = \begin{pmatrix} -\frac{1}{2} & -\frac{\sqrt{3}}{2} \\ \frac{\sqrt{3}}{2} & -\frac{1}{2} \end{pmatrix} \Rightarrow T'(A) = \begin{pmatrix} 1 & 0 & 0 \\ 0 & -\frac{1}{2} & -\frac{\sqrt{3}}{2} \\ 0 & \frac{\sqrt{3}}{2} & -\frac{1}{2} \end{pmatrix}, \text{ etc.}$$

This method will be generalized in the following section.

(2) Representation with smaller dimension.
   We will see later that from the regular representation, we can obtain others of smaller dimensions.
(3) Representation of the same dimension. We distinguish two possibilities:

   (a) The *complex conjugate* representation $(T(X))^* = (T^*(X))$.
      Indeed, when a representation $T(X)$ is not real, it can easily be verified that $T^*(X) = (T(X))^*$ is also a representation:

$$(T(XY))^* = (T(X)T(Y))^* = (T(X)))^* \, (T(Y))^* = T^*(X)T^*(Y).$$

   In many cases, a complex representation can be equivalent to a real one in the sense described in the following (see Section 4.6). Then, the representation is called real. When this occurs for all representations of a group, we say that this group does not admit complex representations.

   (b) The *adjoined representation* $\bar{T}(X)$.
      This is defined as follows:

$$\bar{T}(X) = \left(\widetilde{T(X)}\right)^{-1}, \quad X \in G, \quad \widetilde{T(X)} = \text{transpose of } T(X),$$

$$(\widetilde{T(X)})_{ij} = (T(X))_{ji}. \tag{4.4}$$

   Obviously, if a representation consists of orthogonal matrices, its adjoined coincides with it. Indeed,

$$\bar{T}(XY) = \left(\widetilde{T(XY)}\right)^{-1} = \left(\widetilde{T(Y)}\widetilde{T(X)}\right)^{-1}$$

$$= (\widetilde{T(X)})^{-1}(\widetilde{T(Y)})^{-1} = \bar{T}(X)\bar{T}(Y).$$

   Otherwise, it may be different.

4.5   Direct product of representations

In Section 3.7, we defined the direct product of two groups. In this section, we define the **direct product of two representations of the same group**.

Suppose that we are given two representations, $T^a(X), X \in G$ and $T^b(X), X \in G$. From these, we can obtain another representation as the direct product of two matrices, as follows:

$$T(X) = T^a(X) \otimes T^b(X), \quad X \in G. \tag{4.5}$$

Suppose that two matrices (α), of dimension $m \times n$, and (β), of dimension $m' \times n'$ are given. Their **direct product** is defined as a new matrix γ, of dimension $mm' \times nn'$, written as

$$(\alpha) \otimes (\beta) = (\gamma), \tag{4.6}$$

with the understanding that

$$(\gamma) = \begin{pmatrix} (A)_{11} & (A)_{12} & \cdots & (A)_{1n} \\ (A)_{21} & (A)_{22} & \cdots & (A)_{2,n} \\ \cdots\cdots & \cdots\cdots & \cdots & \cdots\cdots \\ (A)_{m,1} & (A)_{m,2} & \cdots & (A)_{m,n} \end{pmatrix},$$

where (A) has elements made of sub-matrices:

$$(A)_{ij} = \alpha_{ij}(\beta) \text{ a matrix } m' \times n'. \tag{4.7}$$

As an example, let us consider the matrices

$$\alpha = \begin{pmatrix} \alpha_{11} & \alpha_{12} & \alpha_{13} \\ \alpha_{21} & \alpha_{22} & \alpha_{23} \\ \alpha_{31} & \alpha_{32} & \alpha_{33} \end{pmatrix}, \quad \beta = \begin{pmatrix} \beta_{11} & \beta_{12} \\ \beta_{21}, & \beta_{22} \end{pmatrix},$$

$$\alpha \otimes \beta = \begin{pmatrix} \alpha_{11}\beta_{11} & \alpha_{11}\beta_{12} & \alpha_{12}\beta_{11} & \alpha_{12}\beta_{12} & \alpha_{13}\beta_{11} & \alpha_{13}\beta_{12} \\ \alpha_{11}\beta_{21} & \alpha_{11}\beta_{22} & \alpha_{12}\beta_{21} & \alpha_{12}\beta_{22} & \alpha_{13}\beta_{21} & \alpha_{13}\beta_{22} \\ \alpha_{21}\beta_{11} & \alpha_{21}\beta_{12} & \alpha_{22}\beta_{11} & \alpha_{22}\beta_{12} & \alpha_{23}\beta_{11} & \alpha_{23}\beta_{12} \\ \alpha_{21}\beta_{21} & \alpha_{21}\beta_{22} & \alpha_{22}\beta_{21} & \alpha_{22}\beta_{22} & \alpha_{23}\beta_{21} & \alpha_{23}\beta_{22} \\ \alpha_{31}\beta_{11} & \alpha_{31}\beta_{12} & \alpha_{32}\beta_{11} & \alpha_{32}\beta_{12} & \alpha_{33}\beta_{11} & \alpha_{33}\beta_{12} \\ \alpha_{31}\beta_{21} & \alpha_{31}\beta_{22} & \alpha_{32}\beta_{21} & \alpha_{32}\beta_{22} & \alpha_{33}\beta_{21} & \alpha_{33}\beta_{22} \end{pmatrix}, \tag{4.8}$$

$$\beta \otimes \alpha = \begin{pmatrix} \alpha_{11}\beta_{11} & \alpha_{12}\beta_{11} & \alpha_{13}\beta_{11} & \alpha_{11}\beta_{12} & \alpha_{12}\beta_{12} & \alpha_{13}\beta_{12} \\ \alpha_{21}\beta_{11} & \alpha_{22}\beta_{11} & \alpha_{23}\beta_{11} & \alpha_{21}\beta_{12} & \alpha_{22}\beta_{12} & \alpha_{23}\beta_{12} \\ \alpha_{31}\beta_{11} & \alpha_{32}\beta_{11} & \alpha_{33}\beta_{11} & \alpha_{31}\beta_{12} & \alpha_{32}\beta_{12} & \alpha_{33}\beta_{12} \\ \alpha_{11}\beta_{21} & \alpha_{12}\beta_{21} & \alpha_{13}\beta_{21} & \alpha_{11}\beta_{22} & \alpha_{12}\beta_{22} & \alpha_{13}\beta_{22} \\ \alpha_{21}\beta_{21} & \alpha_{22}\beta_{21} & \alpha_{23}\beta_{21} & \alpha_{21}\beta_{22} & \alpha_{22}\beta_{22} & \alpha_{23}\beta_{22} \\ \alpha_{31}\beta_{21} & \alpha_{32}\beta_{21} & \alpha_{33}\beta_{21} & \alpha_{31}\beta_{22} & \alpha_{32}\beta_{22} & \alpha_{33}\beta_{22} \end{pmatrix}. \tag{4.9}$$

The reader can benefit from the fact the direct product of two matrices is obtainable through suitable packages, e.g. Mathematica.

One can easily find that the trace of $(\alpha) \otimes (\beta)$ is the product of the traces of (α) and (β). Indeed,

$$tr((\alpha) \otimes (\beta)) = tr(A_{11}) + tr(A_{22}) + \cdots + tr(A_{nn})$$

$$= \alpha_{11} tr(\beta) + \alpha_{22} tr(\beta) + \cdots + \alpha_{nn} tr(\beta)$$

$$= (\alpha_{11} + \alpha_{22} \cdots + \alpha_{nn}) tr(\beta) = tr(\alpha) tr(\beta).$$

Combining the usual and direct products of matrices, one obtains the following theorem.

Theorem: *For given matrices (α), (β), (α') and (β') with dimensions $m \times n$, $p \times q$, $m' \times n'$ and $p' \times q'$, respectively, with $m' = n$ and $p' = q$, the following holds:*

$$((\alpha) \otimes (\beta))((\alpha') \otimes (\beta')) = ((\alpha)(\alpha')) \otimes ((\beta)(\beta')). \qquad (4.10)$$

Proof: We consider[1] the products $(\gamma) = (\alpha)(\alpha')$ and $(\delta) = (\beta)(\beta')$, which exist not only for square matrices but also for any matrices whose dimensions are chosen as above. Then,

$$((\gamma) \otimes (\delta)) = \begin{pmatrix} (\Delta_{11}) & (\Delta_{12}) & \cdots & (\Delta_{1n'}) \\ (\Delta_{21}) & (\Delta_{22}) & \cdots & (\Delta_{2n'}) \\ \ldots\ldots & \ldots\ldots & \cdots & \ldots\ldots \\ (\Delta_{m1}) & (\Delta_{m2}) & \cdots & (\Delta_{mn'}) \end{pmatrix},$$

$$(\Delta_{ij}) = ((\alpha)(\alpha'))_{ij}(\beta)(\beta') = \sum_k (\alpha)_{ik}(\alpha')_{kj}(\beta)(\beta') = \sum_k (\alpha)_{ik}(\beta)(\alpha')_{kj}(\beta'),$$

In other words,

$$(\Delta_{ij}) = \sum_k (A_{ik})(A'_{kj}), \ (A_{ik}) = (\alpha)_{ik}(\beta), \ (A'_{kj}) = (\alpha')_{ik}(\beta). \qquad (4.11)$$

---

[1] The proof can be omitted without consequences if the reader is not interested in mathematical rigor.

One can easily see that

$$(\alpha) \otimes (\beta) = \begin{pmatrix} (A_{11}) & (A_{12}) & \cdots & (A_{1n}) \\ (A)_{21} & (A_{22}) & \cdots & (A_{2n}) \\ \cdots\cdots & \cdots\cdots & \cdots & \cdots\cdots \\ (A_{m1}) & (A_{m2}) & \cdots & (A_{mn}) \end{pmatrix},$$

$$(\alpha') \otimes (\beta') = \begin{pmatrix} (A'_{11}) & (A'_{12}) & \cdots & (A'_{1q}) \\ (A'_{21}) & (A'_{22}) & \cdots & (A'_{2q}) \\ \cdots\cdots & \cdots\cdots & \cdots & \cdots\cdots \\ (A_{n1}) & (A_{n2}) & \cdots & (A_{nq}) \end{pmatrix}.$$

Comparing this with Eq. (4.11), we get

$$[((\alpha) \otimes (\beta))((\alpha') \otimes (\beta'))]_{ij} = \sum_k A_{ik} A'_{kj} = (\Delta)_{ij}, \qquad (4.12)$$

$$[((\alpha) \otimes (\beta))((\alpha') \otimes (\beta'))]_{ij} = [((\alpha)(\alpha')) \otimes ((\beta)(\beta'))]_{ij}.$$

Thus, finally,

$$((\alpha) \otimes (\beta))((\alpha') \otimes (\beta')) = ((\alpha)(\alpha')) \otimes ((\beta)(\beta')). \qquad (4.13)$$

Consider now three elements $A, B, C \in G$ such that $AB = C$. Equation (4.5) implies that

$$
\begin{aligned}
T(A)T(B) &= (T^a(A) \otimes T^b(A))(T^a(B) \otimes T^b(B)) \\
&= (T^a(A)T^a(B)) \otimes (T^b(A)T^b(B)) \\
&\Rightarrow T(A)T(B) = T^a(AB) \otimes T^b(AB) = T^a(C) \otimes T^b(C),
\end{aligned}
$$

i.e.

$$AB = C \Rightarrow T(A)T(B) = T(C). \qquad (4.14)$$

This shows that the direct product of two representations of a group is another representation.

The direct product has applications in constructing many particle wave functions. In this case, we often have to consider two different functions ψ_i, $i = 1, 2, \ldots, n$ and χ_j, $j = 1, 2, \ldots, n'$ and two operators A and B acting on the functions ψ_i and χ_j, respectively:

$$A\psi_i = \sum_k \alpha_{ki}\psi_k, \ B\chi_j = \sum_k \beta_{mj}\chi_m \Rightarrow AB(\psi_i\chi_j) = \sum_{km} \alpha_{ki}\beta_{mj}(\psi_k\chi_m).$$

Making the correspondence

$$(\psi_k\chi_m) \Leftrightarrow |q\rangle,$$

one finds

$$(AB)|q\rangle = \sum_p \gamma_{pq}|p\rangle,$$

where the matrix (γ) is the direct product of (α) and (β) (see Eq. (4.6)).

The proof is not very hard, but we prefer to illustrate it with a simple example with two functions, $\psi_1, \psi_2\ \chi_1, \chi_2$:

$$|1\rangle = (A\psi_1)(B\chi_1) = \sum_k \alpha_{ki}\psi_k \sum_m \beta_{mj}\chi_m$$

$$= \alpha_{11}\beta_{11}(\psi_1\chi_1) + \alpha_{11}\beta_{21}(\psi_1\chi_2) + \alpha_{21}\beta_{11}(\psi_2\chi_1) + \alpha_{21}\beta_{21}(\psi_2\chi_2)$$

$$= \alpha_{11}\beta_{11}|1\rangle + \alpha_{11}\beta_{21}|2\rangle + \alpha_{21}\beta_{11}|3\rangle + \alpha_{21}\beta_{21}|4\rangle.$$

These coefficients, however, are nothing but the first column of the table γ of Eq. (4.8). The remaining columns are constructed similarly.

4.6   Equivalent representations

Let us consider the group

$$G = \{A_1 = E, A_2, A_3, \ldots, A_g\}$$

and two of its representations, namely

$$T_1 = \{T_1(A_1), T_1(A_2), T_1(A_3), \ldots, T_1(A_g)\},$$
$$T_2 = \{T_2(A_1), T_2(A_2), T_2(A_3), \ldots, T_2(A_g)\}.$$

We say that they are **equivalent** if a non-singular matrix S exists such that

$$T_2(X) = S^{-1}T_1(X)S \quad \text{for every } X \in G.$$

It is not necessary for the two representations to be of the same dimension. We can always increase the dimension of the one with the lowest dimension by adding suitable rows and columns with ones along the diagonal and zeros everywhere else, see case 1 in the previous section.

4.7 Representation construction in the space of functions

Consider an invertible transformation A acting on the coordinates, $\mathbf{r}' = A\mathbf{r}$. Which is the transformation O_A induced by A on the space of functions $\psi(\mathbf{r})$, which depend on the coordinates? We demand

$$\mathbf{r} \to \mathbf{r}', \psi \to \psi' \text{ such that } \psi'(\mathbf{r}') = \psi(\mathbf{r}) \qquad (4.15)$$

i.e.

$$O_A\psi(\mathbf{r}') = \psi(\mathbf{r}) \Rightarrow O_A\psi(A\mathbf{r}) = \psi(\mathbf{r}) \qquad (4.16)$$

or

$$O_A\psi(\mathbf{r}) = \psi(A^{-1}\mathbf{r}), \qquad (4.17)$$

which defines the operator O_A. Obviously, if $AB = C$, we get

$$O_{AB}\psi(\mathbf{r}) = \psi((AB)^{-1}\mathbf{r}) = \psi(B^{-1}A^{-1}\mathbf{r}) = \psi(C^{-1}\mathbf{r}) = O_C\psi(\mathbf{r}).$$

In other words,

$$AB = C \Rightarrow O_A O_B = O_C.$$

Similarly,

$$(O_A)^{-1} = O_{A^{-1}},$$

i.e. a representation of G.

We apply this in some simple cases of practical interest.

4.7.1 *Some examples*

Example 1: Let us consider the group $G = \{E, I\}$ consisting of two elements, the identity and the inversion of coordinates. Then,

$$T_1(E) = \begin{pmatrix} 1 & 0 & 0 \\ 0 & 1 & 0 \\ 0 & 0 & 1 \end{pmatrix}, \quad T_1(I) = \begin{pmatrix} -1 & 0 & 0 \\ 0 & -1 & 0 \\ 0 & 0 & -1 \end{pmatrix},$$

and according to Eq. (4.17),

$$O_E\psi(\mathbf{r}) = \psi(\mathbf{r}), \quad O_I\psi(\mathbf{r}) = \psi(-\mathbf{r}).$$

In other words, the action of the two operators is determined by the values of the functions $\psi(\mathbf{r})$ and $\psi(-\mathbf{r})$. We can now select as a basis,

$$|1\rangle = \psi(\mathbf{r}), \quad |2\rangle = \psi(-\mathbf{r}) \Rightarrow O_E|1\rangle = |1\rangle,$$
$$O_I|1\rangle = |2\rangle, \quad O_E|2\rangle = |2\rangle, \quad O_I|2\rangle = |1\rangle.$$

Then, in this basis, we have the representation

$$T_2'(E) = \begin{pmatrix} 1 & 0 \\ 0 & 1 \end{pmatrix}, \quad T_2'(I) = \begin{pmatrix} 0 & 1 \\ 1 & 0 \end{pmatrix}. \tag{4.18}$$

We can, however, select as a basis the set of the even and odd functions, respectively,

$$|1\rangle = \frac{1}{\sqrt{2}} (\psi(\mathbf{r}) + \psi(-\mathbf{r})), \quad |2\rangle = \frac{1}{\sqrt{2}} (\psi(\mathbf{r}) - \psi(-\mathbf{r}))$$
$$\Rightarrow O_E|1\rangle = O_I|1\rangle = |1\rangle, O_E|2\rangle = O_I|1\rangle = |2\rangle,$$

In such a basis, we find the representation

$$T_2(E) = \begin{pmatrix} 1 & 0 \\ 0 & 1 \end{pmatrix}, \quad T_2(I) = \begin{pmatrix} 1 & 0 \\ 0 & -1 \end{pmatrix}. \tag{4.19}$$

The experienced reader can easily see that the basis of Eq. (4.19) can be obtained after a diagonalization of the matrix $T_2'(I)$. Hence, these two representations are not only isomorphic but equivalent as well. On the other hand, the representations T_1 and T_2 are isomorphic but not equivalent (even if the latter is extended to 3D, as we have indicated above). There is no transformation S such that $S^{-1}T_2(I)S = T_1(I)$).

Example 2: Consider the transformation consisting of the translation of coordinates

$$A: \quad x \to x - a, a = \text{constant}. \tag{4.20}$$

We now find the corresponding transformation O_a induced in the space of analytic functions $f(x)$.

The desired equation is

$$O_A f(x) = f(A^{-1}x) = f(x+a).$$

However,

$$f(x+a) = f(x) + \frac{a}{1!}\frac{df(x)}{dx} + \frac{a^2}{2!}\frac{d^2 f(x)}{dx^2} + \cdots$$

$$= \left(1 + \frac{a}{1!}\frac{d}{dx} + \frac{a^2}{2!}\frac{d^2}{dx^2} + \cdots\right) f(x),$$

Hence,

$$f(x+a) = e^{a\frac{d}{dx}} f(x) \Rightarrow O_A = O_a = e^{a\frac{d}{dx}}.$$

We verify that

$$O_a f = e^{a\frac{d}{dx}} f = \sum_n^{\infty} a^n \frac{d^n}{dx^n} f = f(x+a).$$

Ordinarily in quantum mechanics, instead of the operator $\frac{d}{dx}$, the momentum operator $p_x = \frac{\hbar}{i}\frac{d}{dx}$ is employed, the latter being Hermitian. Thus,

$$O_a = e^{ia\frac{p_x}{\hbar}}. \tag{4.21}$$

Equation (4.20) can easily be extended in 3D:

$$A: \quad \mathbf{r} \to \mathbf{r} - \mathbf{a}, \ \mathbf{a} = \text{constant}. \tag{4.22}$$

Proceeding as above, we find that the analog of Eq. (4.21) is

$$O_A = O_{\mathbf{a}} = e^{i\mathbf{a}\cdot\mathbf{p}/\hbar}, \quad \mathbf{p} = \frac{\hbar}{i}\nabla. \tag{4.23}$$

Example 3: Consider the transformation[2] (rotation) $A = R(-\theta)$ in the coordinate space in 2D, e.g. in the (x,y) plane[3]:

$$\begin{pmatrix} x' \\ y' \end{pmatrix} = \begin{pmatrix} \cos\theta & -\sin\theta \\ \sin\theta & \cos\theta \end{pmatrix}\begin{pmatrix} x \\ y \end{pmatrix} \tag{4.24}$$

(see, also, Section 4.7.2). We want to find the transformation in the space of functions $f(x,y)$, i.e.

$$O_\theta f(x,y) = f(\tilde{x},\tilde{y}), \ \begin{pmatrix} \tilde{x} \\ \tilde{y} \end{pmatrix} = R(\theta)\begin{pmatrix} x \\ y \end{pmatrix}, \quad R(\theta) = \begin{pmatrix} \cos\theta & \sin\theta \\ -\sin\theta & \cos\theta \end{pmatrix}$$

---

[2] The derivations of this example can be omitted on first reading.
[3] Intentionally, we write in the plane rather than around an axis perpendicular to it. In 3D, it makes no difference, since there exists a one-to-one correspondence between the two, i.e. three axes and three planes. This correspondence does not exist in higher dimensions. In 4D, for example, we have four axes but six planes, i.e. the number of possible rotations is six.

(note the sign change in the variable θ). Thus,

$$O_\theta = f\left(x\cos\theta + y\sin\theta, x(-)\sin\theta + y\cos\theta\right).$$

The corresponding infinitesimal transformation is

$$D_\epsilon f(x,y) = f(x,y) + \epsilon\left(\frac{\partial f}{\partial\tilde{x}}\frac{d\tilde{x}}{d\theta} + \frac{\partial f}{\partial\tilde{y}}\frac{d\tilde{y}}{d\theta}\right)_{\theta=0}$$

$$= f(x,y) + \epsilon\left(y\frac{\partial f}{\partial x} + (-x)\frac{\partial f}{\partial y}\right)$$

or

$$D_\epsilon f(x,y) = (1 - \epsilon X)\,f(x,y) = \left(1 - i\epsilon\frac{1}{\hbar}L_{xy}\right)f(x,y),$$

with

$$X = x\frac{\partial}{\partial y} - y\frac{\partial}{\partial x}, \quad L_{xy} = \frac{\hbar}{i}\left(x\frac{\partial}{\partial y} \quad y\frac{\partial}{\partial x}\right).$$

The finite transformation becomes

$$O_\theta = R_{xy} = e^{-i\theta L_{xy}/\hbar}.$$

The operator L_{xy} is known in quantum mechanics as the angular momentum operator in 2D x,y. It can easily be generalized in 3D as follows:

$$L_{jk} = \frac{\hbar}{i}\left(x_j\frac{\partial}{\partial x_k} - x_k\frac{\partial}{\partial x_j}\right). \tag{4.25}$$

The corresponding transformations in the group of rotations in 3D are

$$O_{\theta_{jk}} \Leftrightarrow R_{jk} = e^{-i\theta_{jk}L_{jk}/\hbar}. \tag{4.26}$$

We note in the case of an axis of second order, C_2, we find

$$T(C_2) = R(\pi) = \begin{pmatrix} -1 & 0 \\ 0 & -1 \end{pmatrix}, \tag{4.27}$$

see Eq. (4.24).

Example 4: We examine the case of reflection representations.

Consider a point in the plane $P \Leftrightarrow (x,y)$ and its mirror point $P' \Leftrightarrow (x',y')$ with respect to a plane perpendicular to the plane defined by the two lines OP and OP' with a trace OQ', forming an angle θ with the x-axis, which can be judiciously chosen (see Fig. 4.1(b)). Then, $\vec{PP'} = d\hat{d}$ with $\hat{d}$ a unit vector perpendicular to OQ':

$$\hat{d} = -\sin\theta\hat{e}_1 + \cos\theta\hat{e}_2,$$

$$\vec{OR} = \frac{d}{2}\hat{d} + \mathbf{r}, \quad \hat{d}\cdot\vec{OR} = 0 \Rightarrow \frac{d}{2} + \hat{d}\cdot\mathbf{r} = 0 \Rightarrow d = -2\mathbf{r}\cdot\hat{d}.$$

Since, however, $\mathbf{r}' = \mathbf{r} + d\hat{d}$, we find

$$\mathbf{r}' = \mathbf{r} - 2\left(\mathbf{r}\cdot\hat{d}\right)\hat{d}. \tag{4.28}$$

The required relation follows by successively applying Eq. (4.28) for $\mathbf{r} = \hat{e}_1, \mathbf{r}' = \hat{e}'_1$ and $\mathbf{r} = \hat{e}_2, \mathbf{r}' = \hat{e}'_2$. Then, considering $\hat{e}'_1 = T\hat{e}_1$ and $\hat{e}'_2 = T\hat{e}_2$, we find

$$T\hat{e}_1 = \hat{e}_1 - 2\left(\hat{e}_1\cdot\hat{d}\right)\hat{d} = \hat{e}_1 - 2(-\sin\theta)\left(-\sin\theta\hat{e}_1 + \cos\theta\hat{e}_2\right)$$

$$= \cos 2\theta\hat{e}_1 + \sin 2\theta\hat{e}_2.$$

Similarly,

$$T\hat{e}_2 = \hat{e}_2 - 2\left(\hat{e}_2\cdot\hat{d}\right)\hat{d} = \hat{e}_2 - 2(\cos\theta)\left(-\sin\theta\hat{e}_1 + \cos\theta\hat{e}_2\right)$$

$$= \sin 2\theta\hat{e}_1 - \cos 2\theta\hat{e}_2.$$

Consequently,

$$T_X = \begin{pmatrix} \cos 2\theta & \sin 2\theta \\ \sin 2\theta & -\cos 2\theta \end{pmatrix}. \tag{4.29}$$

In the case of C_{2v} we have two zero trace reflections with angles $\theta = \pi/2$ and $\theta = 2\pi$ corresponding to the reflections σ_{1v} and σ_{2v}. Thus,

$$T(\sigma_1) = \begin{pmatrix} -1 & 0 \\ 0 & 1 \end{pmatrix}, \quad T(\sigma_2) = \begin{pmatrix} 1 & 0 \\ 0 & -1 \end{pmatrix}, \quad T(\sigma_1\sigma_2) = T(I) = \begin{pmatrix} -1 & 0 \\ 0 & -1 \end{pmatrix}. \tag{4.30}$$

The last equation coincides with the representation of rotation through π given by Eq. (4.27) and the representation of the inversion operation. Through Eq. (4.30) and the identity matrix, we construct the 2D representation of C_{2v}. This symmetry involves a second-order axis of rotation and reflections through two planes perpendicular to each other containing that axis.

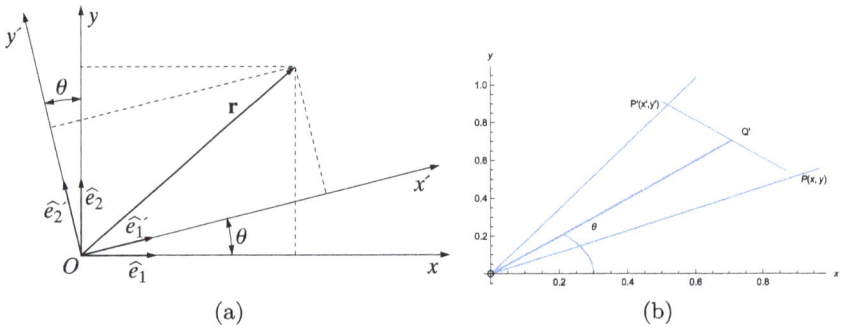

Fig. 4.1. (a) Rotation of the coordinate axis by an angle θ from $\hat{e}_1, \hat{e}_2$ to $\hat{e}'_1, \hat{e}'_2$. (b) Reflection (mirror image) of a point $P(x, y)$ with respect to a plane perpendicular to the plane of the figure forming a trace OQ' with an angle θ with the x-axis. The line PP' is perpendicular to OQ'.

4.7.2 *Application in the case of C_{3v}*

We now construct the faithful 2×2 representation of C_{3v}. We encounter two types of operators, rotations with respect to an axis of third order and reflections with respect to three planes containing this axis.

- Rotations in a plane (see Fig. 4.1(a)).
  Suppose a rotation by an angle θ of the coordinate axis from e_1, e_2 to e'_1, e'_2. We have

$$R|e_1\rangle = \langle e_1|e'_1\rangle|e_1\rangle + \langle e_2|e'_1\rangle|e_2\rangle = \cos\theta|e_1\rangle + \sin\theta|e_2\rangle \Leftrightarrow \begin{pmatrix} \cos\theta \\ \sin\theta \end{pmatrix}.$$

This is the first column. Similarly,

$$R|e_2\rangle = \langle e_1|e'_2\rangle|e_1\rangle + \langle e_2|e'_2\rangle|e_2\rangle = \cos\left(\theta + \frac{\pi}{2}\right)|e_1\rangle + \cos\theta|e_2\rangle \Leftrightarrow \begin{pmatrix} -\sin\theta \\ \cos\theta \end{pmatrix}.$$

This is the second column. Thus,

$$T(\theta) = \begin{pmatrix} \cos\theta & -\sin\theta \\ \sin\theta & \cos\theta \end{pmatrix}. \tag{4.31}$$

We are interested in $(\theta = 2\pi/3, 4\pi/3, 2\pi) \Leftrightarrow (C_3, (C_3)^2, (C_3)^3)$ with $(C_3)^3 = E$. Hence,

$$E = \begin{pmatrix} 1 & 0 \\ 0 & 1 \end{pmatrix}, \quad C_3 = T\left(\frac{2\pi}{3}\right) = \begin{pmatrix} -\frac{1}{2} & -\frac{\sqrt{3}}{2} \\ \frac{\sqrt{3}}{2} & -\frac{1}{2} \end{pmatrix},$$

$$C_3^2 = T\left(\frac{4\pi}{3}\right) = \begin{pmatrix} -\frac{1}{2} & \frac{\sqrt{3}}{2} \\ -\frac{\sqrt{3}}{2} & -\frac{1}{2} \end{pmatrix}.$$

- Reflections. In the case of C_{3v}, we indicate by P_i the traces NH of the planes of symmetry, ordered in some fashion as NH_1, NH_2, NH_3, see Fig. 2.4. Suppose we now select the axes (x, y) so that the axis of y coincides with P_1. Then, $P_1 \Leftrightarrow \theta = \pi/2, P_2 \Leftrightarrow \theta = \pi/2 + 2\pi/3, P_3 \Leftrightarrow \theta = \pi/2 + 4\pi/3$, and we can make the correspondences

$$\theta = \frac{\pi}{2} \Leftrightarrow \begin{pmatrix} -1 & 0 \\ 0 & 1 \end{pmatrix} = T(\sigma_1), \theta = \pi/2 + 2\pi/3 \Leftrightarrow \begin{pmatrix} \frac{1}{2} & \frac{\sqrt{3}}{2} \\ \frac{\sqrt{3}}{2} & -\frac{1}{2} \end{pmatrix} = T(\sigma_2),$$

$$\theta = \pi/2 + 4\pi/3 \Leftrightarrow \begin{pmatrix} \frac{1}{2} & -\frac{\sqrt{3}}{2} \\ -\frac{\sqrt{3}}{2} & -\frac{1}{2} \end{pmatrix} = T(\sigma_3).$$

The above representation of C_{3v} is the faithful representation with the lowest dimension. One can clearly construct faithful representations of higher dimension, in addition to the 6×6 regular representation constructed from the multiplication table. This can be seen from the following example.

Example 5: We construct a 3×3 faithful representation of C_{3v}. Obviously, the identity element is the diagonal matrix with elements 1. Considering the above arrangement NH_1, NH_2, NH_3, C_3 brings the system to the position NH_3, NH_1, NH_2, corresponding to the permutation (132). Then,

$$C_3|1\rangle = |3\rangle, \quad C_3|2\rangle = |1\rangle, \quad C_3|3\rangle = |2\rangle \Rightarrow T_{C_3} = \begin{pmatrix} 0 & 1 & 0 \\ 0 & 0 & 1 \\ 1 & 0 & 0 \end{pmatrix}.$$

Similarly, the rotation $(C_3)^2$ brings the system to NH_2, NH_3, NH_1, corresponding to the permutation (123). Then,

$$(C_3)^2|1\rangle = |2\rangle, \quad (C_3)^2|2\rangle = |3\rangle, \quad (C_3)^2|3\rangle = |1\rangle \Rightarrow T_{C_3^2} = \begin{pmatrix} 0 & 0 & 1 \\ 1 & 0 & 0 \\ 0 & 1 & 0 \end{pmatrix}.$$

Reflection σ_1 corresponds to the permutation (23), σ_2 to (13) and σ_3 to (12). Thus,

$$
T_{\sigma_1} = \begin{pmatrix} 1 & 0 & 0 \\ 0 & 0 & 1 \\ 0 & 1 & 0 \end{pmatrix}, \quad
T_{\sigma_2} = \begin{pmatrix} 0 & 0 & 1 \\ 0 & 1 & 0 \\ 1 & 0 & 0 \end{pmatrix}, \quad
T_{\sigma_3} = \begin{pmatrix} 0 & 1 & 0 \\ 1 & 0 & 0 \\ 0 & 0 & 1 \end{pmatrix}.
$$

The reader is encouraged to check whether this representation is equivalent to the one obtained by enlarging the above faithful 2D representation in the usual way, i.e. by adding one more row and one column with zeros everywhere except along the diagonal, which is filled with ones.

4.7.3 *Application in the case of S_6*

Example 6: We consider the faithful $3 \otimes 3$ representation of the symmetry S_6 we encountered in Section 2.4 (not to be confused with the symmetric group S_n). This representation can be generated by the matrix (s_6):

$$
(s_6) = \begin{pmatrix} -1 & 0 & 0 \\ 0 & -\frac{1}{2} & -\frac{\sqrt{3}}{2} \\ 0 & \frac{\sqrt{3}}{2} & -\frac{1}{2} \end{pmatrix}, (s_6)^2, (s_6)^3, (s_6)^4, (s_6)^5, (s_6)^6
$$

$$
= E = \begin{pmatrix} 1 & 0 & 0 \\ 0 & 1 & 0 \\ 0 & 0 & 1 \end{pmatrix}. \tag{4.32}
$$

This symmetry can characterize a molecule like that[4] of C_2H_6. The three hydrogen atoms connected to each carbon atom lie at the angles of an equilateral triangle. The two triangles thus formed are parallel, and their centers are superimposed over each other. However, the triangles do not form a right prism, as they are rotated at an angle of $\pi/3$ toward each other. The two carbon atoms lie between the triangles on the x axis, which

---

[4]In the lowest energy configuration, the hydrogen atoms connected to each carbon atom will tend to be as far away from each other, and the other carbon atom, as possible while still maintaining the correct distance from the carbon atom they are connected to. This is because the hydrogen atoms repel each other and are also repelled by the other carbon atom. They will, therefore, spread themselves into a configuration similar to that seen in methane, i.e. (approximately) tetrahedral rather than planar. It is for this same reason (hydrogen atoms repelling each other) that the molecule takes on the staggered conformation, though in this case, the repulsion is between the H atoms connected to one carbon atom and the H atoms connected to the other carbon atom.

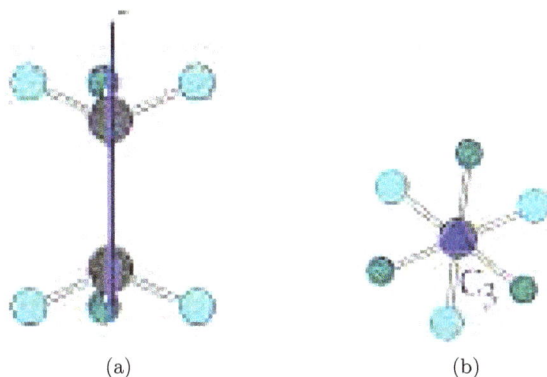

(a)   (b)

Fig. 4.2. (a) A molecule of C_2H_6 in the staggered conformation. The hydrogen atoms of each CH_3 lie at the corners of an equilateral triangle, and the carbon atoms lie on the central axis (x-axis, in blue), which intersects the equilateral triangles at their centers. The lower part is exactly like the upper part but rotated by $\pi/3$ with respect to the x-axis. (b) The system as viewed from the top, the axis now appearing as the sixth order.

is the line connecting the centers of the two triangles, see Fig. 4.2(a). Clearly, this arrangement has a symmetry specified by the third-order axis x with elements C_3 and C_3^2. Then, it has a symmetry with a rotation of $\pi/3$, C_6, and a simultaneous reflection σ_h with respect to a plane perpendicular to the x-axis and in the middle of the distance between the carbon atoms. The result is an improper rotation (orthogonal transformation with determinant -1). Viewed from the top, we see that the set of the symmetry operations, see Fig.4.2(b), are improper orthogonal rotations $(S_6)^k$, $k = 1, 2, \ldots, 6$, $(S_6)^6 = E$. In fact,

$$(S_6)^1 = C_6\sigma_h = \begin{pmatrix} 1 & 0 & 0 \\ 0 & \frac{1}{2} & -\frac{\sqrt{3}}{2} \\ 0 & \frac{\sqrt{3}}{2} & \frac{1}{2} \end{pmatrix} \begin{pmatrix} -1 & 0 & 0 \\ 0 & 1 & 0 \\ 0 & 0 & 1 \end{pmatrix} = \begin{pmatrix} -1 & 0 & 0 \\ 0 & -\frac{1}{2} & -\frac{\sqrt{3}}{2} \\ 0 & \frac{\sqrt{3}}{2} & -\frac{1}{2} \end{pmatrix},$$

$$(S_6)^2 = (C_6\sigma_h)^2 = C_6^2 = C_3 = \begin{pmatrix} 1 & 0 & 0 \\ 0 & -\frac{1}{2} & -\frac{\sqrt{3}}{2} \\ 0 & \frac{\sqrt{3}}{2} & -\frac{1}{2} \end{pmatrix},$$

$$(S_6)^4 = (C_6\sigma_h)^4 = C_6^4 = C_3^2 = \begin{pmatrix} 1 & 0 & 0 \\ 0 & -\frac{1}{2} & \frac{\sqrt{3}}{2} \\ 0 & -\frac{\sqrt{3}}{2} & -\frac{1}{2} \end{pmatrix}.$$

The rotations and the reflection commute, since the reflection plane is perpendicular to the axes of rotation. Proceeding similarly, we find two improper rotations $(S_6)^3$ and $(S_6)^5$, completing the operations of S_6, as given by Eq. (4.32). So, this group is Abelian.

Such configurations, known as staggered conformations, are common not only in individual molecules but also in crystallography, see Tables 8.1 and 17.1.

The above group is isomorphic to the group

$$G = \{E, C_3, C_3^2, \sigma_h, C_3\sigma_h, C_3^2\sigma_h\}.$$

It is also isomorphic to a similar symmetry involving inversion symmetry instead of reflection, see problem 3.7.7.

4.8   Problems

4.1.1 Examine whether from the representation $\Gamma(X)$ of a group G, it can be concluded that $(\Gamma(X))^{-1}$ and $(\Gamma(X))^{+}$ can also be representations of G.

4.1.2 (i) Show that a representation of the group C_{2h} (problem 2.5.7) is of the form

$$T(E) = \begin{pmatrix} 1 & 0 & 0 \\ 0 & 1 & 0 \\ 0 & 0 & 1 \end{pmatrix}, \quad T(C_2) = \begin{pmatrix} -1 & 0 & 0 \\ 0 & -1 & 0 \\ 0 & 0 & 1 \end{pmatrix},$$

$$T(I) = \begin{pmatrix} -1 & 0 & 0 \\ 0 & -1 & 0 \\ 0 & 0 & -1 \end{pmatrix}, \quad \sigma_h = \begin{pmatrix} 1 & 0 & 0 \\ 0 & 1 & 0 \\ 0 & 0 & -1 \end{pmatrix}.$$

The reader should not be satisfied by a mere confirmation that this is indeed a representation. It should be understood that the rotation is a mere extension in 3D of Eq. (4.27). Is the representation of σ_h and I more or less obvious?

(ii) Based on the above results, construct the 2D representation of C_{2v} (problem 2.5.7).

4.1.3 You are given the functions $\psi(\mathbf{r})$, $\psi(-\mathbf{r})$, $\phi(\mathbf{r})$, $\phi(-\mathbf{r})$. Find the matrix T_I corresponding to the inversion operator I. Find a basis in which this operator is diagonal.

4.3.1 Construct the regular representation of the group C_{4v}.

4.3.2 Find the regular representation of the group generated by the matrices

$$\begin{pmatrix} 0 & 1 \\ -1 & 0 \end{pmatrix}, \quad \begin{pmatrix} 0 & 1 \\ 1 & 0 \end{pmatrix}.$$

Is it isomorphic to that of the previous problem?

4.3.3 Do the same for the group $G_1 \otimes G_2$ of problem 3.7.2.

4.3.4 Do the same for the group of problem 3.7.3.

4.3.5 Do the same for the group of problem 3.7.4.

4.3.6 Do the same for the group of problem 3.7.5.

4.3.7 Do the same for the group of problem 3.7.6.

4.3.8 Do the same for the group of problem 3.7.7.

4.7.1 Construct the representation of C_{3v}, defined in the space of functions x^2, xy and y^2, when the symmetry axis is perpendicular to the (x, y) plane, passing through the origin.

4.7.2 Consider the group $G = \{E, C_4^2, \sigma_x, \sigma_y\}$. Show that one of its representations can be defined starting from the function $\phi(x, y) = e^{i(ax+by)}$, with a and b constants. Which are the other functions of this space? Do you recognize this representation?

4.7.3 Find the representation of C_{4v}, defined in the space of functions $e^{\pm imx}$, $e^{\pm imy}$, for a given $m =$ integer > 0. The axis of symmetry is assumed to coincide with z.

4.7.4 Do the same in the space of functions $\cos mx$, $\cos ny$, $m, n =$ integers.

4.7.5 Do the same in the space of the function $\phi(x) = x$.

4.7.6 Do the same starting with functions indicated in the following cases: (i) $\phi(x) = x^2$, (ii) $\phi(x) = x^3$, (iii) $\phi(x) = x^2y$ and (iv) $\phi(x, y) = e^{i(ax+by)}$, with a and b constants, $a \neq b$.

Chapter 5

Representation Reduction — Schur's Lemmas — Character Tables

We have seen in Chapter 4 that there exist many representations of a given group. Not all of them are equally important. In this chapter, after introducing the notion of irreducibility of a given set of representations of a given group, we will see that a selected set of them, called irreducible, play an important role. We will also see that an arbitrary representation can be reduced to irreducible ones, which are sufficient for the study of the applications of a discrete symmetry. We will also devise techniques that can be used to accomplish this reduction.

5.1 Representation reduction

We have seen that there exist many representations of a given group. Are they all important? To provide an answer to this, we introduce an important concept called the reducibility of a given representation.

We say that a representation $T(X), X \in G$ of dimension $n \times n$ is **reducible**, if there exists a non-singular matrix S and two representations $T_1(X)$ and $T_2(X)$, $X \in G$, of dimensions $n_1 \times n_1$ and $n_2 \times n_2$, respectively, such that

$$S^{-1}T(X)S = \begin{pmatrix} T_1(X) & 0 \\ 0 & T_2(X) \end{pmatrix}, \quad \text{for every } X \in G, \ n_1 + n_2 = n.$$

(5.1)

Then, we say that $T(X)$ is reduced to two representations of smaller dimensions. Otherwise, the representation is said to be **irreducible**. The transformation S is called a similarity transformation, and the two representations related this way are called **similar representations**.

It can be shown that the similarity transformations have some interesting properties:

(a) They preserve the trace, namely the sum of the diagonal elements;
(b) they preserve the value of the determinant of a matrix; and
(c) they do not change the eigenvalues of the matrix.

For details, the reader is referred to the literature, e.g. (Bishop, 2017).

The above process can be continued, if $T_1(X)$ and/or $T_2(X)$ are reducible. Continuing this way, one can finally obtain

$$S^{-1}T(X)S = \begin{pmatrix} T_1(X) & 0 & 0 & 0 \\ 0 & T_2(X) & 0 & 0 \\ \cdots & \cdots & \cdots & 0 \\ 0 & 0 & 0 & T_k(X) \end{pmatrix}, \quad \text{for every } X \in G, \quad (5.2)$$

where $n_1 + n_2 + \cdots + n_k = n$. The process ends when all representations along the diagonal are irreducible or 1D.

The representation of C_{2h}, problem 4.1.2 is fully reduced. No further reduction is possible.

Example 1: We consider the simple case of an Abelian group formed from the elements of algebra $1, -1, i -i$. From the multiplication table, we obtain the regular representation

$$T(A_1 = E) = \begin{pmatrix} 1 & 0 & 0 & 0 \\ 0 & 1 & 0 & 0 \\ 0 & 0 & 1 & 0 \\ 0 & 0 & 0 & 1 \end{pmatrix}, \quad T(A_2) = \begin{pmatrix} 0 & 1 & 0 & 0 \\ 1 & 0 & 0 & 0 \\ 0 & 0 & 0 & 1 \\ 0 & 0 & 1 & 0 \end{pmatrix},$$

$$T(A_3) = \begin{pmatrix} 0 & 0 & 1 & 0 \\ 0 & 0 & 0 & 1 \\ 0 & 1 & 0 & 0 \\ 1 & 0 & 0 & 0 \end{pmatrix}, \quad T(A_4) = \begin{pmatrix} 0 & 0 & 0 & 1 \\ 0 & 0 & 1 & 0 \\ 1 & 0 & 0 & 0 \\ 0 & 1 & 0 & 0 \end{pmatrix}.$$

Since these matrices commute, they can be simultaneously diagonalized, Let us choose to diagonalize the matrix $T(A_3)$. Its eigenvalues are $(-1, i, -i, 1)$, and the diagonalizing matrix S is

$$S = \frac{1}{2}\begin{pmatrix} -1 & -1 & -i & i \\ -1 & -1 & i & -i \\ 1 & -1 & 1 & 1 \\ 1 & -1 & -1 & -1 \end{pmatrix}, \quad S^{-1} = \frac{1}{2}\begin{pmatrix} -1 & -1 & 1 & 1 \\ -1 & -1 & -1 & -1 \\ i & -i & 1 & -1 \\ -i & i & 1 & -1 \end{pmatrix},$$

$$\Gamma_1 = \Gamma(E) = \begin{pmatrix} 1 & 0 & 0 & 0 \\ 0 & 1 & 0 & 0 \\ 0 & 0 & 1 & 0 \\ 0 & 0 & 0 & 1 \end{pmatrix}, \quad S^{-1}A_2S = \Gamma_2 = \begin{pmatrix} 1 & 0 & 0 & 0 \\ 0 & 1 & 0 & 0 \\ 0 & 0 & -1 & 0 \\ 0 & 0 & 0 & -1 \end{pmatrix},$$

$$S^{-1}A_3S = \Gamma_3 = \begin{pmatrix} -1 & 0 & 0 & 0 \\ 0 & i & 0 & 0 \\ 0 & 0 & -i & 0 \\ 0 & 0 & 0 & 1 \end{pmatrix}, \quad S^{-1}A_4S = \Gamma_4 = \begin{pmatrix} -1 & 0 & 0 & 0 \\ 0 & 1 & 0 & 0 \\ 0 & 0 & -i & 0 \\ 0 & 0 & 0 & i \end{pmatrix}.$$

The regular representation has been fully reduced to $k = n = 4$ 1×1 submatrices. Something analogous is true for any Abelian group (see Example 3).

It is adequate to know the elements of every irreducible representation (in this example, the diagonal elements). We thus write:

	E	A_2	A_3	A_4
Γ_1	1	1	1	1
Γ_2	1	1	-1	-1
Γ_3	-1	i	$-i$	1
Γ_4	-1	1	$-i$	i

(5.3)

Example 2: We reduce the regular representation of a cyclic group with g elements $\{A, A^2, A^3, \ldots, A^{g-1}, A^g = E\}$. Since $A^g = E$, $\Gamma(A)$ has as eigenvalues the g roots of $\xi^g = 1$, i.e. $1, \xi, \xi^2, \ldots, \xi^{g-1}$. Thus, the matrix $\Gamma(A)$, when diagonalized, takes the form $(S^{-1}\Gamma(A)S)_{\text{diag}} = 1, \xi, \xi^2, \ldots, \xi^{g-1}$, where $\xi = e^{2\pi i/g}$. Since the group is Abelian, all matrices can be simultaneously diagonalized, and in fact, $(S^{-1}\Gamma(A^k)S)_{\text{diag}} = 1, \xi^k, \xi^{2k}, \ldots, \xi^{(g-1)k}$. Proceeding as above, we place the diagonal elements in the columns of the matrix, obtaining Table 5.1.

Table 5.1. The reduction of the regular representation of a cyclic group of order g. The powers of the parameter $\xi = e^{(2\pi i)/g}$ are not independent, since for integers $\lambda > 0$, $\mu > 0$ and $0 < \nu < g$, one finds $\xi^{\lambda g^2 + \mu g + \nu} = \xi^\nu$. Furthermore, since the irreducible representations are all 1D, the character matrix coincides with that of the table given here.

	E	A	A^2	$\cdots$	A^{g-2}	A^{g-1}
E	1	1	1	$\cdots$	1	1
A	1	ξ	ξ^2	$\cdots$	ξ^{g-2}	ξ^{g-1}
A^2	1	ξ^2	ξ^4	$\cdots$	$\xi^{2(g-2)}$	$\xi^{2(g-1)}$
A^3	1	ξ^3	ξ^6	$\cdots$	$\xi^{3(g-2)}$	$\xi^{3(g-1)}$
$\cdots$	$\cdots$	$\cdots$	$\cdots$		$\cdots$	$\cdots$
A^{g-2}	1	$\xi^{(g-2)}$	$\xi^{2(g-2)}$	$\cdots$	$\xi^{(g-2)(g-2)}$	$\xi^{(g-2)(g-1)}$
A^{g-1}	1	$\xi^{(g-1)}$	$\xi^{2(g-1)}$	$\cdots$	$\xi^{(g-1)(g-2)}$	$\xi^{(g-1)(g-1)}$

The orthogonality of the first line (column) with the rest is obvious, since $1 + \xi + \xi^2 + \cdots + \xi^{g-1} = 0$. The rest of the lines (columns) are also mutually orthogonal. In the case, of, for example, columns 2 and 3, noting that $(\xi^k)^* \xi^{(k)} = 1$, we find

$$1 + \xi^* \xi^2 + (\xi^2)^* \xi^4 + \cdots + (\xi^{g-2})^* \xi^{2(g-2)} + (\xi^{g-1})^* \xi^{2(g-1)}$$

$$= 1 + \xi + \xi^2 + \cdots + \xi^{(g-2)} + \xi^{(g-1)} = 0,$$

etc.

The reader who has difficulty understanding the above matrix is encouraged to return to it after studying Section 5.6.

Example 3: We consider now a more involved problem, i.e. the reduction of the regular representation of C_{3v}. This group is non-Abelian, and so, in this case, only one element can be diagonalized. Let us choose it to be the reflection σ_1, i.e. $\Gamma(\sigma_1)$ (see Application 1 of Chapter 4). The corresponding eigenvalues are 1 and -1, both of them triply degenerate. As a result, the associated eigenvectors cannot be determined uniquely. Mathematica is not adequate for the determination of S. The solution to this problem is not easy. To this end, some techniques have been developed based on group theory discussed in Chapter 16.

Returning to C_{3v}, after some luck or repeated attempts, one finds that the desired matrix S is of the form

$$S = \begin{pmatrix} \frac{1}{\sqrt{6}} & \frac{1}{\sqrt{6}} & \frac{1}{\sqrt{3}} & 0 & 0 & \frac{1}{\sqrt{3}} \\ \frac{1}{\sqrt{6}} & \frac{1}{\sqrt{6}} & -\frac{1}{2\sqrt{3}} & \frac{1}{2} & -\frac{1}{2} & \frac{1}{2\sqrt{3}} \\ \frac{1}{\sqrt{6}} & \frac{1}{\sqrt{6}} & -\frac{1}{2\sqrt{3}} & -\frac{1}{2} & \frac{1}{2} & -\frac{1}{2\sqrt{3}} \\ \frac{1}{\sqrt{6}} & -\frac{1}{\sqrt{6}} & -\frac{1}{\sqrt{3}} & 0 & 0 & \frac{1}{\sqrt{3}} \\ \frac{1}{\sqrt{6}} & -\frac{1}{\sqrt{6}} & \frac{1}{2\sqrt{3}} & \frac{1}{2} & \frac{1}{2} & -\frac{1}{2\sqrt{3}} \\ \frac{1}{\sqrt{6}} & -\frac{1}{\sqrt{6}} & \frac{1}{2\sqrt{3}} & -\frac{1}{2} & -\frac{1}{2} & -\frac{1}{2\sqrt{3}} \end{pmatrix}.$$

The eigenvectors are the columns of this matrix ordered according to the eigenvalues $1, -1, -1, 1, -1, 1$, from left to right. We write $\tilde{\Gamma}(X) = S^{-1}\Gamma(X)S$, i.e.

$$\left(\begin{array}{cc|cc|cc} \multicolumn{6}{c}{\tilde{\Gamma}(E)} \\ \hline 1 & 0 & 0 & 0 & 0 & 0 \\ 0 & 1 & 0 & 0 & 0 & 0 \\ \hline 0 & 0 & 1 & 0 & 0 & 0 \\ 0 & 0 & 0 & 1 & 0 & 0 \\ \hline 0 & 0 & 0 & 0 & 1 & 0 \\ 0 & 0 & 0 & 0 & 0 & 1 \end{array} \right) \quad \left(\begin{array}{cc|cc|cc} \multicolumn{6}{c}{\tilde{\Gamma}(\sigma_1)} \\ \hline 1 & 0 & 0 & 0 & 0 & 0 \\ 0 & -1 & 0 & 0 & 0 & 0 \\ \hline 0 & 0 & -1 & 0 & 0 & 0 \\ 0 & 0 & 0 & 1 & 0 & 0 \\ \hline 0 & 0 & 0 & 0 & -1 & 0 \\ 0 & 0 & 0 & 0 & 0 & 1 \end{array} \right),$$

$$\left(\begin{array}{cc|cc|cc} \multicolumn{6}{c}{\tilde{\Gamma}(\sigma_2)} \\ \hline 1 & 0 & 0 & 0 & 0 & 0 \\ 0 & -1 & 0 & 0 & 0 & 0 \\ \hline 0 & 0 & \frac{1}{2} & \frac{\sqrt{3}}{2} & 0 & 0 \\ 0 & 0 & \frac{\sqrt{3}}{2} & -\frac{1}{2} & 0 & 0 \\ \hline 0 & 0 & 0 & 0 & \frac{1}{2} & \frac{\sqrt{3}}{2} \\ 0 & 0 & 0 & 0 & \frac{\sqrt{3}}{2} & -\frac{1}{2} \end{array} \right), \quad \left(\begin{array}{cc|cc|cc} \multicolumn{6}{c}{\tilde{\Gamma}(\sigma_3)} \\ \hline 1 & 0 & 0 & 0 & 0 & 0 \\ 0 & -1 & 0 & 0 & 0 & 0 \\ \hline 0 & 0 & \frac{1}{2} & -\frac{\sqrt{3}}{2} & 0 & 0 \\ 0 & 0 & -\frac{\sqrt{3}}{2} & -\frac{1}{2} & 0 & 0 \\ \hline 0 & 0 & 0 & 0 & \frac{1}{2} & -\frac{\sqrt{3}}{2} \\ 0 & 0 & 0 & 0 & -\frac{\sqrt{3}}{2} & -\frac{1}{2} \end{array} \right),$$

$$\left(\begin{array}{c|cc|cc}
\multicolumn{5}{c}{\tilde{\Gamma}((C_3)|} \\
\hline
1 & 0 & 0 & 0 & 0 & 0 \\
0 & 1 & 0 & 0 & 0 & 0 \\
\cdots\cdots & \cdots & \cdots & \cdots & \cdots \\
0 & 0 & -\frac{1}{2} & -\frac{\sqrt{3}}{2} & 0 & 0 \\
0 & 0 & \frac{\sqrt{3}}{2} & -\frac{1}{2} & 0 & 0 \\
\cdots\cdots & \cdots & \cdots & \cdots & \cdots \\
0 & 0 & 0 & 0 & -\frac{1}{2} & -\frac{\sqrt{3}}{2} \\
0 & 0 & 0 & 0 & \frac{\sqrt{3}}{2} & -\frac{1}{2}
\end{array}\right),\quad
\left(\begin{array}{c|cc|cc}
\multicolumn{5}{c}{\tilde{\Gamma}(C_3^2)|} \\
\hline
1 & 0 & 0 & 0 & 0 & 0 \\
0 & 1 & 0 & 0 & 0 & 0 \\
\cdots\cdots & \cdots & \cdots & \cdots & \cdots \\
0 & 0 & -\frac{1}{2} & \frac{\sqrt{3}}{2} & 0 & 0 \\
0 & 0 & -\frac{\sqrt{3}}{2} & -\frac{1}{2} & 0 & 0 \\
\cdots\cdots & \cdots & \cdots & \cdots & \cdots \\
0 & 0 & 0 & 0 & -\frac{1}{2} & \frac{\sqrt{3}}{2} \\
0 & 0 & 0 & 0 & -\frac{\sqrt{3}}{2} & -\frac{1}{2}
\end{array}\right).$$

$$(5.4)$$

Note that in the upper-left block, we encounter the 1D representations. The other two blocks contain the 2×2 representations, which are nothing more than the representation of Example 2 of Chapter 2. This representation occurs in the regular representation twice.

5.2   Unitary representations

In the case of finite group representations, the following theorem holds.

Theorem 1: *Any given representation $T(X)$ of a finite group G is equivalent to a unitary representation $\Gamma(X), X \in G$. This means that there exists a non-singular matrix S such that*

$$\Gamma(X) = S^{-1}T(X)S, (\Gamma(X))^{-1} = (\Gamma(X))^+ \text{ for every } X \in G \text{ or}$$
$$(\Gamma(X))^+ \Gamma(X) = \Gamma(X)(\Gamma(X))^+ = E, (\Gamma(X))^+_{ij} = \Gamma^*(X)_{ji}. \quad (5.5)$$

For the proof, we construct the quantity

$$H = \sum_{X \in G} T(X)T(X)^+. \quad (5.6)$$

The matrix H is Hermitian, $H^+ = H$, and as a result, there exists a matrix U, which brings it into diagonal form, i.e.

$$U^+ H U = D, \ D_{ij} = d_i \delta_{ij}, \delta_{ij} = \begin{cases} 1 & i = j \\ 0 & i \neq j. \end{cases}$$

We know that the eigenvalues of a Hermitian operator are real. Furthermore, since the operator is positive definite,[1] its eigenvalues are positive,[2] $d_i > 0$.

Consequently, the expressions

$$V = UD^{1/2}, \quad V^{-1} = D^{-1/2}U \tag{5.7}$$

make sense, where

$$(D^{1/2})_{ij} = \delta_{ij}\sqrt{d_i}, \tag{5.8}$$

with the matrix

$$\Gamma(X) = V^{-1}T(X)V \tag{5.9}$$

being unitary. Indeed,

$$\Gamma(X) = D^{-1/2}U^+T(X)UD^{1/2} = D^{-1/2}T'(X)D^{1/2},$$

$$\Gamma^+(X) = D^{1/2}(T'(X))^+D^{-1/2},$$

$$\Gamma(X)\Gamma^+(X) = D^{-1/2}T'(X)D(T'(X))^+D^{-1/2}$$

$$= D^{-1/2}T'(X)\left(\sum_{Y\in G}T'(Y)(T'(Y))^+\right)(T'(X))^+D^{-1/2}$$

---

[1]A real symmetric $n \times n$ matrix M is said to be positive define if the scalar $x^T Mx$ is positive for every non-zero column vector x of n real numbers and x^T its transpose. It follows that every eigenvalue of such a matrix is positive.

[2]In the present case,

$$D = U^{-1}\left(\sum_{X\in G}T(X)T(X)^+\right)U = \sum_{X\in G}\{(U^{-1}T(X)U)(U^{-1}T^+(X)U)\} \Rightarrow$$

$$D = \sum_{X\in G}T'(X)(T'(X))^+, \quad T'(X) = U^{-1}T(X)U, \quad (T'(X))^+ = U^{-1}T^+(X)U$$

$$d_i = D_{ii} = \sum_{X\in G}(T'(X)(T'(X))^+)_{ii}$$

$$= \sum_{X\in G}\sum_j T'(X)_{ij}(T'(X))^+_{ji} = \sum_{X\in G}\sum_j T'(X)_{ij}(T'(X))^*_{ij} \Rightarrow$$

$$d_i = \sum_{X\in G}\sum_j |T'(X)_{ij}|^2 \geq 0.$$

$$= D^{-1/2} \left(\sum_{Y \in G} T'(X) T'(Y) (T'(Y))^+ (T'(X))^+ \right) D^{-1/2}$$

$$= D^{-1/2} \left(\sum_{Y \in G} T'(XY) (T'(XY))^+ \right) D^{-1/2}$$

$$= D^{-1/2} \left(\sum_{Z \in G} T'(Z) (T'(Z))^+ \right) D^{-1/2} \text{ whenever } Z = XY.$$

Thus, finally, noting that the above expression in parenthesis is equal to D (see previous footnote) we obtain:

$$\Gamma(X) \Gamma^+(X) = D^{-1/2} D D^{-1/2} = E.$$

The proof may look complicated, but the conclusion is almost obvious. It is simply saying that in a finite-dimensional space, we can always obtain an orthonormal basis. In this basis, our operators are represented by unitary matrices. This way, we understand why all representations we have encountered so far are unitary, even orthogonal.

The above conclusion does not hold in the case of discrete groups of non-finite order or continuous groups.

Theorem 2: *The regular representation is unitary.*

We have seen that the regular representation satisfies the relation (4.1) and it is real. Hence,

$$\Gamma^+(A_m)_{kj} = \Gamma^*(A_m)_{jk} = \begin{cases} 1, & A_m A_j = A_k \\ 0, & \text{otherwise} \end{cases}$$

$$\Rightarrow \Gamma^+(E)_{kj} = \delta_{jk} = \begin{cases} 1, & k = j \\ 0, & k \neq j. \end{cases}$$

But

$$A_m A_j = A_k \Rightarrow A_m = A_k A_j^{-1} \Rightarrow A_m^{-1} = A_j A_k^{-1} \Rightarrow A_m^{-1} A_k = A_j,$$

and thus,

$$\Gamma(A_m^{-1})_{kj} = \begin{cases} 1, & A_m^{-1} A_k = A_j \\ 0 & \text{otherwise} \end{cases},$$

i.e.

$$\Gamma^+(A_m) = \Gamma(A_m^{-1}) \Rightarrow \Gamma^+(X) = \Gamma(X^{-1}) \quad \text{for every } X \in G.$$

5.3 Characters of representations

We have seen that the representations of a given group are not uniquely defined. Even if the space on which the elements of the group act is selected, the representation depends on the basis chosen. A mere change in basis, leads, of course, to equivalent representations, related by a similarity transformation. Such transformations do not change the character of a matrix. Thus, the characters $\chi(X)$ of a representation $\Gamma(X)$,

$$\chi(X) = tr(X) = \sum_i \Gamma(X)_{ii}, X \in G, \tag{5.10}$$

are numbers characteristic of a representation, not affected by similarity transformations, called **characters of a representation**. Obviously, for a 1D representation, the character coincides with the representation itself. It is easy to see that if two elements A and B are conjugate, i.e. there exists an element $Z \in G$ such that $Z^{-1}AZ = B$, the representations $\Gamma(A)$ and $\Gamma(B)$ have the same character. Indeed, then,

$$\Gamma^{-1}(Z)\Gamma(A)\Gamma(Z) = \Gamma(B) \Rightarrow \chi(A) = \chi(B).$$

Therefore, the representations that correspond to elements of the same class, have the same character. Thus, the number of different traces of representations of a given group cannot exceed that of different classes of the group. Thus, for the representation $\{E, A, B, C, D, F\}$ of C_{3V}, Example 2, Chapter 2, we have three different traces, equal to the number of classes, one for the identity $\chi(E) = 2$, one for the rotations $\chi(A = C_3) = \chi(B = C_3^2) = -1$ and one for the reflections $\chi(C = \sigma_1) = \chi(D = \sigma_2) = \chi(F = \sigma_3)) = 0$. The regular representation of C_{3V} also has the same number of characters, i.e. six for the identity, zero for rotations and zero for reflections. By mere coincidence, the trace of rotations is the same as that of reflections.

The traces of the irreducible representations Γ_1 and Γ_2 (1D) as well as Γ_3 appearing twice (2D) occurring in the regular representation of C_{3V}, are, see Eq. (5.4), as follows: (a) $(1, 1, 2, 2)$ for the identity, (b) $(1, 1, -1, -1)$ for the class of rotations and (c) $(1, -1, 0, 0)$ for the class of reflections.

Before proceeding further into the question of representation reduction, we need the theorem of orthogonality, whose proof depends on the two Schur's lemmas.

5.4   The two lemmas of Schur

First Schur's lemma: *Let $\Gamma^i(X)$ be an irreducible representation of a group G and (ρ) a matrix such that*

$$\Gamma^i(X)(\rho) = (\rho)\Gamma^i(X) \quad \text{for } X \in G. \tag{5.11}$$

Then, $(\rho) = \lambda(\epsilon)$, where ϵ the identity matrix. In other words, (ρ) is a multiple of the identity, provided that the representation is irreducible. Otherwise, the representation is necessarily reducible.

This lemma is not directly a consequence of the group aspects. It depends on notions related to finite-dimensional vector spaces, see, for example, (Vergados, 2004), Section 6.4.

For the proof of the lemma, see Appendix, Chapter 15.

Second Schur's lemma: *Given two representations of a group $\Gamma^i(X)$ and $\Gamma^j(X)$ of dimension $k_i \times k_i$ and $k_j \times k_j$, respectively, and supposing further that there exists a matrix (ρ) with dimension $k_i \times k_j$ which satisfies the relation*

$$\Gamma^i(X)(\rho) = (\rho)\Gamma^j(X). \tag{5.12}$$

Then, the matrix (ρ) is identically zero or it is a square matrix with $\det(\rho) \neq 0$. In other words, if there exists a non-singular matrix (ρ) satisfying Eq. (5.12), the representations $\Gamma^i(X)$ and $\Gamma^j(X)$ are equivalent, i.e. practically the same. Indeed, since $\det(\rho) \neq 0$, the inverse of (ρ) exists and Eq. (5.12) can be written as

$$(\rho)^{-1}\Gamma^i(X)(\rho) = \Gamma^j(X).$$

For the proof of the lemma, see Appendix, Chapter 15.

We are now in position to state a theorem, which is very useful in applications.

5.5   Orthogonality theorem

Theorem 3: *Let* $\Gamma^i(X), X \in G$ *be one irreducible representation of dimension* ℓ_i *of a group* G *of order* g. *Then, one can show the following*:

$$\sum_{X \in G} \Gamma^i(X)_{\alpha\beta} \left(\Gamma^j(X)_{\gamma\delta}\right)^* = \frac{g}{\ell_i} \delta_{ij} \delta_{\alpha\gamma} \delta_{\beta\delta}. \qquad (5.13)$$

This is known as the great orthogonality theorem (GOT).

Indeed, this is an orthogonality in the usual sense for vectors. Let us consider the set of vectors $|i, \alpha\beta\rangle$, defined over a space of g dimensions with projections $|i, \alpha\beta\rangle_k$ corresponding to the elements A_k of the group in such a way that $|i, \alpha\beta\rangle_k \Leftrightarrow \Gamma^i_{\alpha\beta}(A_k)$. Orthogonality means

$$\langle j, \gamma\delta | i, \alpha\beta \rangle = \sum_k (|j, \gamma\delta\rangle_k)^* |i, \alpha\beta\rangle_k = \sum_{A_k \in G} \Gamma^i(A_k)_{\alpha\beta} \left(\Gamma^j(A_k)_{\gamma\delta}\right)^*.$$

$$(5.14)$$

For the proof of the GOT, see Appendix, Chapter 15.

We note that we encounter orthogonality of two types:

- $j = i$ but $(\alpha, \beta) \neq (\gamma, \delta)$,
- $(\alpha, \beta) = (\gamma, \delta)$ but $j \neq i$.

This means that if we have n irreducible representations, with dimensions $\ell_1, \ell_2, \ldots, \ell_n$, there exist $\ell_1^2 + \ell_2^2 + \cdots + \ell_n^2$ orthogonal vectors. Thus, in the g-dimensional group space, we have the condition:

$$\ell_1^2 + \ell_2^2 + \cdots + \ell_n^2 \leq g. \qquad (5.15)$$

In other words, a discrete group can have a finite number of irreducible representations with dimensions satisfying the constraint:

$$\ell_i \leq \sqrt{g}. \qquad (5.16)$$

This condition is quite restrictive. We see, for example, that the regular representation is always reducible.

Example 4: We apply the above in the special case of C_{3v}.

The irreducible representation with the largest dimension is 2×2, which we have encountered many times. In addition, the irreducible representations contained in the regular representation can take the form $|i, \alpha, \beta\rangle_X$ as follows:

$$
\begin{array}{cc|cccccc}
i & \alpha,\beta & E & C_3 & C_3^2 & \sigma_1 & \sigma_2 & \sigma_3 \\
\hline
1 & 1,1 & 1 & 1 & 1 & 1 & 1 & 1 \\
\cdots & \cdots & \cdots & \cdots & \cdots & \cdots & \cdots & \cdots \\
2 & 1,1 & 1 & 1 & 1 & -1 & -1 & -1 \\
\cdots & \cdots & \cdots & \cdots & \cdots & \cdots & \cdots & \cdots \\
3 & 1,1 & 1 & -\frac{1}{2} & -\frac{1}{2} & -1 & \frac{1}{2} & \frac{1}{2} \\
3 & 1,2 & 0 & -\frac{\sqrt{3}}{2} & \frac{\sqrt{3}}{2} & 0 & \frac{\sqrt{3}}{2} & -\frac{\sqrt{3}}{2} \\
3 & 2,1 & 0 & \frac{\sqrt{3}}{2} & -\frac{\sqrt{3}}{2} & 0 & \frac{\sqrt{3}}{2} & -\frac{\sqrt{3}}{2} \\
3 & 2,2 & 1 & -\frac{1}{2} & -\frac{1}{2} & 1 & -\frac{1}{2} & -\frac{1}{2}
\end{array}
\tag{5.17}
$$

The above matrix can be obtained from the reduction of the regular representation of C_{3v} given in Eq. (5.4). Other approaches can be given as we will discuss in the following.[3] In the above expressions of Eq. (5.17), the vector $|i, \alpha, \beta\rangle$ is labeled as follows: The number in the first column denotes the representation. The labels α, β in the second column denote the matrix element of the representation. In the case of 1D representations, this is redundant, but one can put the numbers $\alpha = 1$, $\beta = 1$, viewing it as a 1×1 matrix. The ordering is arbitrary, but we have selected to start from the upper-left corner $\alpha = 1$, $\beta = 1$ and proceed along the first row. At the end of the row, we move to the second one and continue until all the labels of the matrix representation are exhausted. Note that the vectors $|i, \alpha, \beta\rangle$ are not normalized.

5.5.1 *Character orthogonality relations*

A very useful relation is obtained from Eq. (5.13) for $\alpha = \beta$ and $\gamma = \delta$:

$$
\sum_{X \in G} \Gamma^i(X)_{\alpha,\alpha} \Gamma^{*j}(X)_{\gamma\gamma} = \frac{g}{\ell_i} \delta_{ij} \delta_{\alpha,\gamma}.
$$

---

[3]In this construction, one takes into account the fact that the first row, corresponding to the identity representation, is obvious. Also, the orthogonality condition between rows (Eq. (5.14)) and the fact that, in the case of the 1D representations, all elements of a given class have the same value, etc.

Summing with respect to α and γ, we find

$$\sum_{X \in G} \left(\sum_{\alpha} \Gamma^i(X)_{\alpha,\alpha} \right) \left(\sum_{\gamma} \Gamma^{*j}(X)_{\gamma\gamma} \right) = \frac{g}{\ell_i} \delta_{ij} \ell_i \Rightarrow$$

$$\sum_{X \in G} \chi^i(X) \chi^{*j}(X) = g \delta_{ij}. \qquad (5.18)$$

Let us now suppose that the group has n classes $C_1, C_2, \ldots, C_n$, each with elements $g_1, g_2, \ldots, g_n$, respectively. Then, since the elements of a given class have the same character, the previous equation can be written as:

$$\sum_{k=1}^{n} \sqrt{\frac{g_k}{g}} \chi^i(C_k) \sqrt{\frac{g_k}{g}} \chi^{*j}(C_k) = \delta_{ij} \Rightarrow$$

$$\sum_{k=1}^{n} e_k^i e_k^{*j} = \delta_{ij}, \text{ where } e_k^i = \sqrt{\frac{g_k}{g}} \chi^i(C_k), \ e_k^{*j} = \sqrt{\frac{g_k}{g}} \chi^{*j}(C_k). \quad (5.19)$$

The orthogonality is between lines i and j. The unit vectors $e_k^i, (e_k^{*j})$, $k = 1, 2, \ldots, n$, form *a basis in the space of classes of the group*. This means that *the number of irreducible representations of a group is equal to the number of its classes*.

5.5.2 *A criterion of irreducibility of group representations*

Irreducibility criterion: Let C_α be a class of a discrete group G. We construct the matrix

$$M_{C_\alpha} = \sum_{\beta} \Gamma^i(X_\beta), \quad X_\beta \in C_\alpha,$$

where $\Gamma^i(X_\beta)$ is a representation of G. This representation is irreducible if the matrix M_{C_α} is a multiple[4] of the identity for every class C_α of the group G.

We consider the matrix M_{C_α} and a representation $\Gamma^j(A), A \in G$. We then construct the expression

$$\Gamma^j(A) M_{C_\alpha} \left(\Gamma^j(A) \right)^{-1} = \sum_{\beta} \Gamma^j(A) \Gamma^i(X_\beta) \left(\Gamma^j(A) \right)^{-1} = \sum_{\beta} \Gamma^i(Y_\beta) = M_{C_\alpha},$$

hence

$$\Gamma^j(A) M_{C_\alpha} = M_{C_\alpha} \Gamma^j(A),$$

---

[4]The proportionality factor could be zero, that is, the matrix could be 0ϵ.

and according to the first Schur's lemma,

$$M_{C_\alpha} = \lambda(\epsilon), \lambda = \text{constant} \Rightarrow tr\,(M_{C_\alpha}) = \lambda \ell_i$$

$$tr\,(M_{C_\alpha}) = \sum_\beta tr\,(\Gamma^i(X_\beta)) = g_\alpha \chi^j(C_\alpha) \Rightarrow$$

$$g_\alpha \chi^j(C_\alpha) = \lambda \ell_i \Rightarrow \lambda = \frac{g_\alpha}{\ell_i}\chi^j(C_\alpha)$$

or

$$M_{C_\alpha} = \frac{g_\alpha}{\ell_i}\chi^j(C_\alpha)(\epsilon). \tag{5.20}$$

In other words, the sum of the matrices of the representation over all elements of a class is a multiple of the identity matrix *if the representation is irreducible.*

Let us consider a 2×2 representation of C_{3v}. We have:

- The class of the identity:

$$C_1 = \{E\} \Rightarrow M_1 = \begin{pmatrix} 1 & 0 \\ 0 & 1 \end{pmatrix} = (\epsilon).$$

- The class of rotations:

$$C_2 = \{C_3, C_3^2\} \Rightarrow M_2 = \begin{pmatrix} -\frac{1}{2} & -\frac{\sqrt{3}}{2} \\ \frac{\sqrt{3}}{2} & -\frac{1}{2} \end{pmatrix} + \begin{pmatrix} -\frac{1}{2} & \frac{\sqrt{3}}{2} \\ -\frac{\sqrt{3}}{2} & -\frac{1}{2} \end{pmatrix}$$

$$= -\begin{pmatrix} 1 & 0 \\ 0 & 1 \end{pmatrix} = -(\epsilon).$$

We verify that $\frac{g_2}{2}\chi(X \in C_2) = \frac{2}{2}(-1) = -1$.

- The class of reflections:

$$C_3 = \{\sigma_1, \sigma_2, \sigma_3\} \Rightarrow M = \begin{pmatrix} -1 & 0 \\ 0 & 1 \end{pmatrix} + \begin{pmatrix} \frac{1}{2} & \frac{\sqrt{3}}{2} \\ \frac{\sqrt{3}}{2} & -\frac{1}{2} \end{pmatrix} + \begin{pmatrix} \frac{1}{2} & -\frac{\sqrt{3}}{2} \\ -\frac{\sqrt{3}}{2} & -\frac{1}{2} \end{pmatrix}$$

$$= \begin{pmatrix} 0 & 0 \\ 0 & 0 \end{pmatrix} = 0(\epsilon).$$

Indeed, $\frac{g_3}{2}\chi(X \in C_3) = \frac{3}{2}(0) = 0$.

So, this representation is irreducible.

On the other hand, for the representation of C_{3v} discussed in Example 4 of Chapter 4, we find

$$M_1 = \begin{pmatrix} 1 & 0 & 0 \\ 0 & 1 & 0 \\ 0 & 0 & 1 \end{pmatrix},$$

$$M_2 = \begin{pmatrix} 0 & 1 & 0 \\ 0 & 0 & 1 \\ 1 & 0 & 0 \end{pmatrix} + \begin{pmatrix} 0 & 0 & 1 \\ 1 & 0 & 0 \\ 0 & 1 & 0 \end{pmatrix} = \begin{pmatrix} 0 & 1 & 1 \\ 1 & 0 & 1 \\ 1 & 1 & 0 \end{pmatrix},$$

$$M_3 = \begin{pmatrix} 1 & 0 & 0 \\ 0 & 0 & 1 \\ 0 & 1 & 0 \end{pmatrix} + \begin{pmatrix} 0 & 0 & 1 \\ 0 & 1 & 0 \\ 1 & 0 & 0 \end{pmatrix} + \begin{pmatrix} 0 & 1 & 0 \\ 1 & 0 & 0 \\ 0 & 0 & 1 \end{pmatrix} = \begin{pmatrix} 1 & 1 & 1 \\ 1 & 1 & 1 \\ 1 & 1 & 1 \end{pmatrix}.$$

Since the matrices M_2 and M_3 are not multiples of the identity, the representation is reducible.

5.6 More on representation reduction

We have seen that, in general, a representation can be reduced to others of smaller dimension, see Eq. (5.2). When the reduction is complete, i.e. all resulting representations are irreducible, the representation is written as

$$\Gamma(X) = \alpha_1 \Gamma^1(X) \oplus \alpha_2 \Gamma^2(X) \oplus \cdots \oplus \alpha_M \Gamma^M(X), \quad \sum_{k=1}^{M} \alpha_k n_k = d_\Gamma, \quad (5.21)$$

where d_Γ is the dimension of $\Gamma(X)$ and α_k, $1, 2, \ldots, M$ are integers specifying how many times the representation $\Gamma^k(X)$ is found in the above expansion.

Taking the trace of Eq. (5.21), we get

$$\chi(X) = \sum_{k}^{\ell} \alpha_k \chi^k(X). \quad (5.22)$$

Since, however, $\chi^k(X)$ are orthogonal, we obtain

$$\sum_{X \in G} \chi(X) \chi^{*j}(X) = \sum_{k}^{\ell} \alpha_k \sum_{X \in G} \chi^k(X) \chi^{*j}(X) = \sum_{k}^{\ell} \alpha_k g \delta_{jl} = g \alpha_j. \quad (5.23)$$

Hence, the expansion coefficients can be found via the relation:

$$\alpha_j = \frac{1}{g} \sum_{X \in G} \chi(X)\chi^{*j}(X).$$
(5.24)

The quantities $\chi^j(X)$ are called **simple characters**.

Of special interest is the reduction of the regular representation $\Gamma^{\text{reg}}(X)$. We find

$$\Gamma^{\text{reg}}(E) = (\epsilon(n)), \;\; \Rightarrow \;\; \chi^{\text{reg}}(E) = n.$$

$\Gamma^{\text{reg}}(X \neq E)$ does not have non-zero diagonal elements. Thus,

$$\chi^{\text{reg}}(E) = n, \quad \chi^{\text{reg}}(X \neq E) = 0.$$
(5.25)

As a result, Eq. (5.24) in the case of the regular representation, which is a real $g \times g$ one, yields:

$$\alpha_j = \frac{1}{g} \sum_{X \in G} \chi(X)\chi^j(X) = \frac{1}{g}\chi(E)\chi^j(E) = \chi^j(E) = \ell_j,$$
(5.26)

$$\Gamma^{\text{reg}}(X) = \ell_1\Gamma_1(X) \oplus \ell_2\Gamma_2(X) \oplus \cdots \oplus \ell_M\Gamma_M(X).$$

Taking the trace of the last expression for $X = E$ and making use of Eq. (5.25), we find

$$\sum_{i=1}^{M} \ell_i^2 = g.$$
(5.27)

This expression is very useful for two reasons:

- In the case of groups of low order g, often adequate for most applications, it allows the determination of all the irreducible representations of the group, see, for example, the case of C_{3v}, Section 5.4 .
- It allows the determination of another orthogonality relation, beyond that of Eq. (5.13), Section 5.4.

Indeed, in Section 5.4, we found that in the g-space of $\Gamma_{\alpha\beta}(X)$, we can define a basis made of $\sum_i \ell_i^2$ orthogonal vectors $|i, \alpha, \beta\rangle_X$. Equation (5.27) tells us that the number of elements of the basis is g. We can therefore determine another basis $|i, X\rangle_{\alpha,\beta} \in g$ as follows:

$$\sqrt{\frac{\ell_i}{g}}\Gamma(A_1), \sqrt{\frac{\ell_i}{g}}\Gamma(A_2), \ldots, \sqrt{\frac{\ell_i}{g}}\Gamma(A_g),$$

$$\sum_{i=1}^{M}\sum_{\alpha=1}^{\ell_i}\sum_{\beta=1}^{\ell_i} \sqrt{\frac{\ell_i}{g}}\Gamma^i_{\alpha,\beta}(A_j)\sqrt{\frac{\ell_i}{g}}\Gamma^{*i}_{\alpha,\beta}(A_k) = \delta_{jk}.$$

(5.28)

This relation implies that, for example, in the case of Eq. (5.17), we have not only orthogonality with respect to the rows but with respect to the columns as well. In general, for every discrete group, **the reduction matrix of the regular representation is unitary.** The analog of matrix (5.17) using the now normalized vectors becomes

i	α, β	E	C_3	C_3^2	σ_1	σ_2	σ_3
1	1,1	$\frac{1}{\sqrt{6}}$	$\frac{1}{\sqrt{6}}$	$\frac{1}{\sqrt{6}}$	$\frac{1}{\sqrt{6}}$	$\frac{1}{\sqrt{6}}$	$\frac{1}{\sqrt{6}}$
...	...	...	...	...	...	...	...
2	1,1	$\frac{1}{\sqrt{6}}$	$\frac{1}{\sqrt{6}}$	$\frac{1}{\sqrt{6}}$	$-\frac{1}{\sqrt{6}}$	$-\frac{1}{\sqrt{6}}$	$-\frac{1}{\sqrt{6}}$
...	...	...	...	...	...	...	...
3	1,1	$\frac{1}{\sqrt{3}}$	$-\frac{1}{2\sqrt{3}}$	$-\frac{1}{2\sqrt{3}}$	$-\frac{1}{\sqrt{3}}$	$\frac{1}{2\sqrt{3}}$	$\frac{1}{2\sqrt{3}}$
3	1,2	0	$-\frac{1}{2}$	$\frac{1}{2}$	0	$\frac{1}{2}$	$-\frac{1}{2}$
3	2,1	0	$\frac{1}{2}$	$-\frac{1}{2}$	0	$\frac{1}{2}$	$-\frac{1}{2}$
3	2,2	$\frac{1}{\sqrt{3}}$	$-\frac{1}{2\sqrt{3}}$	$-\frac{1}{2\sqrt{3}}$	$\frac{1}{\sqrt{3}}$	$-\frac{1}{2\sqrt{3}}$	$-\frac{1}{2\sqrt{3}}$

(5.29)

The expression (5.28) leads to a new orthogonality relation involving the characters

$$\sum_{i=1}^{n}e^i_\alpha e^{*i}_\beta = \delta_{\alpha\beta}, \text{ where } e^i_\alpha = \sqrt{\frac{g_\alpha}{g}}\chi^i(C_\alpha), e^{*j}_\beta = \sqrt{\frac{g_\beta}{g}}\chi^{*j}(C_\beta). \quad (5.30)$$

In other words, the orthogonality is not only with respect to the rows, Eq. (5.19), but with respect to the columns α and β, that is, the matrix e^i_α is unitary.

5.7 Some examples of representation reduction

The representation reduction will be performed according to the conclusions drawn in the previous sections. Special attention will be paid to the regular representation, which contains all the irreducible representations of the group. In summary, we mention the following:

- The number of the irreducible representations of a group equals that of its classes n, Eq. (5.21), i.e.

$$\Gamma(X) = \ell_1\Gamma_1(X) \oplus \ell_2\Gamma_2(X) \oplus \cdots \oplus \ell_n\Gamma_n(X),$$

where ℓ_i is the dimension of $\Gamma_i(X)$.
- The dimensions ℓ_i of the irreducible representations are restricted by the condition $\sum_i^n \ell_i^2 = g$, where g, the order of the group.
- The irreducible representations $\Gamma_i(X)$ satisfy two orthogonality conditions, given by the relations (5.13) and (5.28), that is, the matrices that have the elements of the group as columns and the labels i, α, β of the representation as rows are unitary.
- The characters of the irreducible representations also obey two conditions of orthogonality, namely Eqs. (5.19) and (5.30); the character matrices e_α^i, with columns the classes C_α of the group and rows the index i of the representation, are unitary.

Example 5: Abelian groups of small order.

We have already seen that in the case of Abelian groups, all elements of the regular representation can be simultaneously diagonalized with a similarity transformation, which makes the reduction a simple matter. In most cases, the orthogonality relations are adequate. We simply begin by selecting the characters of the first row and the first column equal to 1:

- $g = 2$, $G = \{E, A\}$. The character matrix takes the form

$$\begin{array}{c|cc} & E & A \\ \hline i=1 & 1 & 1 \\ i=2 & 1 & a \end{array} \Leftrightarrow \text{orthogonality condition} \Rightarrow a = -1 \Rightarrow \begin{array}{c|cc} & E & A \\ \hline i=1 & 1 & 1 \\ i=2 & 1 & -1 \end{array}.$$

- $g = 3$, $G = \{E, A, A^2\}$. Since $A^3 = 1$, there exist eigenvalues satisfying $\lambda^3 = 1$, i.e. with the roots $1, \xi, \xi^2$, $\xi = e^{2\pi i/3}$. These roots satisfy the condition $1 + \xi + \xi^2 = 0$. The character matrix, with the first line and column made of 1's, takes the form

$$\begin{array}{c|ccc} & E & A & A^2 \\ \hline i=1 & 1 & 1 & 1 \\ i=2 & 1 & \xi & \xi^2 \\ i=3 & 1 & \xi^2 & \xi \end{array}.$$

All orthogonality conditions are satisfied. The orthogonal matrix e_α^i is obtained by a mere multiplication with $1/\sqrt{3}$.

- $g = 4$, $G = \{E, A, A^2, A^3\}$. This case has been considered in Example 2. We find

	E	A	A^2	A^3
$\Gamma(E)$	1	1	1	1
$\Gamma(A)$	1	ξ	ξ^2	ξ^3
$\Gamma(A^2)$	1	ξ^2	ξ^4	ξ^6
$\Gamma(A^3)$	1	ξ^3	ξ^6	ξ^9

$\xi = e^{i\pi/2} \Rightarrow$

	E	A	A^2	A^3
$\Gamma(E)$	1	1	1	1
$\Gamma(A)$	1	i	-1	$-i$
$\Gamma(A^2)$	1	-1	1	-1
$\Gamma(A^3)$	1	$-i$	-1	i

The other possibility for $g = 4$ with a multiplication table given in Table 3.1(a) can also be treated as in Example 6, Eq. (5.31).

- $g = 6$. A somewhat realistic example is C_{3h}.

$C_{3h} = \{E, A, A^2, \sigma_h, A\sigma_h, A^2\sigma_h\}$, where A is an axis of third order and σ_h is a reflection with respect to a plane perpendicular to this axis.

One can show that this group is Abelian. The first character row is $1, 1, 1, 1, 1, 1$. An orthogonal combination corresponding to the element A is of the form $1, \xi, \xi^2, a, a\xi, a\xi^2$, $\xi = e^{(2\pi i)/3}$, $|a| = 1$. Two solutions can be found by electing $a = 1$ and $a = -1$. The rows corresponding to A^2 will have the form $1, \xi^2, \xi, a, a\xi^2, a\xi$, and the two solutions correspond to $a = 1$ and $a = -1$. Finally, the line corresponding to σ_h can be selected to be $1, 1, 1, -1, -1, -1$. Thus, we get

C_{3h}	E	C_3	C_3^2	σ_h	$C_2\sigma_h$	$C_2^2\sigma_h$	$\epsilon = e^{(2\pi i)/3}$
$i = 1$	1	1	1	1	1	1	
$i = 2$	1	ϵ	ϵ^2	1	ϵ	ϵ^2	
$i = 3$	1	ϵ^2	ϵ	1	ϵ^2	ϵ	
$i = 4$	1	1	1	-1	-1	-1	
$i = 5$	1	ϵ	ϵ^2	-1	$-\epsilon$	$-\epsilon^2$	
$i = 6$	1	ϵ^2	ϵ	-1	$-\epsilon^2$	$-\epsilon$	

Example 6: We reduce the representations of a group with four elements. We distinguish two cases:

(i) The cyclic group. This has already been examined in the Example 5 above.

(ii) The elements of the group obey the multiplication in Table 3.1(a). We order the elements of the group as

$\{E, A, B, AB\}$. We know that $\chi_1^1 = \chi_2^1, \chi_3^1 = \chi_4^1 = 1$ and $\chi_1^2 = \chi_1^3, \chi_1^4 = 1$. For the evaluation of the remaining characters, we may proceed as above. We however, employ a more elegant method. There exist three subgroups $\{E, A\}$, $\{E, B\}$ and $\{E, AB\}$ and 3 $\{H_1, K_1\}$ with the multiplication table $H_1H_1 = H_1$, $H_1K_1 = K_1$, $K_1K_1 = H_1$, which is nothing but the Abelian of second order with character matrix

$$
\begin{array}{c|cc}
 & H_1 & H_1 \; K_1 \\
\hline
i = 1 & 1 & 1 \\
i = 2 & 1 & -1
\end{array}.
$$

The possible cosets are

(1) $H_1 = \{E, A\}$, $K_1 = \{B, AB\}$. Then, from the above character table, we obtain

$$\chi_2^2 = \chi^2(H_1) = \chi^2(A) = 1, \chi_3^2 = \chi^2(K_1) = \chi^2(B) = -1,$$
$$\chi_4^2 = \chi^2(AB) = \chi^2(K_1) = -1.$$

(2) $H_1 = \{E, B\}$, $K_1 = \{B, AB\}$. Now,

$$\chi_2^3 = \chi^2(K_1) = \chi^2(A) = -1, \chi_3^3 = \chi^2(K_1) = \chi^2(B) = 1,$$
$$\chi_4^3 = \chi^2(AB) = \chi^2(K_1) = -1.$$

(3) $H_1 = \{E, AB\}$, $K_1 = \{A, B\}$. Now,

$$\chi_2^4 = \chi^2(K_1) = \chi^2(A) = -1, \chi_3^4 = \chi^2(K_1) = \chi^3(B) = -1,$$
$$\chi_4^4 = \chi^2(AB) = \chi^2(K_1) = 1.$$

Finally,

$$
\begin{array}{c|cccc}
 & E & A & B & AB \\
\hline
i = 1 & 1 & 1 & 1 & 1 \\
i = 2 & 1 & 1 & -1 & -1 \\
i = 3 & 1 & -1 & 1 & -1 \\
i = 4 & 1 & -1 & -1 & 1
\end{array}. \tag{5.31}
$$

Note that this table is different from that of case (i) above.

In both cases, the unitary matrix e_α^i is obtained after multiplication of the above with $\frac{1}{2}$.

Example 7: One more review of the group C_{3v}.

The reduction of the regular representation has already been accomplished (see Eq. (5.17)). Here, we follow a different procedure in the spirit of the previous section.

C_{3v} is characterized by three classes, $K_1 = \{E\}$, $K_2 = \{C_3, C_3^2\}$, $K_3 = \{\sigma_1, \sigma_2, \sigma_3\}$ and, as a result, three irreducible representations with dimensions satisfying the condition $\ell_1^2 + \ell_2^2 + \ell_3^2 = 6$, $\ell_i \leq 2$. This admits only one solution: $\ell_1 = 1$, $\ell_2 = 1$, $\ell_3 = 2$. Thus,

$$\Gamma(X) = \Gamma^1(X) \oplus \Gamma^2(X) \oplus 2\Gamma^3(X).$$

Let us begin with the determination of the characters. Clearly, $\chi_1(E) = 1$, $\chi_2(E) = 1$, $\chi_3(E) = 2$. We can select $\chi_{K_1}^1 = 1, \chi_{K_2}^1 = 1$. Also, $\chi_{K_1}^2 = 1$, $\chi^2(C_3) = \chi^2(C_3^2) = \left(\chi^{(}C_3)\right)^2 = 1$. As a result,

	K_1	K_2	K_3
	$\{E\}$	$\{C_3, C_3^2\}$	$\{\sigma_1, \sigma, \sigma_3\}$
$i = 1$	1	1	1
$i = 2$	1	1	α
$i = 3$	2	β	γ

The number of elements in each class is $n(K_1) = 1$, $n(K_2) = 2$ and $n(K_3) = 3$. As a result, the condition of orthogonality of the first two rows yields $1 + 1 \times n(K_2) + n(K_3) \times \alpha = 1 + 2 + 3\alpha = 0 \Rightarrow \alpha = -1$. Similarly, the orthogonality between the first and third rows gives $2 + 2\beta + 3\gamma = 0$, while that between the second and third rows yields $2 + 2\beta - 3\gamma = 0$. So, the solution is $\gamma = 0$ and $\beta = -1$. Hence,

	K_1	K_2	K_3
	$\{E\}$	$\{C_3, C_3^2\}$	$\{\sigma_1, \sigma, \sigma_3\}$
$i = 1$	1	1	1
$i = 2$	1	1	-1
$i = 3$	2	-1	0

$\Rightarrow e_\alpha^i \Leftrightarrow$

	$\alpha = 1$	$\alpha = 2$	$\alpha = 3$
$i = 1$	$\frac{1}{\sqrt{6}}$	$\frac{1}{\sqrt{3}}$	$\frac{1}{\sqrt{2}}$
$i = 2$	$\frac{1}{\sqrt{6}}$	$\frac{1}{\sqrt{3}}$	$-\frac{1}{\sqrt{2}}$
$i = 3$	$\frac{2}{\sqrt{6}}$	$-\frac{1}{\sqrt{3}}$	0

From the character table, we can determine the 2D representation; this has already been determined on general grounds in Section 4.7.2. This can also be obtained using the symmetric group S_3, since S_3 is isomorphic to C_{3v}. For this purpose, it is adequate to consider the basis of functions

$$|1\rangle = \psi(1) + \psi(2) - 2\psi(3), \quad |2\rangle = \psi(1) - \psi(2).$$

We prefer to evaluate it from the character table, since this can be employed in other symmetries a well.

The representation of the identity element is obvious. Another element can be chosen to be diagonal. We select this to be the element σ_1 of the C_3. This must also be traceless. Thus,

$$\Gamma_3(\sigma_1) = \begin{pmatrix} \alpha & 0 \\ 0 & -\alpha \end{pmatrix}, \quad |\alpha|^2 = 1 \Rightarrow \alpha = -1 \text{ (sign choice)}.$$

Though $\sigma_i^2 = E$ and $\Gamma_3(\sigma_i)$ can be chosen to be unitary, in this case, they can also be Hermitian and traceless, i.e.

$$\Gamma_3(\sigma_2) = \begin{pmatrix} \alpha & \beta \\ \beta^* & -\alpha \end{pmatrix}, \quad \Gamma_3(\sigma_3) = \begin{pmatrix} \gamma & \delta \\ \delta^* & -\gamma \end{pmatrix},$$

$$|\alpha|^2 + |\beta|^2 = 1, \ |\gamma|^2 + |\delta|^2 = 1,$$

with α and γ real. According to the criterion of irreducibility, their sum must be a multiple of the identity

$$\Gamma_3(\sigma_1) + \Gamma_3(\sigma_2) + \Gamma_3(\sigma_3) = 0 \Rightarrow -1 + \alpha + \gamma = 0,$$

$$\beta + \delta = 0 \Rightarrow \delta = -\beta, \ \gamma = 1 - \delta.$$

Now,

$$\alpha^2 + |\beta|^2 = 1, \quad \gamma^2 + |\delta|^2 = 1 \Rightarrow \alpha^2 + |\beta|^2 = 1,$$

$$(1 - \alpha)^2 + |\beta|^2 = 1 \Rightarrow \alpha = \tfrac{1}{2}, \beta = e^{i\phi} \tfrac{\sqrt{3}}{2}.$$

The angle ϕ remains arbitrary. The matrices, however, can be chosen so as to be real ($\phi = 0$). Thus,

$$\Gamma_3(\sigma_2) = \begin{pmatrix} \frac{1}{2} & \frac{\sqrt{3}}{2} \\ \frac{\sqrt{3}}{2} & -\frac{1}{2} \end{pmatrix}, \quad \Gamma_3(\sigma_3) = \begin{pmatrix} \frac{1}{2} & -\frac{\sqrt{3}}{2} \\ -\frac{\sqrt{3}}{2} & -\frac{1}{2} \end{pmatrix}.$$

Now, however, the relations $C_3 = \sigma_2\sigma_3$ and $C_3^2 = C_3C_3$ imply

$$\Gamma_3(C_3) = \Gamma_3(\sigma_2)\Gamma_3(\sigma_3) = \begin{pmatrix} \frac{1}{2} & \frac{\sqrt{3}}{2} \\ \frac{\sqrt{3}}{2} & -\frac{1}{2} \end{pmatrix} \begin{pmatrix} \frac{1}{2} & -\frac{\sqrt{3}}{2} \\ -\frac{\sqrt{3}}{2} & -\frac{1}{2} \end{pmatrix} = \begin{pmatrix} -\frac{1}{2} & -\frac{\sqrt{3}}{2} \\ \frac{\sqrt{3}}{2} & -\frac{1}{2} \end{pmatrix},$$

$$\Gamma_3(C_3^2) = \Gamma_3(C_3)\Gamma_3(C_3) = \begin{pmatrix} -\frac{1}{2} & -\frac{\sqrt{3}}{2} \\ \frac{\sqrt{3}}{2} & -\frac{1}{2} \end{pmatrix} \begin{pmatrix} -\frac{1}{2} & -\frac{\sqrt{3}}{2} \\ \frac{\sqrt{3}}{2} & -\frac{1}{2} \end{pmatrix} = \begin{pmatrix} -\frac{1}{2} & \frac{\sqrt{3}}{2} \\ -\frac{\sqrt{3}}{2} & -\frac{1}{2} \end{pmatrix},$$

and the construction is complete.

Example 8: The reduction of C_{4V}, see Section 2.4.2.

This group is simple and, at the same time, very useful. It describes a pyramid with a square base vertical to its symmetry axis of fourth order. Its elements are:

$$G = C_{4V} = \{E, C_4, C_4^2, C_4^3, \sigma_x, \sigma_y, \sigma_a, \sigma_b\},$$

where C_4 is the fourth-order axis, perpendicular to the square, σ_x, σ_y are reflections with respect to planes which are perpendicular to the square passing through the middle of the opposite sides of the square and σ_a, σ_b are reflections perpendicular to the plane of the square passing through the two diagonals of the square.

One faithful 2×2 representation can be constructed easily on the basis of the discussion in Section 4.7. In fact, one can show that the construction of the representations of C_4 and σ_a suffices, since those two elements are generators of the group. This is obvious for the rotations (and the identity). Furthermore, $C_4\sigma_a = \sigma_y$, $C_4^2\sigma_a = \sigma_b$, $C_4\sigma_b = \sigma_x$. Thus,

$$C_4 = \begin{pmatrix} 0 & 1 \\ -1 & 0 \end{pmatrix}, \quad \sigma_a = \begin{pmatrix} 0 & 1 \\ 1 & 0 \end{pmatrix} \Rightarrow C_4^2 = \begin{pmatrix} -1 & 0 \\ 0 & -1 \end{pmatrix}, \quad C_4^3 = \begin{pmatrix} 0 & -1 \\ 1 & 0 \end{pmatrix},$$

$$E = C_4^4 = \begin{pmatrix} 1 & 0 \\ 0 & 1 \end{pmatrix}, \quad \sigma_x = \begin{pmatrix} -1 & 0 \\ 0 & 1 \end{pmatrix}, \quad \sigma_y = \begin{pmatrix} 1 & 0 \\ 0 & -1 \end{pmatrix}, \quad \sigma_b = \begin{pmatrix} 0 & -1 \\ -1 & 0 \end{pmatrix}.$$

From the resulting multiplication table, one obtains the 8×8 regular representation of the group, which we will try to reduce.

We first find that this group has five classes:

$$\{E\}, \{C_4^2\}, \{C_4, C_4^3\}, \{\sigma_x, \sigma_y\}, \{\sigma_a, \sigma_b\}$$

and, as a result, five irreducible representations. From the discussion in Section 5.5, the largest dimension an irreducible representation can have is 2. Furthermore, since $\sum_i \ell_i^2 = 8$, the only solution is $\ell_i = 1$, $i = 1, 2, 3, 4$ and $\ell_5 = 2$.

The 1-dimensional, or 1D for short, representations $\Gamma^{(i)}(X)$, $i = 1, 2, 3, 4$, coincide with the characters $\chi^i(X)$, and they can be constructed from the orthogonality conditions, as discussed above. The 2D representation $\Gamma^5(X)$ has just been constructed. The obtained results are given in

Table 5.32. We have chosen to present the results in the form $i; \alpha, \beta \leftrightarrow \Gamma^i_{\alpha,\beta}(X)$ so that they can easily be used in Chapter 16.

i	α, β	E	C_4	C_4^2	C_4^3	σ_x	σ_y	σ_a	σ_a
1	1,1	1	1	1	1	1	1	1	1
2	1,1	1	−1	1	−1	−1	−1	1	1
3	1,1	1	−1	1	−1	1	1	−1	−1
4	1,1	1	1	1	1	−1	−1	−1	−1
5	1,1	1	0	−1	0	−1	1	0	0
5	1,2	0	1	0	−1	0	0	1	−1
5	2,1	0	−1	0	1	0	0	1	−1
5	2,2	1	0	−1	0	1	−1	0	0

$$(5.32)$$

Equation (5.26) yields

$$\Gamma(X) = \Gamma^{(1)}(X) \oplus \Gamma^{(2)}(X) \oplus \Gamma^{(3)}(X) \oplus \Gamma^{(4)}(X) \oplus 2\Gamma^{(5)}(X). \qquad (5.33)$$

The character reduction of the representation of some groups of interest to crystal structure are given in an appendix, see Chapter 17.

5.8   Reduction of the direct product of group representations

In Section 4.5, we defined the direct product or Kronecker product of two representations. This product has many applications in many particle theories. Here, we consider the reduction of the direct product of two irreducible representations $\Gamma^a(X)$ and $\Gamma^b(X)$ of a group G. To this end, we construct the matrices $\Gamma(X)$:

$$\Gamma(X) = \Gamma^a(X) \otimes \Gamma^b(X), \ X \in G. \qquad (5.34)$$

We know already, Section 4.5, that, if the elements of the group satisfy the relation $AB = C$, one must have $\Gamma(A)\Gamma(B) = \Gamma(C)$. Indeed,

$$\Gamma(A)\Gamma(B) = \left(\Gamma^a(A) \otimes \Gamma^b(A)\right)\left(\Gamma^a(B) \otimes \Gamma^b(B)\right)$$

$$= \left(\Gamma^a(A)\Gamma^b(B)\right) \otimes \left(\Gamma^a(A)\Gamma^b(B)\right)$$

$$= \Gamma^a(AB) \otimes \Gamma^b(AB) = \Gamma^a(C) \otimes \Gamma^b(C) = \Gamma(C).$$

In other words, the **direct product of two representations of a group constitutes another representation of the group**. We have also seen in Section 4.5 that the direct product preserves the trace in a multiplicative way, i.e.

$$\chi(X) = \chi^a(X)\chi^b(X). \tag{5.35}$$

It is sufficient to study the direct product of irreducible representations. The direct product of two irreducible representations is, however, in general reducible. Thus, we write

$$\Gamma_i(X) \otimes \Gamma_j(X) = \alpha_1^{i,j}\Gamma_1(X) \oplus \alpha_2^{i,j}\Gamma_2(X) \oplus \cdots \alpha_n^{i,j}\Gamma_n(X), \tag{5.36}$$

with $\alpha_k^{i,j}$, $k = 1, 2, \ldots, n$ non-negative integers. This expression is often called Glebsch–Gordan expansion. In general, this problem is rather complicated, and we only explain it with a simple example.

Example 9: We study the reduction of the direct product of the irreducible representations of C_{3v} for all possible arrangements:

(1) The 1D

$$\Gamma^1(X) \otimes \Gamma^1(X) = \Gamma_1(X), \quad \Gamma^2(X) \otimes \Gamma^2(X) = \Gamma_1(X),$$

$$\Gamma^1(X) \otimes \Gamma^2(X) = \begin{cases} 1, & X = E, C_3, C_3^2 \\ -1, & X = \sigma_1, \sigma_2, \sigma_3 \end{cases}$$

$$\Rightarrow \Gamma^1(X) \otimes \Gamma^2(X) = \Gamma^2(X).$$

(2) $\Gamma^i(X) \otimes \Gamma^3(X)$, $i = 1, 2$.

$$\Gamma^1(X) \otimes \Gamma^3(X) = \Gamma^3(X),$$

$$\Gamma^2(X) \otimes \Gamma^3(X) = \begin{cases} \Gamma^3(X), & X = E, C_3, C_3^2 \\ -\Gamma^3(X), & X = \sigma_1, \sigma_2 \sigma_3 \end{cases},$$

$$\alpha_3^{23} = 1, X = E, C_3, C_3^2; \quad \alpha_3^{23} = -1, \quad X = \sigma_1, \sigma_2, \sigma_3.$$

(3) $\Gamma^3(X) \otimes \Gamma^3(X)$. In this case, the product representation consists of 4×4 matrices. The two diagonal ones are $\Gamma_3(E) \otimes \Gamma_3(E) = \mathrm{diag}(1,1,1,1)$, $\Gamma^3(\sigma_1) \otimes \Gamma^3(\sigma_1) = \mathrm{diag}(1,-1,-1,1)$. In addition,

$$\Gamma^3(C_3) \otimes \Gamma^3(C_3) = \begin{pmatrix} -\frac{1}{2}C_3 & -\frac{\sqrt{3}}{2}C_3 \\ \frac{\sqrt{3}}{2}C_3 & -\frac{1}{2}C_3 \end{pmatrix} = \begin{pmatrix} \frac{1}{4} & \frac{\sqrt{3}}{4} & \frac{\sqrt{3}}{4} & \frac{3}{4} \\ -\frac{\sqrt{3}}{4} & \frac{1}{4} & -\frac{3}{4} & \frac{\sqrt{3}}{4} \\ -\frac{\sqrt{3}}{4} & -\frac{3}{4} & \frac{1}{4} & \frac{\sqrt{3}}{4} \\ \frac{3}{4} & -\frac{\sqrt{3}}{4} & -\frac{\sqrt{3}}{4} & \frac{1}{4} \end{pmatrix},$$

$$\Gamma^3(C_3^2) \otimes \Gamma^3(C_3^2) = \begin{pmatrix} \frac{1}{4} & -\frac{\sqrt{3}}{4} & -\frac{\sqrt{3}}{4} & \frac{3}{4} \\ \frac{\sqrt{3}}{4} & \frac{1}{4} & -\frac{3}{4} & -\frac{\sqrt{3}}{4} \\ \frac{\sqrt{3}}{4} & -\frac{3}{4} & \frac{1}{4} & -\frac{\sqrt{3}}{4} \\ \frac{3}{4} & \frac{\sqrt{3}}{4} & \frac{\sqrt{3}}{4} & \frac{1}{4} \end{pmatrix},$$

$$\Gamma^3(\sigma_2) \otimes \Gamma^3(\sigma_2) = \begin{pmatrix} \frac{1}{4} & -\frac{\sqrt{3}}{4} & -\frac{\sqrt{3}}{4} & \frac{3}{4} \\ -\frac{\sqrt{3}}{4} & \frac{1}{4} & \frac{3}{4} & -\frac{\sqrt{3}}{4} \\ -\frac{\sqrt{3}}{4} & \frac{3}{4} & \frac{1}{4} & -\frac{\sqrt{3}}{4} \\ \frac{3}{4} & -\frac{\sqrt{3}}{4} & -\frac{\sqrt{3}}{4} & \frac{1}{4} \end{pmatrix},$$

$$\Gamma^3(\sigma_3) \otimes \Gamma^3(\sigma_3) = \begin{pmatrix} \frac{1}{4} & \frac{\sqrt{3}}{4} & \frac{\sqrt{3}}{4} & \frac{3}{4} \\ \frac{\sqrt{3}}{4} & \frac{1}{4} & \frac{3}{4} & \frac{\sqrt{3}}{4} \\ \frac{\sqrt{3}}{4} & \frac{3}{4} & \frac{1}{4} & \frac{\sqrt{3}}{4} \\ \frac{3}{4} & \frac{\sqrt{3}}{4} & \frac{\sqrt{3}}{4} & \frac{1}{4} \end{pmatrix}.$$

From the structure of the matrices, one can see that these representations are reducible. For their reduction, the corresponding character relations (5.35) are adequate, that is, one element for each class is sufficient. Thus,

$$a_1^{3,3} + a_2^{3,3} + a_3^{3,3} = 4, \quad a_1^{3,3} + a_2^{3,3} - a_3^{3,3} = 1, \quad a_1^{3,3} - a_2^{3,3} = 0$$

for the elements $X = E, C_3, \sigma_1$, respectively. The non-negative integer solutions are $\alpha_1^{3,3} = \alpha_2^{3,3} = \alpha_3^{3,3} = 1$. In other words,

$$\Gamma^3(X) \otimes \Gamma^3(X) = \Gamma^1(X) \oplus \Gamma^2(X) \oplus \Gamma^3(X), \tag{5.37}$$

a really neat result.

5.9 Reduction of the direct product of groups

The elements of the direct product of two groups $G \times G'$, see Section 3.7, can take the form

$$A_1 A_1', A_1 A_2', \ldots, A_1 A_{g'}', A_2 A_1', A_2 A_2', \ldots, A_2 A_{g'}', A_3 A_1', A_3 A_2', \ldots,$$

$$A_3 A_{g'}', \ldots, A_g A_g', A_g A_2', \ldots, A_g A_{g'}'.$$

Let us consider the case

$$A = A_k A_{k'}', \quad B = A_\ell A_{\ell'}',$$

$$A_k A_\ell = A_m, A_{k'}' A_{\ell'}' = A_{m'}' \Rightarrow C = A_m A_{m'}' \Leftrightarrow AB = C$$

$$\Gamma(A) = \Gamma_{A_k} \otimes \Gamma(A_{k'}'), \Gamma(B) = \Gamma(A_\ell) \otimes \Gamma(A_{\ell'}'), \Gamma(C) = \Gamma(A_m) \otimes \Gamma(A_{m'}').$$

Making use of Eq. (4.10), we find

$$\Gamma(A)\Gamma(B) = (\Gamma_{A_k} \otimes \Gamma(A_{k'}'))\,(\Gamma(A_\ell) \otimes \Gamma(A_{\ell'}'))$$

$$= (\Gamma(A_k)\Gamma(A_\ell)) \otimes (\Gamma(A_{k'}')\Gamma(A_{\ell'}'))$$

$$= \Gamma(A_m) \otimes \Gamma(A_{m'}') = \Gamma(C) \Rightarrow \Gamma(A)\Gamma(B) = \Gamma(C).$$

In other words, through the direct product of representations of G and G', a representation of $G \times G'$ is obtained.

We illustrate this by considering the group in Example 6 of Section 3.7 to be $G = C_I = \{E, I\}$ and $G' = C_3 = \{E, \sigma_x\}$, which are found in crystal structure. In this case, we have

$$P = \begin{pmatrix} 0 & 1 \\ 1 & 0 \end{pmatrix}, \quad Q = \begin{pmatrix} -1 & 0 \\ 0 & -1 \end{pmatrix}, \quad PQ = \begin{pmatrix} 0 & -1 \\ -1 & 0 \end{pmatrix}.$$

We write this in the "canonical form", i.e. $G \times G' = \{EE, EP, QE, PQ\}$ and get:

$$\Gamma(E) = E \otimes E = \begin{pmatrix} 1 & 0 & 0 & 0 \\ 0 & 1 & 0 & 0 \\ 0 & 0 & 1 & 0 \\ 0 & 0 & 0 & 1 \end{pmatrix},$$

$$\Gamma(P) = E \otimes P = \begin{pmatrix} 0 & 1 & 0 & 0 \\ 1 & 0 & 0 & 0 \\ 0 & 0 & 0 & 1 \\ 0 & 0 & 1 & 0 \end{pmatrix},$$

$$\Gamma(Q) = Q \otimes E = \begin{pmatrix} -1 & 0 & 0 & 0 \\ 0 & -1 & 0 & 0 \\ 0 & 0 & -1 & 0 \\ 0 & 0 & 0 & -1 \end{pmatrix},$$

$$\Gamma(PQ) = Q \otimes P = \begin{pmatrix} 0 & -1 & 0 & 0 \\ -1 & 0 & 0 & 0 \\ 0 & 0 & 0 & -1 \\ 0 & 0 & -1 & 0 \end{pmatrix}.$$

One can easily verify that this representation obeys the same multiplication table with that of the group, Eq. (3.9). It is clearly reducible, which is also true of those of the groups G and G', and is reduced as above, Eq. (5.31).

We will show that whenever any representations of G and G' are irreducible, the representation of $G \times G'$ is also irreducible.

Before proceeding with the proof, let us consider the reduction $C_{3h} = (E, I) \times C_{3v}$. This group has 12 elements:

$$\{E, C_3, C_3^2, \sigma_1, \sigma_2, \sigma_3, I, IC_3, IC_3^2, I\sigma_1, I\sigma_2, I\sigma_3\}.$$

The reduction of the regular representation can proceed as usual, but with the method of direct product, this can be accomplished in a much simpler manner. Let us consider, for example, the representation $\tilde{\Gamma}(\sigma_2) = \{E, I\} \otimes \Gamma(\sigma_2)$, making use of the reduced forms of $\{E, I\}$ and $\Gamma(\sigma_2)$, that is, $\{E, I\} = \mathrm{diag}\{1, -1\}$ and $\Gamma(\sigma_2)$, as given in the Example 3 above. Making use of packages, such as Mathematica, we find

$$
\tilde{\Gamma}(\sigma_2) =
\begin{pmatrix}
1 & 0 & 0 & 0 & 0 & 0 & 0 & 0 & 0 & 0 & 0 & 0 \\
0 & -1 & 0 & 0 & 0 & 0 & 0 & 0 & 0 & 0 & 0 & 0 \\
0 & 0 & \frac{1}{2} & \frac{\sqrt{3}}{2} & 0 & 0 & 0 & 0 & 0 & 0 & 0 & 0 \\
0 & 0 & \frac{\sqrt{3}}{2} & -\frac{1}{2} & 0 & 0 & 0 & 0 & 0 & 0 & 0 & 0 \\
0 & 0 & 0 & 0 & \frac{1}{2} & \frac{\sqrt{3}}{2} & 0 & 0 & 0 & 0 & 0 & 0 \\
0 & 0 & 0 & 0 & \frac{\sqrt{3}}{2} & -\frac{1}{2} & 0 & 0 & 0 & 0 & 0 & 0 \\
0 & 0 & 0 & 0 & 0 & 0 & -1 & 0 & 0 & 0 & 0 & 0 \\
0 & 0 & 0 & 0 & 0 & 0 & 0 & 1 & 0 & 0 & 0 & 0 \\
0 & 0 & 0 & 0 & 0 & 0 & 0 & 0 & -\frac{1}{2} & -\frac{\sqrt{3}}{2} & 0 & 0 \\
0 & 0 & 0 & 0 & 0 & 0 & 0 & 0 & -\frac{\sqrt{3}}{2} & \frac{1}{2} & 0 & 0 \\
0 & 0 & 0 & 0 & 0 & 0 & 0 & 0 & 0 & 0 & -\frac{1}{2} & -\frac{\sqrt{3}}{2} \\
0 & 0 & 0 & 0 & 0 & 0 & 0 & 0 & 0 & 0 & -\frac{\sqrt{3}}{2} & \frac{1}{2}
\end{pmatrix}.
$$

Thus, it breaks into two pairs of 1D and two pairs of 2D, with the members of one pair opposite to those of the other pair. The same is true for the other elements of the group. Thus,

$$
\Gamma^{\mathrm{reg}}(C_{3h}) = \Gamma^1 + \Gamma^2 + (\Gamma')^1 + (\Gamma')^2 + 2\Gamma^3 + 2(\Gamma')^3, \quad (\Gamma')^i = -\Gamma^i, \; i = 1, 2, 3.
$$

Theorem: *Given the representations of G and G', which are irreducible, then the resulting representations of $G \times G'$ are also irreducible.*

Proof: The representation $\Gamma^i(X) \otimes \Gamma^j(X')$ is a matrix of dimension $d_i d_j$, where $d_i = \dim\left(\Gamma^i(X)\right)$, $d_j = \dim\left(\Gamma^j(X')\right)$. A given matrix ρ can be considered a matrix $d_i \times d_j$, i.e. composed of d_i rows and d_j columns, indicated as $(\rho)^{(km)}$:

$$
(\rho) =
\begin{pmatrix}
(\rho)^{(11)} & (\rho)^{(12)} & \cdots & (\rho)^{(1d_j)} \\
(\rho)^{(21)} & (\rho)^{(22)} & \cdots & (\rho)^{(2d_j)} \\
\cdots & \cdots & \cdots & \cdots \\
(\rho)^{(d_i 1)} & (\rho)^{(d_i 2)} & \cdots & (\rho)^{(d_i d_j)}
\end{pmatrix}.
$$

We consider the representation

$$\Gamma(EX') = \Gamma^i(E) \otimes \Gamma^j(X') = \begin{pmatrix} \Gamma^j(X') & 0 & \cdots & 0 \\ 0 & \Gamma^j(X') & \cdots & 0 \\ \cdots & \cdots & \cdots & \cdots \\ 0 & 0 & 0 & \Gamma^j(X') \end{pmatrix}.$$

Suppose now that $(\rho)\Gamma(EX') = \Gamma(EX')(\rho)$. Then,

$$(\rho) = \begin{pmatrix} (\rho)^{(11)} & (\rho)^{(12)} & \cdots & (\rho)^{(1d_j)} \\ (\rho)^{(21)} & (\rho)^{(22)} & \cdots & (\rho)^{(2d_j)} \\ \cdots & \cdots & \cdots & \cdots \\ (\rho)^{(d_i 1)} & (\rho)^{(d_i 2)} & \cdots & (\rho)^{(d_i d_j)} \end{pmatrix} \begin{pmatrix} \Gamma^j(X') & 0 & \cdots & 0 \\ 0 & \Gamma^j(X') & \cdots & 0 \\ \cdots & \cdots & \cdots & \cdots \\ 0 & 0 & 0 & \Gamma^j(X') \end{pmatrix}$$

$$= \begin{pmatrix} \Gamma^j(X') & 0 & \cdots & 0 \\ 0 & \Gamma^j(X') & \cdots & 0 \\ \cdots & \cdots & \cdots & \cdots \\ 0 & 0 & 0 & \Gamma^j(X') \end{pmatrix} \begin{pmatrix} (\rho)^{(11)} & (\rho)^{(12)} & \cdots & (\rho)^{(1d_j)} \\ (\rho)^{(21)} & (\rho)^{(22)} & \cdots & (\rho)^{(2d_j)} \\ \cdots & \cdots & \cdots & \cdots \\ (\rho)^{(d_i 1)} & (\rho)^{(d_i 2)} & \cdots & (\rho)^{(d_i d_j)} \end{pmatrix}$$

$$\Rightarrow (\rho)^{(km)}\Gamma^j(X') = \Gamma^j(X')(\rho)^{(km)}, \ k = 1, 2, \ldots, d_i, \ m = 1, 2, \ldots, d_j.$$

Since the representations are irreducible, Schur's second lemma implies that

$$(\rho)^{(km)} = \xi_{km}\epsilon_{d_i}, \tag{5.38}$$

with $\epsilon_{d_i} =$ the identity matrix in d_i dimensions and ξ_{km} constants dependent on the matrix $(\rho)^{(km)}$.

We consider now the representation $\Gamma(XE)$:

$$\Gamma(XE) = \Gamma^i(X) \otimes \Gamma^j(E) = \begin{pmatrix} \Gamma^i(X)_{11}(\epsilon_{d_j}) & \cdots & \cdots & \Gamma^i(X)_{1m}(\epsilon_{d_j}) \\ \Gamma^i(X)_{21}(\epsilon_{d_j}) & \cdots & \cdots & \Gamma^i(X)_{2m}(\epsilon_{d_j}) \\ \cdots & \cdots & \cdots & \cdots \\ \Gamma^i(X)_{m1}(\epsilon_{d_j}) & \cdots & \cdots & \Gamma^i(X)_{mm}(\epsilon_{d_j}) \end{pmatrix}.$$

Suppose further that $(\rho)\Gamma(XE) = \Gamma(XE)(\rho)$, then

$$\sum_n (\rho)^{(kn)}\Gamma^i(X)_{nm}(\epsilon_{d_j}) = \sum_n \Gamma^i(X)_{mn}(\epsilon_{d_j})(\rho)^{(nk)} \Rightarrow$$

$$\sum_n (\rho)^{(kn)}(\epsilon_{d_j})\Gamma^i(X)_{nm} = \sum_n \Gamma^i(X)_{mn}(\rho)^{(nk)}(\epsilon_{d_j}),$$

and then, making use of Eq. (5.38), we find

$$\sum_n \xi_{kn}\Gamma^i(X)_{nm} = \sum_n \Gamma^i(X)_{mn}\xi_{nk} \Rightarrow (\xi)\Gamma^i(X) = \Gamma^i(X)(\xi),$$

i.e. the matrix (ξ) commutes with all the matrices $\Gamma^i(X)$. As a result, according to the first Schur lemma, the matrix (ρ) is a multiple of the identity, that is, the representation $\Gamma(X)$ is irreducible.

We will adopt the notation

$$\Gamma^{(i,j)}(X) = \Gamma^i(X) \otimes \Gamma^j(X), \tag{5.39}$$

hence the characters obey the relation

$$\chi^{(i,j)}(X) = \chi^i(X)\chi^j(X). \tag{5.40}$$

Using the orthogonality of characters, we find that the representations of Eq. (5.39) are all different, i.e. non-equivalent. Their number is $k \times k'$, with k the irreducible representations of G and k' the irreducible representations of G'. On the other hand, $k \times k'$ is the number of classes of $G \times G'$.

It should be stressed that Eq. (5.39) provides all the irreducible representations of $G \times G'$.

Example 10: We now study the reduction of the group direct product of the irreducible representations of the group with $g = 6$ of Example 5 above. We have:

$$C_{3h} = \{E, A, A^2, \sigma_h, A\sigma_h, A^2\sigma_h\} = \{E, A, A^2\} \otimes \{E, \sigma_h\}.$$

From the known representations of $g = 2$ and $g = 3$, we obtain

$$\begin{pmatrix} 1 & 1 \\ 1 & -1 \end{pmatrix} \otimes \begin{pmatrix} 1 & 1 & 1 \\ 1 & \epsilon & \epsilon^2 \\ 1 & \epsilon^2 & \epsilon \end{pmatrix} = \begin{pmatrix} 1 & 1 & 1 & 1 & 1 & 1 \\ 1 & \epsilon & \epsilon^2 & 1 & \epsilon & \epsilon^2 \\ 1 & \epsilon^2 & \epsilon & 1 & \epsilon^2 & \epsilon \\ 1 & 1 & 1 & -1 & -1 & -1 \\ 1 & \epsilon & \epsilon^2 & -1 & -\epsilon & -\epsilon^2 \\ 1 & \epsilon^2 & \epsilon & -1 & -\epsilon^2 & -\epsilon \end{pmatrix}.$$

There exists, of course, the equivalent expression

$$
\begin{pmatrix} 1 & 1 & 1 \\ 1 & \epsilon & \epsilon^2 \\ 1 & \epsilon^2 & \epsilon \end{pmatrix} \otimes \begin{pmatrix} 1 & 1 \\ 1 & -1 \end{pmatrix} = \begin{pmatrix} 1 & 1 & 1 & 1 & 1 & 1 \\ 1 & -1 & 1 & -1 & 1 & -1 \\ 1 & 1 & \epsilon & \epsilon & \epsilon^2 & \epsilon^2 \\ 1 & -1 & \epsilon & -\epsilon & \epsilon^2 & -\epsilon^2 \\ 1 & 1 & \epsilon^2 & \epsilon^2 & \epsilon & \epsilon \\ 1 & -1 & \epsilon^2 & -\epsilon^2 & \epsilon & -\epsilon \end{pmatrix},
$$

which is nothing but a reordering of the elements of the group C_{3h}.

The conclusions of this section find extensive applications in the study of crystal structure and solid-state physics, since many of the groups encountered are such direct products (D_{nh}, D_{nd}, T_n, O_n, etc.). The most common case is whenever a group G can be extended to include point inversion, see Problems 3.7.4–3.7.7 and 5.9.1–5.9.10. The resulting group is often called improper, the name arising from that of orthogonal transformations with a determinant -1.

5.10   Double covering groups of discrete groups

Any given discrete group with elements $G : \{E, A_1, A_2, \ldots, A_{g-1}\}$ can be extended by adding the elements $A_i I$ and I, where I is the opposite of the identity, yielding a group G' given by

$$G' : \{E, A_1, A_2, \ldots, A_{g-1}, I, A_1 I, A_2 I, \ldots, A_{g-1} I\}. \tag{5.41}$$

Indeed, the set G' is a group:

$$A_i' A_j' = A_i I A_j I = A_i A_j II = A_i A_j, \quad A_i A_j' = A_i A_j I = (A_i A_j)'.$$

The group G' is often called double covering of G.

If the group G is a point group encountered in crystallography, the group G' does not belong to the same category unless G contains the symmetry of point inversion.

Popular groups are $S_4, S_4', A_4, A_4', S_5, A_5, A_5'$, where A_n indicates the subgroup formed by the set of even permutations of S_n. Note that A_n' has the same number of elements as S_n, but the two groups are not isomorphic. The group A_4 will be discussed in detail in Chapter 13. It has two generators, $T = (123)$ and $S = (14)(23)$, with properties

$S^2 = T^3 = (ST)^3 = (TS)^3$. We will see that this group has four irreducible representations, three which are 1D, indicated by $\Gamma^{(1)} = \underline{1}, \Gamma^{(2)} = \underline{1}', \Gamma^{(3)} = \underline{1}''$, and one which is 3D, $\Gamma^{(4)} = \underline{3}$. The double covering group of A_4 is the group A'_4 or T_h. This group has seven conjugate classes, $1C_1, 1C_2, 6C_4, 4C_6, 4C_3, 4C'_3, 4C'_6$. It has, therefore, seven irreducible representations, four of which are the irreducible representations of A_4. The remaining three must satisfy the condition $\ell_1^2 + \ell_2^2 + \ell_3^2 = 24 - 12 = 12$, see Eq. (5.27), which has the only solution $\ell_1 = \ell_2 = \ell_3 = 2$. Thus, the irreducible representations are

$$\Gamma^{(1)} = \underline{1}, \Gamma^{(2)} = \underline{1}', \Gamma^{(3)} = \underline{1}'', \Gamma^{(4)} = \underline{2}, \Gamma^{(5)} = \underline{2}', \Gamma^{(6)} = \underline{2}'', \Gamma^{(7)} = \underline{3}.$$

For the group A_5, one finds that it has two generators S and T with properties $S^2 = T^3 = (ST)^5 = E$, see Section 14.5.2. Its elements are distributed into five classes and the five irreducible representations are

$$\Gamma^{(1)} = \underline{1}, \Gamma^{(2)} = \underline{3}, \Gamma^{(3)} = \underline{3}', \Gamma^{(4)} = \underline{5}, \Gamma^{(5)} = \underline{6}.$$

A_5 is isomorphic to the icosahedral rotation group, related to the icosahedron studied by the ancient Greeks, see Fig. 2.12(b), and is of order 60. The group A'_5 has 120 elements (Wang et al., 2021). These 120 elements are categorized into nine conjugacy classes, i.e. four additional classes compared to A_5, and the irreducible representations are

$$\Gamma^{(1)} = \underline{1}, \quad \Gamma^{(2)} = \underline{2}, \quad \Gamma^{(3)} = \underline{2}', \quad \Gamma^{(4)} = \underline{3},$$
$$\Gamma^{(5)} = \underline{3}', \quad \Gamma^{(6)} = \underline{4}, \quad \Gamma^{(7)} = \underline{4}', \quad \Gamma^{(8)} = \underline{5}, \quad \Gamma^{(9)} = \underline{6}.$$

The group A'_5 is isomorphic to I_h, the full icosahedral group, recently applied in particle physics (Everett and Stuart, 2011).

5.11 Problems

5.1.1 Consider the representation

$$\begin{pmatrix} 1 & 0 \\ 0 & 1 \end{pmatrix}, \quad \begin{pmatrix} 0 & 1 \\ 1 & 0 \end{pmatrix}$$

of $G = \{E, I\}$. Is it reducible? If yes, find the matrix S reducing it.

5.1.2 Do the same for

$$
\begin{pmatrix} 1 & 0 & 0 & 0 \\ 0 & 1 & 0 & 0 \\ 0 & 0 & 1 & 0 \\ 0 & 0 & 0 & 1 \end{pmatrix}, \quad
\begin{pmatrix} 0 & 1 & 0 & 0 \\ 1 & 0 & 0 & 0 \\ 0 & 0 & 0 & 1 \\ 0 & 0 & 1 & 0 \end{pmatrix}.
$$

Is S uniquely specified?

5.1.3 Find the regular representation of the cyclic group of four elements. Is it reducible? If yes, find a matrix S causing the reduction and the diagonal form of the representation.

5.1.4 Do the same for the group of five elements.

5.2.1 Consider the matrices

$$
\begin{pmatrix} 1 & 0 \\ 0 & 1 \end{pmatrix}, \quad
\begin{pmatrix} 0 & 1 \\ 1 & 0 \end{pmatrix}, \quad
\begin{pmatrix} -1 & -1 \\ 0 & 1 \end{pmatrix},
$$

$$
\begin{pmatrix} 1 & 0 \\ -1 & -1 \end{pmatrix}, \quad
\begin{pmatrix} -1 & -1 \\ -1 & 0 \end{pmatrix}, \quad
\begin{pmatrix} 0 & 1 \\ -1 & -1 \end{pmatrix}.
$$

Show that they constitute a representation of the group S_3. Then, construct an equivalent unitary representation, and find the matrix S connecting them. Is the resulting representation reducible? Hint: Consider the matrix $S = \begin{pmatrix} 1/\sqrt{3} & 1 \\ 1/\sqrt{3} & -1 \end{pmatrix}$, and show that the matrices of Eq. (2.6) are obtained this way.

5.2.2 Find the representation of the group C_{4v} resulting from the basis $|\psi_1\rangle = \frac{1}{\sqrt{2}}(\hat{e}_1 + \hat{e}_2)$, $|\psi_2\rangle = \hat{e}_3$.

5.3.1 Find the characters of a representation of the group with $g = 3$.

5.3.2 Do the same in the case of the group with $g = 4$.

5.3.3 Do the same for the regular representation of the group C_{4v}.

5.3.4 Do the same for the representation of problem 4.3.2.

5.3.5 Do the same for the representation of problem 4.7.1.

5.3.6 Do the same for the representation of problem 4.7.2.

5.3.7 Do the same for the representation of problem 4.7.3.

5.3.8 Do the same for the representation of problems 4.7.4–4.7.5.

5.3.9 Do the same for the representation of problem 4.7.6.

5.4.1 Construct the quantity

$$M_{K_\alpha} = \sum_\beta \Gamma\left(X_\beta \in K_\alpha\right),$$

where K_α are the classes of the group C_{3v} and $\Gamma(X)$ is their regular representation. What conclusions can be drawn?

5.4.2 Do the same for the representation of problem 5.2.1.

5.4.3 Examine whether the representation of problem 4.3.2 satisfies the relations (5.13) and (5.19).

5.4.4 Do the same for the representation of problem 4.7.8.

5.6.1 Find the irreducible representations of the groups C_4 and S_4.

5.6.2 Find the number of the irreducible representations of the symmetric group S_3 and the dimension of each one.

5.6.3 Do the same in the case of the group C_{4v}.

5.6.4 Do the same in the case of the group of problem 4.3.2.

5.6.5 Do the same in the case of the group of problem 3.7.2.

5.6.6 Do the same in the case of the group T (Section 2.4.2).
It may be useful to show first that the 12 elements are distributed in four classes.

5.6.7 Do the same in the case of the group T_d (Section 2.4.2).
It may be useful to show first that its 24 elements are distributed in 5 classes (Section 2.4.2).

5.7.1 Find the irreducible representations of the group C_{3h}, and construct the corresponding character table.

5.7.2 Find the character table of the irreducible representations of problem 5.6.3.

5.7.3 Do the same in the case of the groups C_{2d} and D_4.

5.7.4 Do the same in the case of the group of problem 5.6.4.

5.7.5 Do the same in the case of the group of problem 5.6.6.

5.7.6 Do the same in the case of the group C_6.

5.7.7 Do the same in the case of the group D_6.

5.7.8 Do the same in the case of the group C_{6v}.

5.7.9 Do the same in the case of the group D_{3h}.

5.8.1 Find the direct sum $A \oplus B$ and the direct product $A \otimes B$ with

$$A = \begin{pmatrix} 2 & 5 & 9 \\ 1 & 4 & 7 \\ 3 & 3 & 3 \end{pmatrix}, \quad B = \begin{pmatrix} 6 & 4 \\ 2 & 7 \end{pmatrix}.$$

After that, obtain $B \oplus A$ and $B \otimes A$.

Note: The direct sum of two matrices A and B is a matrix with matrices A and B along the diagonal and zeros elsewhere.

5.8.2 Do the same for

$$A = \begin{pmatrix} -2 & 3 & 4 \\ 8 & 7 & -6 \end{pmatrix}, \quad B = \begin{pmatrix} 9 \\ 6 \\ 3 \end{pmatrix}.$$

5.8.3 Show that $(A \otimes B)^{-1} = A^{-1} \otimes B^{-1}$ (without changing the order of A and B).

5.8.4 Show that $A \otimes B$ is unitary provided that A and B are unitary.

5.8.5 Suppose that $\Gamma_i, i = 1, \ldots, 4$ are the 1D representations and Γ_5 is the irreducible 2D one of C_{4v}. Show that

$$\Gamma_1 \otimes \Gamma_i = \Gamma_i, \quad i = 1, \ldots, 4, \quad \Gamma_i \otimes \Gamma_i = \Gamma_1, \quad i = 1, \ldots, 4,$$

$$\Gamma_2 \otimes \Gamma_3 = \Gamma_4, \Gamma_3 \otimes \Gamma_4 - \Gamma_2,$$

$$\Gamma_2 \otimes \Gamma_4 = \Gamma_3, \Gamma_i \otimes \Gamma_5 = \Gamma_5, \quad i = 1, \ldots, 4,$$

$$\Gamma_5 \otimes \Gamma_5 = \Gamma_1 \otimes \Gamma_2 \otimes \Gamma_3 \otimes \Gamma_4.$$

5.8.6 Write down the analogous relations for the irreducible representations of problem 5.6.4.

5.9.1 Consider the group $C_{2h} = C_2 \otimes C_i$, $C_i = \{E, I\}$. Obtain all the irreducible representations of the part on the left, i.e. of the group C_{2h}.

5.9.2 Do the same in the case of the group $C_{4h} = C_4 \otimes C_i$.

5.9.3 Do the same in the case of the group $C_{6h} = C_6 \otimes C_i$.

5.9.4 Do the same in the case of the group $C_{4h} = D_4 \otimes C_i$.

5.9.5 Do the same in the case of the group $C_{3h} = C_3 \otimes C_i$.

5.9.6 Do the same in the case of the group $D_{2h} = D_2 \otimes C_i$.

5.9.7 Do the same in the case of the group $D_{6h} = D_6 \otimes C_i$.

5.9.8 Do the same in the case of the group $D_{3d} = D_3 \otimes C_i$.

5.9.9 Go as far as you can in obtaining the irreducible representations of the group $T_h = T \otimes C_i$.

5.9.10 Go as far as you can in obtaining the irreducible representations of the group $O_h = O \otimes C_i$.

Chapter 6

Simple Applications in Quantum Mechanics

The arrival of quantum mechanics was the main reason for the familiarity of physicists with group theory, which had already been developed by mathematicians. At this stage, Eugene Wigner played a very crucial role, which is discussed in this chapter. In addition, a great contribution toward the familiarity of physicists with group theory can be attributed to Heisenberg and Weyl, mainly in that they showed the equivalence of Schrödinger's picture of quantum mechanics with that of Heisenberg. As a result of such efforts, many discrete groups were invented through the neat theory of Heisenberg–Wigner–Weyl, which we briefly discuss in Chapter 14.

6.1 Behavior of a quantum system acted upon by group operators

If we denote as x all the coordinates that describe a system, the time-independent Schrödinger differential equation may take the following form:

$$H(x)\psi(x) = E\psi(x), \qquad (6.1)$$

where $H(x)$ is the Hamiltonian describing the system. In this space, we consider a linear transformation, which is an element of a group, $A \in G$:

$$x \to x' = Ax. \qquad (6.2)$$

This implies a transformation T_A in functional space

$$\psi \to \psi' = T_A\psi = \psi(A^{-1}x). \qquad (6.3)$$

see Eq. (4.17). Then, Eq. (6.1) becomes

$$H(A^{-1}x)\psi(A^{-1}x) = E\psi(A^{-1}x). \qquad (6.4)$$

Thus, Eq. (6.1) remains unchanged as long as

$$H(A^{-1}x) = H(x). \qquad (6.5)$$

This is equivalent to

$$HT_A = T_A H. \qquad (6.6)$$

In fact, from Eqs. (6.5) and (6.4), one finds

$$H(x)\psi(A^{-1}x) = E\psi(A^{-1}x) \Rightarrow H(x)T_A\psi(x) = ET_A\psi(x) \Rightarrow \qquad (6.7)$$

$$\left(T_A^{-1}HT_A\right)\psi(x) - E\psi(x) \to T_A^{-1}HT_A - H \to HT_A = T_A H \qquad (6.8)$$

Now, considering a representation of T_A, $T_A \to T(A)$, in some basis, we show that

$$T(A)T(H) = T(H)T(A),$$

where $T(H)$ is the representation of the operator H in the same basis.

Let $\psi_i(x)$ be an orthogonal basis in functional space. Then,

$$T(H)_{ij} = \langle \psi_i(x)|H|\psi_j(x)\rangle \equiv H_{ij}, \ T(A)_{ij} = \langle \psi_i(x)|T_A|\psi_j(x)\rangle,$$

$$H'_{ij} = T'(H)_{ij} = \int dx \psi_i'^*(x)|H|\psi_j'(x) = \int dx \psi_i^*(A^{-1}x)|H|\psi_j(A^{-1}x)$$

$$= \int dx' \psi_i^*(x')|H(x')|\psi_j(x') = H_{ij},$$

$$H'_{ij} = T'(H)_{ij} = \int dx T_A \psi_i(x)|H|T_A\psi_j(x)$$

$$= \sum_k \sum_\ell (T(A))^*_{kj}(T(A))_{\ell j} \int \psi_k^*(x)|H|\psi_\ell$$

$$= \sum_k \sum_\ell (T(A))^*_{kj}(T(A))_{\ell j} H_{k\ell} = \left((T(A))^+ T(H)T(A)\right)_{ij}$$

$$\Rightarrow T(A) = (T(A))^+ T(H)T(A).$$

We suppose that the representation is unique:[1]

$$T(A) \Rightarrow \Gamma(A), \Gamma^{+}(A) = \Gamma^{-1}(A) = \Gamma(A^{-1}) \Leftrightarrow T(H) = \Gamma^{-1}(A)T(H)\Gamma(A)$$
$$\Rightarrow \Gamma(A)T(H) = T(H)\Gamma(A). \tag{6.9}$$

In order to solve Eq. (6.1), the following strategy is used: We expand the function ψ in the basis ψ_i:

$$\psi = \sum_i \alpha_i \psi_i.$$

Equation (6.1) now becomes

$$\sum_i \alpha_i H \psi_i = \sum_i \alpha_i E \psi_i.$$

Taking the internal product of both sides of the equation with $\langle \psi_j |$, we find

$$\sum_i \alpha_i \langle \psi_j | H | \psi_i \rangle = \sum_i \alpha_i E \langle \psi_j \psi_i \rangle = E\alpha_j \Rightarrow \sum_i \alpha_i (H_{ji} - E\delta_{ij}) = 0,$$

or in matrix form,

$$(H)|\alpha\rangle = E|\alpha\rangle. \tag{6.10}$$

Equation (6.10) is equivalent to Eq. (6.1).

It is sufficient, therefore, to solve Eq. (6.10). If, by good luck, we hit the bull's eye when selecting the basis, we may be able to reduce the matrix (H):

$$(H) = \begin{pmatrix} (H^{(1)}) & 0 & 0 & \cdots & \cdots & 0 \\ 0 & (H^{(2)}) & 0 & \cdots & \cdots & 0 \\ 0 & 0 & (H^{(3)}) & \cdots & \cdots & 0 \\ 0 & 0 & 0 & (H^{(4)}) & \cdots & 0 \\ 0 & 0 & 0 & \cdots & \cdots & (H^{(k)}) \end{pmatrix}. \tag{6.11}$$

As we will see in the following section, group theory plays an important part in this reduction. Finally, let us mention for the sake of completeness

---

[1]As we have already mentioned, this is always possible in the case of discrete groups. It is also possible in the case of the usual compact continuous groups.

that in applications, H is written as

$$H = H^{(0)} + V,$$

where V is a small enough perturbation and the eigenfunctions and eigenvalues of $H^{(0)}$ are known, i.e.:

$$H^{(0)}|\psi_i\rangle = E^{(0)}|\psi_i\rangle,$$

so the solution of our unperturbed problem provides the desired basis. Then,

$$H_{ij} = E^{(0)}\delta_{ij} + V_{ij}, \ V_{ij} = \langle\psi_j|V|\psi_i\rangle. \tag{6.12}$$

6.2 Wigner's theorem

Let $\Gamma(X)$, $X \in G$ be a representation in an appropriate basis so as to be fully reduced, i.e.:

$$\Gamma(X) = \alpha_1\Gamma_1(X) \oplus \alpha_2\Gamma_2(X)\cdots\alpha_k\Gamma_k(X), \ \Gamma_i(X),$$

$$i = 1, 2, \ldots, k \text{ is irreducible.} \tag{6.13}$$

Suppose also that

$$(H)\Gamma(X) = \Gamma(X)(H), \tag{6.14}$$

where (H) is the corresponding Hamilton matrix, i.e. that which is constructed in the same basis. Then,

$$(H) = (H^{(1)}) \oplus (H^{(2)}) \oplus \cdots \oplus (H^{(k)}),$$

$$\dim(H^{(i)}) \le \alpha_i\ell_i, \quad i = 1, 2, \ldots, k, \ell_i = \dim\Gamma_i \tag{6.15}$$

Proof: The basis in which $\Gamma(X)$ takes the form of (6.13) may be written as

$$\psi_\alpha^{\rho,i} = |\rho, i, \alpha\rangle,$$

where $\rho = 1, 2, \ldots, \alpha_i$, denoting which of the α_i a specific $\Gamma_i(X)$ belongs to, and α is the specific basis vector in the space $\ell_i \times \ell_i$. As long as $\Gamma_i(X)$ is unique,

$$\langle\psi_\beta^{\rho,j}|\rho, i, \alpha\rangle \propto \delta_{ij}\delta_{\alpha,\beta}.$$

In fact,

$$\langle \rho', j, \beta | \rho, i, \alpha \rangle = \langle \rho', j, \beta | \Gamma^+(X)\Gamma(X) | \rho, i, \alpha \rangle$$

$$= \frac{1}{g} \sum_{X \in G} \langle \rho', j, \beta | \Gamma^+(X)\Gamma(X) | \rho, i, \alpha \rangle$$

$$= \frac{1}{g} \sum_{X \in G, m, k} \Gamma^{*j}_{m\beta}(X)\Gamma^{i}_{k\alpha}(X) \langle \rho', j, m | \rho, i, k \rangle.$$

According to the orthogonality theorem, Eq. (5.13), we have

$$\langle \rho', j, \beta | \rho, i, \alpha \rangle = \frac{1}{\ell_i} \delta_{ij} \delta_{\alpha\beta} \delta_{mk} \sum_k \langle \rho', i, k | \rho, i, k \rangle.$$

Unfortunately, there is no orthogonality with respect to the index ρ. This is a problem with the group theoretical approach. It can, however, be achieved in practice using the inelegant Gram–Schmidt method.

According to the above, the representation $\Gamma(X)$ may take the following form:

$$\Gamma(X) = \begin{pmatrix} \Gamma^{(1)} & 0 & 0 & 0 & 0 & 0 & 0 & 0 & 0 & 0 & 0 & 0 & 0 \\ 0 & \Gamma^{(1)} & 0 & 0 & 0 & 0 & 0 & 0 & 0 & 0 & 0 & 0 & 0 \\ 0 & 0 & \cdots & 0 & 0 & 0 & 0 & 0 & 0 & 0 & 0 & 0 & 0 \\ 0 & 0 & 0 & \Gamma^{(1)} & 0 & 0 & 0 & 0 & 0 & 0 & 0 & 0 & 0 \\ 0 & 0 & 0 & 0 & \Gamma^{(2)} & 0 & 0 & 0 & 0 & 0 & 0 & 0 & 0 \\ 0 & 0 & 0 & 0 & 0 & 0 & \cdots & 0 & 0 & 0 & 0 & 0 & 0 \\ 0 & 0 & 0 & 0 & 0 & 0 & 0 & \Gamma^{(2)} & 0 & 0 & 0 & 0 & 0 \\ 0 & 0 & 0 & 0 & 0 & 0 & 0 & 0 & \cdots & 0 & 0 & 0 & 0 \\ 0 & 0 & 0 & 0 & 0 & 0 & 0 & 0 & 0 & \Gamma^{(k)} & 0 & 0 & 0 \\ 0 & 0 & 0 & 0 & 0 & 0 & 0 & 0 & 0 & 0 & \Gamma^{(k)} & 0 & 0 \\ 0 & 0 & 0 & 0 & 0 & 0 & 0 & 0 & 0 & 0 & 0 & \cdots & 0 \\ 0 & 0 & 0 & 0 & 0 & 0 & 0 & 0 & 0 & 0 & 0 & 0 & \Gamma^{(k)} \end{pmatrix}.$$

Now, let us see what form the matrix (H) takes in this basis. To simplify the symbolism, let us consider two irreducible representations, $\Gamma_1(X)$ and $\Gamma_2(X)$, which are found $\alpha_1 = 3$ and $\alpha_2 = 2$ times, respectively. Then,

$$\Gamma(X) = \begin{pmatrix} \Gamma^{(1)} & 0 & 0 & 0 & 0 \\ 0 & \Gamma^{(1)} & 0 & 0 & 0 \\ 0 & 0 & \Gamma^{(1)} & 0 & 0 \\ 0 & 0 & 0 & \Gamma^{(2)} & 0 \\ 0 & 0 & 0 & 0 & \Gamma^{(2)} \end{pmatrix}.$$

The matrix (H) will take the form

$$(H) = \begin{pmatrix} (H_{11}^{(11)}) & (H_{12}^{(11)}) & (H_{13}^{(11)}) & (H_{11}^{(12)}) & (H_{12}^{(12)}) \\ (H_{21}^{(11)}) & (H_{22}^{(11)}) & (H_{23}^{(11)}) & (H_{21}^{(12)}) & (H_{22}^{(12)}) \\ (H_{31}^{(11)}) & (H_{32}^{(11)}) & (H_{33}^{(11)}) & (H_{31}^{(12)}) & (H_{32}^{(12)}) \\ (H_{11}^{(21)}) & (H_{12}^{(21)}) & (H_{13}^{(21)}) & (H_{11}^{(22)}) & (H_{22}^{(22)}) \\ (H_{21}^{(21)}) & (H_{22}^{(21)}) & (H_{23}^{(21)}) & (H_{21}^{(22)}) & (H_{22}^{(22)}) \end{pmatrix}.$$

The equation $\Gamma(X)(H) = (H)\Gamma(X)$ means that $\Gamma_1(H_{ab}^{(12)}) = (H_{ab}^{(12)})\Gamma_2$, $\Gamma_1(H_{ab}^{(21)}) = (H_{ab}^{(21)})\Gamma_2$ for all the sub-matrices with different upper indices. Because of the orthogonality theorem, these will be equal to 0. The matrix (H) can be written as

$$(H) = \begin{pmatrix} (H_{11}^{(1)}) & (H_{12}^{(1)}) & (H_{13}^{(1)}) & 0 & 0 \\ (H_{21}^{(1)}) & (H_{22}^{(1)}) & (H_{23}^{(1)}) & 0 & 0 \\ (H_{31}^{(1)}) & (H_{32}^{(1)}) & (H_{33}^{(1)}) & 0 & 0 \\ 0 & 0 & 0 & (H_{11}^{(2)}) & (H_{22}^{(2)}) \\ 0 & 0 & 0 & (H_{21}^{(2)}) & (H_{22}^{(2)}) \end{pmatrix}$$

(one upper index is sufficient). The dimensions of the sub-matrices are $\alpha_i \times \dim\Gamma_i$.

$\Gamma(X)(H) = (H)\Gamma(X)$ means that $\Gamma_1(H_{ab}^{(1)}) = (H_{ab}^{(1)})\Gamma_1$, $\Gamma_2(H_{cd}^{(2)}) = (H_{cd}^{(2)})\Gamma_2$. The orthogonality theorem means that $(H_{ab}^{(1)})_{\alpha\beta} = H_{ab}^{(1)}\delta_{\alpha\beta}$, $(H_{cd}^{(2)})_{\alpha\beta} = H_{cd}^{(2)}\delta_{\alpha\beta}$, where α, β are the indices of the representational matrices. Assuming that their dimensions are 2×2, we see that

$$\begin{pmatrix} \Gamma_1 & 0 & 0 \\ 0 & \Gamma_1 & 0 \\ 0 & 0 & \Gamma_1 \end{pmatrix} \Leftrightarrow (H^{(1)}) = \begin{pmatrix} H_{11}^{(1)} & 0 & H_{12}^{(1)} & 0 & H_{13}^{1} & 0 \\ 0 & H_{11}^{(1)} & 0 & H_{12}^{(1)} & 0 & H_{13}^{1} \\ H_{21}^{(1)} & 0 & H_{22}^{(1)} & 0 & H_{23}^{(1)} & 0 \\ 0 & H_{21}^{(1)} & 0 & H_{22}^{(1)} & 0 & H_{23}^{(1)} \\ H_{31}^{(1)} & 0 & H_{32}^{(1)} & 0 & H_{33}^{(1)} & 0 \\ 0 & H_{31}^{(1)} & 0 & H_{32}^{(1)} & 0 & H_{33}^{(1)} \end{pmatrix}.$$

$$(6.16)$$

The zeros do not appear in the case of one-dimensional representations. A large number of zeros is expected in the case of representations of higher dimensions.[2] The presence of many zeros in matrices (sparse matrices) is no longer very useful.

In summary, we can conclude that

$$\langle \rho', j, \beta | (H) \rho, i, \alpha \rangle = \delta_{ij} \delta_{\alpha\beta} H_{\rho\rho'}^{(i)}. \qquad (6.17)$$

The initial matrix (H) is reduced to sub-matrices of dimensions $\alpha_i \times \dim \Gamma_i$ and of the form seen in Eq. (6.16), i.e. generally, of dimensions sufficiently smaller than that of the initial matrix (H). Thus, the problem is simplified significantly.

- As long as $\alpha_i = 1$, the sub-matrix is diagonal. If $\alpha_i > 1$, symmetry alone is not sufficient to diagonalize the matrix $H^{(i)}$.
- In the basis where the matrix $H^{(i)}$ is defined, it is possible using other methods, even numerically if necessary, to find a similarity transformation that diagonalizes it. As long as the basis is orthonormal, the transformation is unitary $U^{(i)}$, of dimensions $\alpha_i \times \dim \Gamma_i$, with no effect on the other subspaces.
- The unitary matrix

$$U = U^{(1)} \oplus U^{(2)} \oplus \cdots \oplus U^{(k)} = \begin{pmatrix} U^{(1)} & 0 & \cdots & 0 \\ 0 & U^{(2)} & \cdots & 0 \\ \cdots & \cdots & \cdots & \cdots \\ 0 & 0 & \cdots & U^{(k)} \end{pmatrix}$$

can diagonalize the entire matrix (H).

Alternatively, we may suppose that the matrix (H) remains unaltered under the transformations of the symmetry group G, i.e. that Eq. (6.14) holds true and that the basis on which it was constructed is complete, i.e. it can generate all the irreducible representations of the group. Let us also suppose that the matrix (H) can by some means be diagonalized. We may then expect the following:

---

[2]Several years ago, it was commonly used in representations of large dimensions, which led to the appearance of many zeros. The primitive computers in use at that time had little storage, and by using this form, one could save only the non-zero elements. In addition, certain numerical processes were more effective in sparse matrices (where there are few non-zero elements) (handling sparse matrices). Clearly, group theory was much more useful as a tool then compared to now.

(i) The resulting eigenvectors must necessarily belong to some representation of G, as long as there is no degeneracy, i.e. as long as no two representations share the same eigenvalue.

(ii) When there is degeneracy, they may be classified as representations of G following the diagonalization of another operator, T, which also remains unaltered under the transformations of the symmetry group G and commutes with the Hamilton function, $TH = HT$. An appropriate such operator will remove the degeneracy. This can always occur if there is "hidden" symmetry greater than G. Such a symmetry can give meaning to the index ρ, changing Eq. (6.17) to one that is also orthogonal with respect to ρ.

(iii) Even in the case of diagonalization using traditional methods, the use of symmetry can give meaning to the resulting eigenvectors.

6.3 Useful examples

Example 1: Study of the nucleus[3] ^{12}C as a system of three α particles in the framework of C_{3V} symmetry.

Clearly, our system is described by a 3×3 representation, which is reducible. In order to reduce it, we consider the basis

$$|1\rangle = \frac{1}{\sqrt{6}} \left(\psi(1) + \psi(2) - 2\psi(3)\right), \quad |2\rangle = \frac{1}{\sqrt{2}} \left(\psi(1) - \psi(2)\right),$$

$$|3\rangle = \frac{1}{\sqrt{3}} \left(\psi(1) + \psi(2) + \psi(3)\right)$$

i.e. through the unitary matrix S, in the order $T(C_3)$, $T((C_3)^2)$, $T(\sigma_1), T(\sigma_2)$, $T(\sigma_3)$, we find

$$S = \begin{pmatrix} \frac{1}{\sqrt{6}} & \frac{1}{\sqrt{2}} & \frac{1}{\sqrt{3}} \\ \frac{1}{\sqrt{6}} & -\frac{1}{\sqrt{2}} & \frac{1}{\sqrt{3}} \\ -\sqrt{\frac{2}{3}} & 0 & \frac{1}{\sqrt{3}} \end{pmatrix} \Rightarrow S^+ \begin{pmatrix} 0 & 1 & 0 \\ 0 & 0 & 1 \\ 1 & 0 & 0 \end{pmatrix} S = \begin{pmatrix} -\frac{1}{2} & -\frac{\sqrt{3}}{2} & 0 \\ \frac{\sqrt{3}}{2} & -\frac{1}{2} & 0 \\ 0 & 0 & 1 \end{pmatrix},$$

$$S^+ \begin{pmatrix} 0 & 0 & 1 \\ 1 & 0 & 0 \\ 0 & 1 & 0 \end{pmatrix} S = \begin{pmatrix} -\frac{1}{2} & \frac{\sqrt{3}}{2} & 0 \\ -\frac{\sqrt{3}}{2} & -\frac{1}{2} & 0 \\ 0 & 0 & 1 \end{pmatrix},$$

---

[3] Other systems, such as the NH_3 molecule, can be similarly studied.

$$S^+ \begin{pmatrix} 1 & 0 & 0 \\ 0 & 0 & 1 \\ 0 & 1 & 0 \end{pmatrix} S = \begin{pmatrix} -\frac{1}{2} & \frac{\sqrt{3}}{2} & 0 \\ \frac{\sqrt{3}}{2} & \frac{1}{2} & 0 \\ 0 & 0 & 1 \end{pmatrix},$$

$$S^+ \begin{pmatrix} 0 & 0 & 1 \\ 0 & 1 & 0 \\ 1 & 0 & 0 \end{pmatrix} S = \begin{pmatrix} -\frac{1}{2} & -\frac{\sqrt{3}}{2} & 0 \\ -\frac{\sqrt{3}}{2} & \frac{1}{2} & 0 \\ 0 & 0 & 1 \end{pmatrix},$$

$$S^+ \begin{pmatrix} 0 & 1 & 0 \\ 1 & 0 & 0 \\ 0 & 0 & 1 \end{pmatrix} S = \begin{pmatrix} 1 & 0 & 0 \\ 0 & -1 & 0 \\ 0 & 0 & 1 \end{pmatrix}.$$

Note the change of roles in the reflections; σ_1, σ_2 and σ_3 of Section 4.7.2 correspond to σ_3, σ_1 and σ_2 in the reduction. This is due to the choice of axes in Example 5 of Section 4.7.2 and of S, and it is of no significance.

In any case, reduction of the three-dimensional representation gives us:

$$\Gamma = \Gamma^3 + \Gamma^1. \tag{6.18}$$

Let us now consider the matrix (H) which in the original basis will have a form such that it is commutable with all the matrices of Example 5 in Section 4.7.2. The only option is

$$(H) = \begin{pmatrix} E_0 & h & h \\ h & E_0 & h \\ h & h & E_0 \end{pmatrix}.$$

As during the reduction, both the two and three-dimensional representations each appear only once ($\alpha_1, \alpha_2 = 1$); according to the above, the vectors $|i\rangle$ are the eigenvectors we were seeking for (H), with eigenvalues that can easily be found to be $\lambda = E_0 - h$, $E_0 - h$, $E_0 = 2h$. The reader can easily verify this with the direct diagonalization of (H). As there clearly is degeneracy, i.e. two of the eigenvalues of (H) are the same, the corresponding eigenvectors are not unique. So, the choice of vectors $|i\rangle$, $i = 1, 2, 3$, is not unique. Any linear combination of the two first leading to an orthonormal system is equally acceptable, for example, a combination of the general form is

$$|1'\rangle = \cos\theta|1\rangle + \sin\theta|2\rangle, \quad |2'\rangle = -\sin\theta|1\rangle + \cos\theta|2\rangle.$$

Example 2: Consider the inversion group $G = \{E, I\}$. A representation in function space

$$\psi(\mathbf{r}), \psi(-\mathbf{r}), \phi(\mathbf{r}), \phi(-\mathbf{r})$$

is the following:

$$\Gamma(E) = \begin{pmatrix} 1 & 0 & 0 & 0 \\ 0 & 1 & 0 & 0 \\ 0 & 0 & 1 & 0 \\ 0 & 0 & 0 & 1 \end{pmatrix}, \quad \Gamma(I) = \begin{pmatrix} 0 & 1 & 0 & 0 \\ 1 & 0 & 0 & 0 \\ 0 & 0 & 0 & 1 \\ 0 & 0 & 1 & 0 \end{pmatrix}.$$

The operator (H) which satisfies the equation $(H)\Gamma(X) = \Gamma(X)(H)$ will be in the form

$$(H) = \begin{pmatrix} a & b & e & f \\ b & a & f & e \\ e & f & c & d \\ f & e & d & c \end{pmatrix}.$$

The above representation is reducible and can be reduced through the transformation

$$\psi_+ = \frac{1}{\sqrt{2}}\left(|\psi(\mathbf{r})\rangle + |\psi(-\mathbf{r})\rangle\right), \quad \phi_+ = \frac{1}{\sqrt{2}}\left(|\phi(\mathbf{r})\rangle + |\phi(-\mathbf{r})\rangle\right),$$

$$\psi_- = \frac{1}{\sqrt{2}}\left(|\psi(\mathbf{r})\rangle - |\psi(-\mathbf{r})\rangle\right), \quad \phi_- = \frac{1}{\sqrt{2}}\left(|\phi(\mathbf{r})\rangle - |\phi(-\mathbf{r})\rangle\right).$$

In fact,

$$S = \begin{pmatrix} \frac{1}{\sqrt{2}} & 0 & \frac{1}{\sqrt{2}} & 0 \\ \frac{1}{\sqrt{2}} & 0 & -\frac{1}{\sqrt{2}} & 0 \\ 0 & \frac{1}{\sqrt{2}} & 0 & \frac{1}{\sqrt{2}} \\ 0 & \frac{1}{\sqrt{2}} & 0 & -\frac{1}{\sqrt{2}} \end{pmatrix} \Rightarrow S^+\Gamma(I)S == \begin{pmatrix} 1 & 0 & 0 & 0 \\ 0 & 1 & 0 & 0 \\ 0 & 0 & -1 & 0 \\ 0 & 0 & 0 & -1 \end{pmatrix}.$$

That is,

$$\Gamma(X) = 2\Gamma_1 \oplus 2\Gamma_2, \quad \Gamma_1 = 1, \quad \Gamma_2 = -1.$$

Also,

$$S^+(H)S = \begin{pmatrix} a+b & e+f & 0 & 0 \\ e+f & c+d & 0 & 0 \\ 0 & 0 & a-b & e-f \\ 0 & 0 & e-f & c-d \end{pmatrix}.$$

This is in the same form as Eq. (6.16): $(\alpha_i \dim \Gamma_i) \times (\alpha_i \dim \Gamma_i) = 2 \times 2$. It has not been automatically diagonalized, which is to be expected, as $\alpha_1 = \alpha_2 = 2$. However, it does not combine functions of different parity (that is, with different eigenvalues of the operator I). The existence of symmetry simplifies the problem, in the sense that it is only necessary to diagonalize a 2×2, rather than a 4×4, matrix.

The larger the number of functions, N, the greater the gain (the dimension of the matrix (H) becomes $N/2$ instead of N).

Example 3: Here, we study a linear molecule comprising three atoms, two of the same type in the positions r_1 and r_3 and a third, different type of atom between them, in r_2 (Fig. 6.1).

This problem is characterized by inversion symmetry, and in fact,

$$I: r_1 \to -r_3, r_3 \to -r_1, r_2 \to -r_2 \Rightarrow \Gamma(I) = \begin{pmatrix} 0 & 0 & -1 \\ 0 & -1 & 0 \\ -1 & 0 & 0 \end{pmatrix}.$$

It can easily be seen that the matrix (H), which satisfies the equation $(H)\Gamma(X) = \Gamma(X)(H)$, is of the form

$$(H) = \begin{pmatrix} a & b & c \\ b & d & b \\ c & b & a \end{pmatrix}.$$

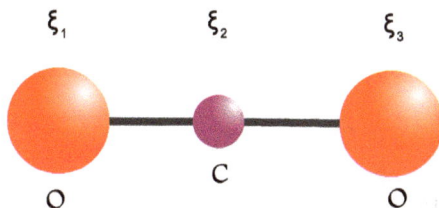

Fig. 6.1. A symmetric linear molecule, such as CO_2.

The above representation $\Gamma(X)$ can be diagonalized in the basis $\mathbf{r}^{\pm} = \frac{1}{\sqrt{2}}(\mathbf{r}_1 \mp \mathbf{r}_3)$, $\mathbf{r}_0 = \mathbf{r}_2$, i.e. with a transformation S such that

$$S = \begin{pmatrix} \frac{1}{\sqrt{2}} & 0 & \frac{1}{\sqrt{2}} \\ 0 & 1 & 0 \\ -\frac{1}{\sqrt{2}} & 0 & \frac{1}{\sqrt{2}} \end{pmatrix} \Rightarrow S^+\Gamma(X)S = \begin{pmatrix} 1 & 0 & 0 \\ 0 & -1 & 0 \\ 0 & 0 & -1 \end{pmatrix} = \Gamma^1 \oplus 2\Gamma^2,$$

$$\Gamma^1 = 1, \quad \Gamma^2 = -1. \tag{6.19}$$

Also,

$$S^+(H)S = \begin{pmatrix} a-c & 0 & 0 \\ 0 & a+c & \sqrt{2}b \\ 0 & \sqrt{2}b & d \end{pmatrix},$$

which means that it has the expected form: diagonal in that part that corresponds to $\alpha_1 = 1$, with a value of $a - c$, and a non-diagonal part that corresponds to $\alpha_2 = 2$.

The reduction could also have been carried out through the character table of the irreducible representations of $G = \{E, I\}$, which is as follows:

	E	I
$\Gamma_1 = \chi_1$	1	1
$\Gamma_2 = \chi_2$	1	−1.

This leads to the equations

$$\chi(E) = \alpha_1 + \alpha_2, \quad \chi(I) = \alpha_1 - \alpha_2 \Rightarrow 3 = \alpha_1 + \alpha_2,$$

$$-1 = \alpha_1 - \alpha_2 \Rightarrow \alpha_1 = 1, \quad \alpha_2 = 2.$$

Example 4: We will study a molecule of orthorhombic symmetry, i.e. symmetry of the group D_{2h}.

One such molecule is described by a system of masses

$$(M, m_1, m_2, m_3, m_4, m_5, m_6)$$

placed at the points

$$(0,0,0), (a,0,0), (-a,0,0), (0,b,0), (0,-b,0), (0,0,c), (0,0,-c),$$

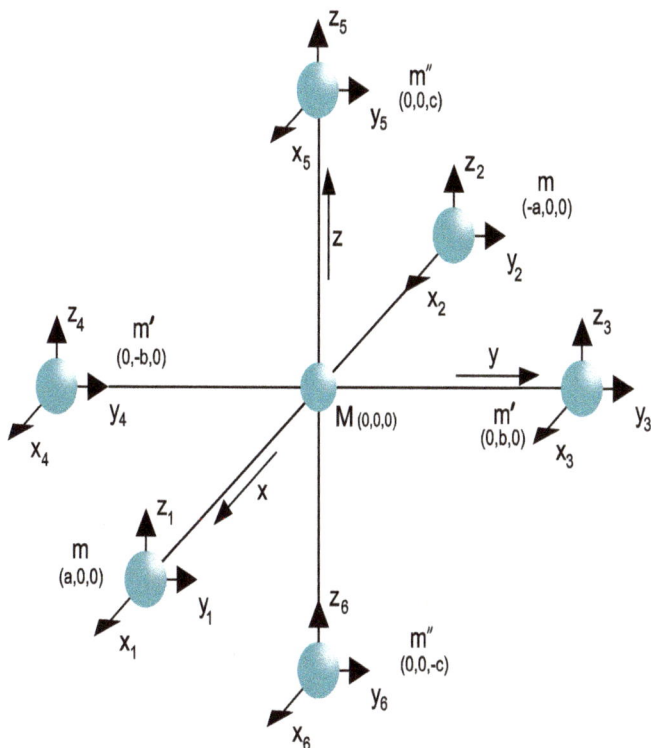

Fig. 6.2. Molecule characterized by orthorhombic ($D_{2}h$) symmetry.

respectively (Fig. 6.2). Suppose that $m_1 = m_2 = m$, $m_3 = m_4 = m'$ and $m_5 = m_6 = m''$. Such a system will have the following symmetry elements:

$$\{E, C_x^2, C_y^2, C_z^2, I\}, \; I \text{ being the inversion operator and}$$

$$C_i^2, i = x, y, z, \text{ being rotations by } \pi$$

about the corresponding axes. We know that this group is D_{2h}, which is an Abelian group with eight elements.

To find the irreducible representations, we use the fact that it is isomorphic to the group $G_1 \otimes G_2$, where $G_1 = \{E, \sigma_x, \sigma_y, \sigma_z\}$ and $G_2 = \{E, I\}$. We know, as seen from Eq. (5.40) and the arguments leading to it, that the characters of these groups are connected through the equation

$$\chi^{i,j}(XY) = \chi^i(X)\chi^i(Y), \quad i = 1, \ldots, 4, \; j = 1, 2, \; X \in G_1, \; Y \in G_2.$$

The characters of G_1 and G_2 are trivial to find; therefore,

	E	C_x^2	C_y^2	C_z^2
χ^1	1	1	1	1
χ^2	1	1	-1	-1
χ^3	1	-1	1	-1
χ^4	1	-1	-1	1

	E	I
χ^1	1	1
χ^2	1	-1.

	E	C_x^2	C_y^2	C_z^2	I	$C_x^2 I$	$C_y^2 I$	$C_z^2 I$
$\chi^1 = \chi^{1,1}$	1	1	1	1	1	1	1	1
$\chi^2 = \chi^{2,1}$	1	1	-1	-1	1	1	-1	-1
$\chi^3 = \chi^{3,1}$	1	-1	1	-1	1	-1	1	-1
$\chi^4 = \chi^{4,1}$	1	-1	-1	1	1	-1	-1	1
$\chi^5 = \chi^{1,2}$	1	1	1	1	-1	-1	-1	-1
$\chi^6 = \chi^{2,2}$	1	1	-1	-1	-1	-1	1	1
$\chi^7 = \chi^{3,2}$	1	-1	1	-1	-1	1	-1	1
$\chi^8 = \chi^{4,2}$	1	-1	-1	1	-1	1	1	-1.

The Hamiltonian which describes the system will depend on the coordinates. We observe that rotation about an axis does not change the coordinate along that axis, but it does change the sign of the other two coordinates, for example,

$$C_x^2 \begin{pmatrix} x \\ y \\ z \end{pmatrix} = \begin{pmatrix} x \\ -y \\ -z \end{pmatrix}.$$

The actions of the group's operators are given in brief in Table 6.1. Based on this table, the 21×21 representation of $\Gamma(X)$ can be constructed. This construction is simple but will not be given here. The reader is encouraged to obtain the representations $\Gamma(C_z^2)$ and $\Gamma(I)$, which are two of the three generators of this group.

This representation is reducible and can be reduced by diagonalizing, for example, $\Gamma(I)$. Thus, we have

$$\Gamma(X) = 3\Gamma^1(X) \oplus +2(\Gamma^2(X) \oplus \Gamma^3(X) \oplus \Gamma_4(X))$$
$$+ 4(\Gamma^5(X) \oplus \Gamma^6(X) \oplus \Gamma^7(X)). \qquad (6.20)$$

Table 6.1. Actions of the elements of D_{2h} on the coordinates (x_i, y_i, z_i), $i = 1, 2, \ldots, 6, 0$. The action of an element, X, on a coordinate, q, can be found as the element (q, X) in this table.

	E	C_2^z	C_2^y	C_2^x	I	$C_2^z I$	$C_2^y I$	$C_2^x I$
x_1	x_1	$-x_2$	$-x_2$	x_1	$-x_2$	x_1	x_1	$-x_2$
y_1	y_1	$-y_2$	y_2	$-y_1$	$-y_2$	y_1	$-y_1$	y_2
z_1	z_1	z_2	$-z_2$	$-z_1$	$-z_2$	$-z_1$	z_1	z_2
x_2	x_2	$-x_1$	$-x_1$	x_2	$-x_1$	x_2	x_2	$-x_1$
y_2	y_2	$-y_1$	y_1	$-y_2$	$-y_1$	y_2	$-y_2$	y_1
z_2	z_2	z_1	$-z_1$	$-z_2$	$-z_1$	$-z_2$	z_2	z_1
x_3	x_3	$-x_4$	$-x_3$	x_4	$-x_4$	x_3	x_4	$-x_3$
y_3	y_3	$-y_4$	y_3	$-y_4$	$-y_4$	y_3	$-y_4$	y_3
z_3	z_3	z_4	$-z_3$	$-z_4$	$-z_4$	$-z_3$	z_4	z_3
x_4	x_4	$-x_3$	$-x_4$	x_3	$-x_3$	x_4	x_3	$-x_4$
y_4	y_4	$-y_3$	y_4	$-y_3$	$-y_3$	y_4	$-y_3$	y_4
z_4	z_4	z_3	$-z_4$	$-z_3$	$-z_3$	$-z_4$	z_3	z_4
x_5	x_5	$-x_5$	$-x_6$	x_6	$-x_6$	x_6	x_5	$-x_5$
y_5	y_5	$-y_5$	y_6	$-y_6$	$-y_6$	y_6	$-y_5$	y_5
z_5	z_5	z_5	$-z_6$	$-z_6$	$-z_6$	$-z_6$	z_5	z_5
x_6	x_6	$-x_6$	$-x_5$	x_5	$-x_5$	x_5	x_6	$-x_6$
y_6	y_6	$-y_6$	y_5	$-y_5$	$-y_5$	y_5	$-y_6$	y_6
z_6	z_6	z_6	$-z_5$	$-z_5$	$-z_5$	$-z_5$	z_6	z_6
x_0	x_0	$-x_0$	$-x_0$	x_0	$-x_0$	x_0	x_0	$-x_0$
y_0	y_0	$-y_0$	y_0	$-y_0$	$-y_0$	y_0	$-y_0$	y_0
z_0	z_0	z_0	$-z_0$	$-z_0$	$-z_0$	$-z_0$	z_0	z_0

The matrix (H) is 21×21 and, as long as it is real, comprises 241 elements. However, as it is characterized by D_{2h} symmetry, i.e. $\Gamma(X)(H) = (H)\Gamma(X), X \in D_{2h}$, only 21 are independent, and they are given by the matrices

$$\left(H^{(1)}\right), 3 \times 3; \ \left(H^{(i)}\right), \quad i = 2, 3, 4, 2 \times 2; \ \left(H^{(i)}\right) \quad i = 5, 6, 7, 4 \times 4,$$

where the dimensions correspond to the multiplicity in the representation in the above reduction (no sparse zeros). We see the great advantage offered by symmetry. Because of symmetry, the largest matrix that needs to be diagonalized is 4×4, while without it, the matrix to be diagonalized would be 21×21.

6.4 Problems

6.3.1 Table 6.1 is a special case of a general transformation: $\hat{x}_i, \hat{y}_i, \hat{z}_i$ of the unit vectors on the a molecule with N atoms:

$$A : \hat{x}_i \to \hat{x}'_i, \ \hat{y}_i \to \hat{y}'_i, \ \hat{z}_i \to \hat{z}'_i$$

associated with a representation, $\Gamma(A)$, of a given group. Show that

$$\chi(A) = n(1 + 2\cos\phi) \quad \text{if } A \text{ rotation,}$$

$$\chi(A) = n(-1 + 2\cos\phi) \quad \text{if } A \text{ rotation} \times \text{ inversion,}$$

where n is the number of molecules that remain motionless.

   In the case of a linear molecule, $\chi(E) = N$. For a rotation, $\chi(C_2) = n$, where $n = 0$ if $N = $ even and $n = 1$ if $N = $ odd.

6.3.2 In the case of the molecule NH_3, one can construct a table analogous to Table 6.1. However, by using the method discussed in the previous problem, show that

$$\chi(E) = 12, \chi_{C_3} =, \chi_{C_3^2} = 0, \chi_{\sigma_1} = \chi_{\sigma_2} = \chi_{\sigma_3} = 0.$$

Then, show that

$$\Gamma = 2\Gamma^{(1)} + 4\Gamma^{(2)} + 3\Gamma^{(3)}.$$

Hint: It may be useful to use the expression $\chi(A) = a\chi_1(A) + b\chi_2(A) + c\chi_3(A)$ and to determine the coefficients a, b, c.

6.3.3 Derive Eq. (6.20).

6.3.4 The molecule BF_6 is characterized by O_h symmetry. Show that the representation that corresponds to the transformation

$$\Gamma : \hat{x}_i \to \hat{x}'_i, \quad \hat{y}_i \to \hat{y}'_i, \quad \hat{z}_i \to \hat{z}'_i$$

is characterized by the following character table:

	E	$8C_3$	$3C_2$	$6C_4$	$6C_2'$	I	$8C_3I$	$3C_2I$	$6C_4I$	$6C_2'I$
χ	21	0	0	-1	1	-3	0	1	-1	1

Then, show that

$$\Gamma(X) = \Gamma_g^{(1)}(X) \oplus \Gamma_g^{(2)}(X) \oplus \Gamma_u^{(3)}(X) \oplus 2\Gamma_u^{(4)}(X) \oplus \Gamma_g^{(5)}(X)$$

$$\oplus \Gamma_u^{(6)}(X) \oplus 3\Gamma_g^{(7)}(X) \oplus \Gamma_u^{(8)}(X).$$

Hint: You may use, even without proof, the following character table:

	E	$8C_3$	$3C_2$	$6C_4$	$6C_2'$	I	$8C_3I$	$3C_2I$	$6C_4I$	$6C_2'I$
A_{1g}	1	1	1	1	1	1	1	1	1	1
A_{2g}	1	1	1	-1	-1	1	1	1	-1	-1
E_g	2	-1	2	0	0	2	-1	2	0	0
F_{1g}	3	0	-1	1	-1	3	0	-1	1	-1
F_{2g}	3	0	-1	-1	1	3	0	-1	-1	1.
A_{1u}	1	1	1	1	1	-1	-1	-1	-1	-1
A_{2u}	1	1	1	-1	-1	-1	-1	-1	1	1
E_u	2	-1	2	0	0	-2	1	-2	0	0
F_{1u}	3	0	-1	1	-1	-3	0	1	-1	1
F_{2u}	3	0	-1	-1	1	-3	0	1	1	-1

In this table, $A_{1g}, A_{2g}, A_{1u}, A_{2u}$ are 1D, E_g, E_u 2D and F_d, F_u 3D. (problem 5.9.10 may help in proving this.) The mathematical symbols used to describe the classes are those found in Section 2.4.2.

6.3.5 The molecule C_2H_2 is linear. (a) Identify the symmetry group that it is characterized by. (b) Find the representation Γ of the operators in problem 6.3.1. (c) Find coefficients α_i such that

$$\Gamma(X) = \sum_i \oplus \alpha_i \Gamma^i,$$

where Γ^i are the irreducible representations of the group.

6.3.6 Do the same for the molecule CH_4 assuming that it is on a plane.[4]

6.3.7 Do the same for the benzene molecule, C_6H_6, assuming that the carbon atoms are arranged on the vertices of a regular hexagon.

---

[4]In reality, this molecule is characterized by tetrahedral symmetry. One truly plane molecule with square, C_4, symmetry is XeF_4.

Chapter 7

Symmetries and Normal Modes
of Oscillation

The study of normal modes of oscillation is a subject of interest to many branches of physics and other physical sciences, both at the classical as well as quantum levels. In this chapter, we examine whether, under certain interactions of the constituent particles, the system can undergo oscillations as a whole and we determine the corresponding oscillation frequencies, known as normal modes. We consider in particular such systems in the domain of molecular and solid state physics in the context of harmonic interactions.

7.1   Introduction

Let us consider a system of N particles described by the coordinates q_i, $i = 1, 2, \ldots, 3N$. In the case[1] of classical physics, the system is described by a Hamiltonian function of the form

$$H = T + V, \; T = \text{kinetic and } V = \text{potential energy}, \tag{7.1}$$

where

$$T = \frac{1}{2} \sum_i m_i \dot{q}_i^2, \quad \dot{q}_i = \frac{dq_i}{dt}, \quad V = \frac{1}{2} \sum_{i \neq j} V_{i,j}(q_i, q_j). \tag{7.2}$$

---

[1]In some ordering, e.g., in the case of Cartesian coordinates, we write q_1, q_2, q_3, q_4, q_5, q_6 instead of x_1, y_1, z_1, x_2, y_2, z_2. Clearly, the number $3N$ is appropriate in three dimensions. In the case of two dimensions, we have $2N$ parameters and in one dimension, just N of them.

In writing the above expressions, we have assumed that the potential energy is independent of the velocities $\dot{q}_i$ and it does not depend explicitly on time t. Suppose further that

$$V_{i,j}(q_i, q_j) = V_{j,i}(q_j, q_i),\tag{7.3}$$

i.e. $V_{i,j}$ is a symmetric function of q_i, q_j. These assumptions are quite general. Consider now a Taylor expansion around a point, e.g. around the points $q_i = 0, q_j = 0$, i.e.

$$V = \sum_{i,j} v_{i,j}(0,0) + \sum_i q_i \left.\frac{\partial V}{\partial q_i}\right|_{q_i=0} + \frac{1}{2}\sum_{i\neq j} \left.\frac{\partial^2 V}{\partial q_i \partial q_j}\right|_{q_i=0, q_j=0} q_i q_j + \cdots$$

The first term is a constant of no special meaning. The second term becomes zero due to the equilibrium condition:

$$\left.\frac{\partial V}{\partial q_i}\right|_{q_i=0} = 0,$$

and all the terms of order higher than 2 can be considered negligible:

$$V_{i,j} = v'_{i,j} q_i q_j, \quad v'_{i,j} = \left.\frac{\partial^2 V}{\partial q_i \partial q_j}\right|_{q_i=0, q_j=0}.$$

Furthermore for future convenience, let us change the scale of the coordinates via the transformation

$$\eta_i = \sqrt{m_i} q_i,$$

thus we get

$$T = \frac{1}{2}\sum_i \dot{\eta}_i^2 \quad \text{and} \quad V = \frac{1}{2}\sum_{ij} v_{i,j}\eta_i\eta_j, \quad v_{i,j} = \frac{v'_{i,j}}{\sqrt{m_i m_j}}.\tag{7.4}$$

That is to say the matrix T is diagonal and, thanks to Eq. (7.3), the matrix (v) is symmetric. Thus, it can be diagonalized via a similarity transformation S, i.e.

$$S^{-1}(v)S = d, \quad d = \text{diagonal}, \quad S^{-1} = S^T,\tag{7.5}$$

where S^T is the transpose of S and the diagonal matrix d consists of the eigenvalues of the matrix (v), while the kinetic energy matrix remains diagonal.

7.2 Normal modes of oscillation

According to the above discussion, the eigenvectors of (v) are

$$Q_i = \sum_j S_{ij} \eta_j, \tag{7.6}$$

the inverse being

$$\eta_i = \sum_j S_{ij}^{-1} Q_j = \sum_j S_{ji} Q_j. \tag{7.7}$$

The variables S_{ij} and Q_i are real, since the matrix (v) is real. Thus, in the new basis, we have

$$T = \frac{1}{2} \sum_i^{3N} \dot{Q}_i^2, \quad V = \frac{1}{2} \sum_i^{3N} \lambda_i Q_i^2, \tag{7.8}$$

where $\lambda_1, \lambda_2, \ldots, \lambda_{3N}$ are the eigenvalues of (v), being real as well.

Moreover, since the system is stationary, meaning that at equilibrium the potential energy attains a minimum value, the eigenvalues cannot be negative and we can set

$$\lambda_i = \omega_i^2, \quad i = 1, 2, \ldots, 3N. \tag{7.9}$$

At this point, let us consider the Lagrangian function $L = T - V$, which in our case becomes

$$L = T - V = \frac{1}{2} \sum_{i=1}^{3N} \left(\dot{Q}_i^2 - \omega_i^2 Q_i^2 \right). \tag{7.10}$$

while the Hamiltonian is given by

$$H = \frac{1}{2} \sum_{i=1}^{3N} \left(P_i^2 + \omega_i^2 Q_i^2 \right), \tag{7.11}$$

where

$$P_i = \frac{\partial L}{\partial \dot{Q}_i}. \tag{7.12}$$

In other words, we have a superposition of classical harmonic oscillations with equations of motion

$$\frac{d^2 Q_i}{dt^2} + \omega_i^2 Q_i = 0. \tag{7.13}$$

The quantities Q_i are the canonical coordinates and ω_i the canonical frequencies, i.e. the **normal modes** of oscillation.

The quantum mechanical description of the system can be given in the standard form by putting

$$P_i = \frac{\partial}{\partial Q_i}. \tag{7.14}$$

Schrödinger's equation can be written

$$H\psi(Q_1, Q_2, \ldots, Q_{3N}) = E\psi(Q_1, Q_2, \ldots, Q_{3N}), \tag{7.15}$$

where

$$H = \sum_i^{3N} H_i(Q_i), H_i(Q_i) = \frac{1}{2}\left(-\hbar^2 \frac{\partial^2}{\partial Q_i^2} + \omega_i^2 Q_i^2\right). \tag{7.16}$$

This simple form of the Hamiltonian function leads to

$$\psi(Q_1, Q_2, \ldots, Q_{3N}) = \psi_1(Q_1)\psi_2(Q_2), \ldots, \psi(Q_{3N}),$$
$$E = E_1 + E_2 + \cdots + E_{3N} \tag{7.17}$$

$$\frac{1}{2}\left(-\hbar^2 \frac{\partial^2}{\partial Q_i^2} + \omega_i^2 Q_i^2\right)\psi_i(Q_i) = E_i\psi_i(Q_i) \tag{7.18}$$

$$E_i = \left(n_i + \frac{1}{2}\hbar\omega_i\right), \quad n_i = 0, 1, 2, \ldots \tag{7.19}$$

We note that the above description refers to the external coordinates q_i, defined in some absolute reference frame. In many instances, however, it is required to consider the relative (internal) coordinates, that is, in the case of a molecule, the coordinates of the center of mass, which usually coincides with the position of a heavy nucleus, and the relative coordinates with respect to the center of mass. In such a case, the expression of the kinetic energy becomes complicated and is no longer diagonal. In the simple case of a small number of particles, it is possible to write the kinetic energy in terms of the relative coordinates and proceed with the diagonalization of the Hamiltonian operator as above.

7.2.1   *Summary*

The kinetic and potential energy can be cast in the form

$$T = \frac{1}{2}\sum_i m_i \dot{q}_i^2, \quad \dot{q}_i = \frac{dq_i}{dt}, \quad V = \frac{1}{2}\sum_{i \neq j} V_{i,j}(q_i, q_j). \tag{7.20}$$

In the case of coupled harmonic oscillators,

$$V = \frac{1}{2} \sum_{i \neq j} k_{i,j}(q_i - q_j)^2.$$

Putting $\sqrt{m_i}\dot{q}_i = \dot{\eta}_i$, we find

$$T = \frac{1}{2} \sum_i \dot{\eta}_i^2, \quad \dot{\eta}_i = \frac{d\eta_i}{dt}, \quad V = \frac{1}{2} \sum_{i \neq j} \frac{k_{i,j}}{\sqrt{m_i m_j}}(\eta_i - \eta_j)^2. \qquad (7.21)$$

In this basis, the kinetic energy matrix is diagonal and it suffices to diagonalize the Potential energy matrix $\mathcal{R}$. In this way, we find the eigenvectors

$$Q_j = \mathcal{R}_{ij}\eta_j, \quad \eta_i = \mathcal{R}_{ji}Q_j.$$

The eigenvalues of the system cannot become negative, i.e. $\lambda_i \geq w_i^2$. The functions of Lagrange and Hamilton, $\mathcal{L}$ and H, respectively, take the form

$$\mathcal{L} = \frac{1}{2} \sum_{i=1}^{3N}((\dot{Q}_i^2 - w_i^2 Q_i^2)), \quad H = \frac{1}{2} \sum_{i=1}^{3N}(P_i^2 + w_i^2 Q_i^2), \quad P_i = \frac{\partial \mathcal{L}}{\partial \dot{Q}_i}.$$

In other words, we have a superposition of classical harmonic oscillations with equations of motion:

$$\frac{d^2 Q_i}{dt^2} + w_i^2 Q_i = 0, \quad i = 1, \dots, 3N, \qquad (7.22)$$

which characterize the normal modes.

Motivated by the discussion of Chapter 6, we expect that the solution to the problem of finding the normal modes can be simplified if we consider the symmetries of the problem in question. Some applications of this idea will be considered in the following section.

7.3 Some applications involving the use of symmetry

Application 1a. Let us consider the molecule CO_2, with structure familiar to us from Example 2 of Chapter 6.

Here, $q_i = \xi_i$, where ξ_i define the positions of the nuclei along the chain:

$$T = \frac{1}{2}m\dot{\xi}_1^2 + \frac{1}{2}M\dot{\xi}_2^2 + \frac{1}{2}m\dot{\xi}_3^2, \quad V = \frac{1}{2}k\left((\xi_1 - \xi_2)^2 + (\xi_3 - \xi_2)^2\right).$$

In the expression of the potential energy, we assumed that, due to symmetry, the spring constant is the same. Furthermore, due to their large separation, the two end atoms are not dynamically coupled. The potential energy takes the form

$$V = \frac{1}{2}k\left(\xi_1^2 + 2\xi_2^2 + \xi_3^2 - (\xi_1\xi_2 + \xi_2\xi_1 + \xi_1\xi_3 + \xi_3\xi_1 + \xi_2\xi_3 + \xi_3\xi_2)\right).$$

We now make the change of scale $\eta_1 = \sqrt{m}\xi_1$, $\eta_2 = \sqrt{M}\xi_2$, $\eta_3 = \sqrt{m}\xi_3$. As we have seen in the previous section, in the basis η_1, η_2, η_3, the kinetic energy is a multiple of the identity and the matrix of the potential energy is 3×3 and takes the form

$$(v') = \frac{k}{2m}\begin{pmatrix} 1 & -\sqrt{\frac{m}{M}} & 0 \\ -\sqrt{\frac{m}{M}} & \frac{2m}{M} & -\sqrt{\frac{m}{M}} \\ 0 & -\sqrt{\frac{m}{M}} & -1 \end{pmatrix}.$$

We know that this molecule is characterized by the symmetry E, I and the three-dimensional representation is reducible and

$$\Gamma(X) = \Gamma^1(X) + 2\Gamma^2(X).$$

We have also seen in Chapter 6 that the maximum matrix that needs be diagonalized is 2×2. In fact, using the basis of ψ_1, ψ_2, ψ_3 of Example 2 of Chapter 6, we find

$$\xi_1 = \frac{1}{\sqrt{2}}(\psi_1 + \psi_3), \quad \xi_3 = \frac{1}{\sqrt{2}}(\psi_3 - \psi_1), \quad \xi_2 = \psi_2,$$

$$T = \frac{1}{2}m\dot{\psi}_1^2 + \frac{1}{2}M\dot{\psi}_2^2 + \frac{1}{2}m\dot{\psi}_3^2,$$

$$V = k\left(\psi_1^2 + 2\psi_2^2 + \psi_3^2 - 2\sqrt{2}\psi_2\psi_3\right).$$

Using again the canonical coordinates $\eta_i = \sqrt{m}\psi_i$, $i = 1,3$, $\eta_2 = \sqrt{M}\psi_2$, we find

$$(t) = \frac{1}{2}\begin{pmatrix} 1 & 0 & 0 \\ 0 & 1 & 0 \\ 0 & 0 & 1 \end{pmatrix}, \quad (v) = \frac{1}{2}\begin{pmatrix} \frac{k}{m} & 0 & 0 \\ 0 & 2\frac{k}{M} & -\sqrt{2}\frac{k}{\sqrt{mM}} \\ 0 & -\sqrt{2}\frac{k}{\sqrt{mM}} & \frac{k}{m} \end{pmatrix}.$$

One of the eigenvalues is $\lambda_1 = k/m$. The other two can be found by diagonalizing the remaining 2×2 matrix and are given by $\lambda_2 = 0$

$\lambda_3 = k/m + 2k/M$. Hence,

$$Q_1 = \eta_1 = \sqrt{m}\psi_1, \quad Q_2 = \frac{1}{\sqrt{3}}\eta_2 + \frac{\sqrt{2}}{\sqrt{3}}\eta_3, \quad Q_3 = \frac{\sqrt{2}}{\sqrt{3}}\eta_2 - \frac{1}{\sqrt{3}}\eta_3,$$

$$\omega_1 = \sqrt{\frac{k}{m}}, \quad \omega_2 = 0, \quad \omega_3 = \sqrt{\frac{k}{m} + 2\frac{k}{M}} \quad \text{or}$$

$$\omega_3 = \omega_1\sqrt{1 + \frac{2m}{M}},$$

or

$$Q_1 = \sqrt{\frac{m}{2}}(\xi_3 - \xi_1), \quad Q_2 = \frac{1}{\sqrt{3}}\left(\sqrt{m}(\xi_1 + \xi_3) + \sqrt{M}\xi_2\right),$$

$$Q_3 = \frac{1}{\sqrt{6}}\left(2\sqrt{M}\xi_2 - \sqrt{m}(\xi_1 + \xi_3)\right).$$

It is clear that the oscillation mode Q_2 is spurious, since it corresponds to the coordinate of the center of mass and it is nothing but the transfer of the whole molecule. This is made obvious by the fact that it corresponds to zero eigenvalue.

The classical equations of motion are given via the relations (7.13) At the quantum level, $E = E_1 + E_3$ as well as the relations (7.18) and (7.19). The eigenfunctions are known from elementary quantum mechanics:

$$\psi_{n_i} = \left(\frac{\omega_i/\hbar}{\sqrt{\pi}2^{n_i}(n_i)!}\right)^{1/2} e^{-(1/2)((\omega_i Q_i)/\hbar)^2} H_{n_i}\left(\frac{\omega_i Q_i}{\hbar}\right), \quad i = 1, 3,$$

where

$$n_1 = 0, 1, \ldots, n_3 = 0, 1, \ldots, \quad \text{and } H_{n_i}(x) \text{ the Hermite polynomials}$$

The spring constant k can only be determined experimentally. The accepted value is $k = 1600\,\mathrm{Nm^{-1}}$. Thus, $\omega_1 = 2.5 \times 10^{14}\,\mathrm{s^{-1}}$, $\omega_3 = \omega_1\sqrt{1 + \frac{2\times 16}{12}} = 4.7 \times 10^{14}\,\mathrm{s^{-1}}$.

One can, of course, obtain the same results without making use of the symmetry by diagonalizing the 3×3 matrix of Eq. (7.3). The benefit of using symmetry is not so great in this simple case!

Application 1b. We are going to discuss the normal modes of the molecule H_2O and TeO_2.

The H_2O molecule is assumed to be planar with the lines OH_1 and OH_2 forming an angle of $104.5°$. The molecule TeO_2 has similar structure and will be examined here.

Such a system has a symmetry plane perpendicular to the plane of the molecule passing through the bisector of the angle. We consider a coordinate system with its origin at the location of the heavy atom taking the x-axis along the bisector. Then, under the reflection σ, one has

$$\sigma : x_i \to x_i, \quad y_i \to -y_i, \quad i = 1, 2, 3$$
$$\Rightarrow \Gamma(\sigma) = \text{diag}\{1, 1, 1, -1, -1, -1\}.$$

This representation is already reduced. The Hamiltonian operator is $H = T + V$, where

$$T = \frac{1}{2}m\dot{x}_1^2 + \frac{1}{2}M\dot{x}_2^2 + \frac{1}{2}m\dot{x}_3^2 + \frac{1}{2}m\dot{y}_1^2 + \frac{1}{2}M\dot{y}_2^2 + \frac{1}{2}m\dot{y}_3^2.$$

$$V = \frac{1}{2}k\left((x_1 - x_2)^2 + (x_3 - x_2)^2 + (y_1 - y_2)^2 + (y_3 - y_2)^2\right).$$

We observe that the Hamiltonian does not mix the x_i and y_i components, and hence, the solution of the problem amounts to two 3×3 subspaces, which have the form of the previous application. Therefore, we find the solutions we have obtained previously, i.e. one for the x_i and another for the y_i components.

The spring constant of TeO_2 is $2\,\text{Nm}^{-1}$. We thus find

$$\omega_1 = 8.9 \times 10^{12}\,\text{s}^{-1}, \quad \omega_3 = \omega_1\sqrt{1 + \frac{2 \times 16}{130}} = 9.9 \times 10^{12}\,\text{s}^{-1}.$$

Application 2. We will study the normal oscillation modes of the molecule NH_3, which is characterized with the familiar symmetry C_{3V}.

We suppose the three hydrogen atoms occupy the positions $A = (x_1, y_1)$, $B = (x_2, y_2)$ and $C = (x_3, y_3)$, respectively. The axis origin is in the middle of BC with x-axis parallel to it. Considering that the coupling between the heavy nucleus and the hydrogen atoms can be neglected, something verified experimentally, we have oscillations along the sides of the equilateral triangle ABC. Therefore,

$$V_{23} = \frac{1}{2}k(x_3 - x_2)^2, \quad V_{12} = \frac{1}{2}k\left(\frac{1}{2}(x_1 - x_2) + \frac{\sqrt{3}}{2}(y_1 - y_2)\right)^2,$$

$$V_{13} = \frac{1}{2}k\left(\frac{1}{2}(x_1 - x_3) - \frac{\sqrt{3}}{2}(y_1 - y_3)\right)^2$$

(the case whereby the coupling between the heavy nucleus and the rest cannot be neglected is left as an exercise in problem 7.3.4). As a result,

$$V = k\frac{1}{2}\left\{\frac{1}{2}x_1^2 + \frac{3}{2}y_1^2 + \frac{5}{4}x_2^2 + \frac{3}{4}y_2^2 + \frac{5}{4}x_3^2 + \frac{3}{4}y_3^2\right.$$

$$+ \frac{\sqrt{3}}{2}(x_3y_1 + x_2y_2 - x_2y_1 - y_2x_1 - x_3y_3 + y_3x_1)$$

$$\left. -2x_2x_3 - \frac{1}{2}(x_1x_2 + x_1x_3) - \frac{3}{2}(y_1y_2 + y_1y_3)\right\},$$

$$q_1 = x_1,\ q_2 = y_1,\ q_3 = x_2,\ q_4 = y_2,\ q_5 = x_3,\ q_6 = y_3.$$

In the basis, $\eta_i = \sqrt{m}q_i$, we have

$$H = T + V,$$

where

$$T = \frac{1}{2}\left(\dot\eta_1^2 + \dot\eta_2^2 + \dot\eta_3^2 + \dot\eta_4^2 + \dot\eta_5^2 + \dot\eta_6^2\right), \quad V = \frac{1}{2}\sum_{i \neq j} v_{ij}\eta_i\eta_j.$$

$$\frac{1}{2}(v) = m\omega_0^2 \begin{pmatrix} \frac{1}{2} & 0 & -\frac{1}{4} & -\frac{\sqrt{3}}{4} & -\frac{1}{4} & \frac{\sqrt{3}}{4} \\ 0 & \frac{3}{2} & -\frac{\sqrt{3}}{4} & -\frac{3}{4} & \frac{\sqrt{3}}{4} & -\frac{3}{4} \\ -\frac{1}{4} & -\frac{\sqrt{3}}{4} & \frac{5}{4} & \frac{\sqrt{3}}{4} & -1 & 0 \\ -\frac{\sqrt{3}}{4} & -\frac{3}{4} & \frac{\sqrt{3}}{4} & \frac{3}{4} & 0 & 0 \\ -\frac{1}{4} & \frac{\sqrt{3}}{4} & -1 & 0 & \frac{5}{4} & -\frac{\sqrt{3}}{4} \\ \frac{\sqrt{3}}{4} & -\frac{3}{4} & 0 & 0 & -\frac{\sqrt{3}}{4} & \frac{3}{4} \end{pmatrix}.$$

This representation is reducible and, in fact, one symbolically finds

$$\Gamma(X) = 3\Gamma^3(X),$$

$$\Gamma^3(X) = \text{the irreducible } 2 \times 2 \text{ representation of } C_{3V}.$$

In this case, $X = \frac{1}{2}(v)$. This means that by invoking symmetry, the matrix that needs be diagonalized has dimension equal to the number 3 appearing above as the multiplicity times the dimension of $\Gamma^3(X)$, i.e. 6, the same is the dimension of $\Gamma(X)$. So, in this case, group theory does not offer any advantage. By direct diagonalization, we find the eigenvalues

$$\det\left(\left(\frac{1}{2}v\right)/(m\omega^2)\right) = \lambda^3\left(\lambda - \frac{3}{2}\right)^2(\lambda - 3),$$

and the corresponding eigenvectors can be chosen as follows:

$$\lambda = 3 \Rightarrow Q_1 = \left(0, \frac{1}{\sqrt{3}}, -\frac{1}{2}, -\frac{1}{2\sqrt{3}}, \frac{1}{2}, -\frac{1}{2\sqrt{3}}\right),$$

$$\lambda = \frac{3}{2} \Rightarrow Q_2 = \left(\frac{1}{\sqrt{3}}, 0, -\frac{1}{2\sqrt{3}}, -\frac{1}{2}, -\frac{1}{2\sqrt{3}}, \frac{1}{2}\right),$$

$$\lambda = \frac{3}{2} \Rightarrow Q_3 = \left(0, \frac{1}{\sqrt{3}}, \frac{1}{2}, -\frac{1}{2\sqrt{3}}, -\frac{1}{2}, -\frac{1}{2\sqrt{3}}\right),$$

$$\lambda = 0 \Rightarrow Q_4 = \left(\frac{1}{\sqrt{3}}, 0, \frac{1}{\sqrt{3}}, 0, \frac{1}{\sqrt{3}}, 0\right),$$

$$\lambda = 0 \Rightarrow Q_5 = \left(0, \frac{1}{\sqrt{3}}, 0, \frac{1}{\sqrt{3}}, 0, \frac{1}{\sqrt{3}}\right),$$

$$\lambda = 0 \Rightarrow Q_6 = \left(\frac{1}{\sqrt{3}}, 0, -\frac{1}{2\sqrt{3}}, \frac{1}{2}, -\frac{1}{2\sqrt{3}}, -\frac{1}{2}\right)$$

(in the case of degeneracy, the eigenvectors are not determined uniquely).

Not all states are interesting in the spectroscopy of the NH_3 molecule. All those with eigenvalue (frequency) zero are spurious without any physical meaning. Indeed, Q_4 is characterized by $y_1 = y_2 = y_3 = 0$, i.e. it corresponds to a motion of the entire molecule along the x-axis. Q_5 corresponds to $x_1 = x_2 = x_3 = 0$, i.e. translation of the entire molecule along the y-axis. Q_6 corresponds to a rotation around an axis perpendicular to the plane of the molecule. The situation is exhibited in Fig. 7.1.

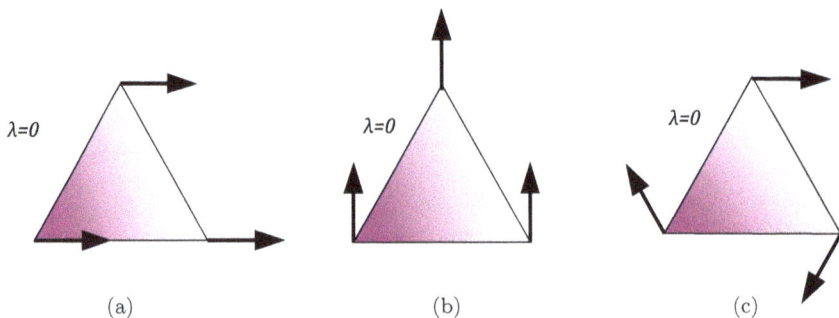

          (a)   (b)   (c)

Fig. 7.1. The three spurious motions of the NH_3 molecule, two translations and one rotation, of the entire molecule, which do not affect its spectrum.

Application 3. We will study the normal oscillation modes of the linear molecule of acetylene, C_2H_2, with masses and coordinates as follows:

$$
\begin{array}{cccc}
m_1 & m_2 & m_2 & m_1 \\
H \;-\; C & \equiv & C \;-\; H. \\
x_1 & x_2 & x_3 & x_4
\end{array}
$$

This molecule has symmetry $G = \{E, I\}$, see Example 3 of Chapter 16. The Hamiltonian, considering only the interaction between neighbors, is of the form

$$
V = \frac{1}{2}\left(k_1\left(x_1 - x_2\right)^2 + (x_3 - x_4)^2\right) + k_2(x_2 - x_3)^2),
$$

$$
T = \frac{1}{2}m_1\dot{x}_1^2 + \frac{1}{2}m_2\dot{x}_2^2 + \frac{1}{2}m_2\dot{x}_3^2 + \frac{1}{2}m_1\dot{x}_4^2,
$$

Again, to make the kinetic energy a multiple of the identity, we write

$$
\eta_1 = \sqrt{m_1}x_1, \eta_2 = \sqrt{m_2}x_2, \eta_3 = \sqrt{m_2}x_3, \eta_4 = \sqrt{m_1}x_4,
$$

$$
T = \frac{1}{2}\dot{\eta}_1^2 + \frac{1}{2}\dot{\eta}_2^2 + \frac{1}{2}\dot{\eta}_3^2 + \frac{1}{2}\dot{\eta}_4^2,
$$

$$
V = \frac{1}{\sqrt{2m_1m_2}}
$$

$$
((\eta_2 - \eta_3)^2\, k_2 m_1 + k_1\left((\eta_2^2 + \eta_3^2)\right)m_1
$$

$$
-2\left(\eta_1\eta_2 + \eta_3\eta_4\right)\sqrt{m_1}\sqrt{m_2} + \left(\eta_1^2 + \eta_4^2\right)m_2)).
$$

The corresponding matrices can be written as

$$
T = \frac{1}{2}\begin{pmatrix} 1 & 0 & 0 \\ 0 & 1 & 0 \\ 0 & 0 & 1 \end{pmatrix} \quad \text{(kinetic energy)},
$$

$$
v = \frac{1}{2}\begin{pmatrix}
\frac{k_1}{m_1} & -\frac{k_1}{\sqrt{m_1 m_2}} & 0 & 0 \\
-\frac{k_1}{\sqrt{m_1 m_2}} & \frac{k_1 + k_2}{m_2} & -\frac{k_2}{m_2} & 0 \\
0 & -\frac{k_2}{m_2} & \frac{k_1 + k_2}{m_2} & -\frac{k_1}{\sqrt{m_1 m_2}} \\
0 & 0 & -\frac{k_1}{\sqrt{m_1 m_2}} & \frac{k_1}{m_1}
\end{pmatrix}.
$$

The Hamiltonian operator is invariant under the symmetry transformations, but, unfortunately, the functions

$$\Psi_1 = \frac{1}{\sqrt{2}}(\eta_1 - \eta_4), \quad \Psi_2 = \frac{1}{\sqrt{2}}(\eta_2 - \eta_3),$$

$$\Psi_3 = \frac{1}{\sqrt{2}}(\eta_1 + \eta_4), \quad \Psi_4 = \frac{1}{\sqrt{2}}(\eta_2 + \eta_3)$$

are not eigenfunctions of the Hamiltonian, since the irreducible representations occur twice in the above representation of the symmetry group E, I of the system.

Clearly, one eigenvector with eigenvalue 0 corresponds to the motion of the center of mass of the molecule, i.e it is analogous to

$$R = m_1 x_1 + m_2 x_2 + m_2 x_3 + m_1 x_4 = \sqrt{m_1}\eta_1 + \sqrt{m_2}\eta_2 + \sqrt{m_2}\eta_3 + \sqrt{m_1}\eta_4.$$

Indeed after normalization it takes the form

$$\Psi_3' = \left(\frac{\sqrt{m_1}}{\sqrt{2(m_1 + m_2)}}, \frac{\sqrt{m_2}}{\sqrt{2(m_1 + m_2)}}, \frac{\sqrt{m_2}}{\sqrt{2(m_1 + m_2)}}, \frac{\sqrt{m_1}}{\sqrt{2(m_1 + m_2)}} \right).$$

Another eigenvector is the orthogonal combination

$$\Psi_4 = \left(\frac{\sqrt{m_2}}{\sqrt{2(m_1 + m_2)}}, -\frac{\sqrt{m_1}}{\sqrt{2(m_1 + m_2)}}, -\frac{\sqrt{m_1}}{\sqrt{2(m_1 + m_2)}}, \frac{\sqrt{m_2}}{\sqrt{2(m_1 + m_2)}} \right).$$

The other two basis vectors remain unchanged. Thus, under the transformation,

$$S = \begin{pmatrix} \frac{1}{\sqrt{2}} & 0 & 0 & \frac{1}{\sqrt{2}} \\ 0 & \frac{1}{\sqrt{2}} & -\frac{1}{\sqrt{2}} & 0 \\ \frac{\sqrt{m_1}}{\sqrt{2(m_1+m_2)}} & \frac{\sqrt{m_2}}{\sqrt{2(m_1+m_2)}} & \frac{\sqrt{m_2}}{\sqrt{2(m_1+m_2)}} & \frac{\sqrt{m_1}}{\sqrt{2(m_1+m_2)}} \\ \frac{\sqrt{m_2}}{\sqrt{2(m_1+m_2)}} & -\frac{\sqrt{m_1}}{\sqrt{2(m_1+m_2)}} & -\frac{\sqrt{m_1}}{\sqrt{2(m_1+m_2)}} & \frac{\sqrt{m_2}}{\sqrt{2}\sqrt{2(m_1+m_2)}} \end{pmatrix},$$

one gets

$$S^T(v)S = \frac{1}{2} \begin{pmatrix} \frac{k_1}{2m_1} & 0 & 0 & \frac{k_1\sqrt{\frac{m_1}{m_2}+1}}{2m_1} \\ 0 & \frac{k_1+2k_2}{2m_2} & 0 & 0 \\ 0 & 0 & 0 & 0 \\ \frac{k_1\sqrt{\frac{m_1}{m_2}+1}}{2m_1} & 0 & 0 & \frac{k_1(m_1+m_2)}{2m_1 m_2} \end{pmatrix}.$$

As expected, one eigenvalue is non-zero, $\omega_1 = \sqrt{\frac{k_1 + 2k_2}{4m_2}}$, while the other is spurious with eigenvalue 0. The remaining two can be found by diagonalizing 2×2 sub-matrix. One eigenvalue is non-zero, $\omega_2 = \sqrt{\frac{k_1(m_1 + 2m_2)}{2(2m_1 m_2)}}$, and the other 0, corresponding to the eignvectors

$$Q_1 = \Psi_2, Q_2 = \sqrt{\frac{m_2}{m_1 + 2m_2}} \Psi_1 + \sqrt{\frac{m_1 + m_2}{m_1 + 2m_2}} \Psi_4$$

or

$$Q_1 = \frac{1}{\sqrt{2}}(\eta_2 - \eta_3),$$

$$Q_2 = \sqrt{\frac{m_2}{2(m_1 + 2m_2)}} \eta_1 + \sqrt{\frac{m_1 + m_2}{2(m_1 + 2m_2)}} \eta_2$$

$$+ \sqrt{\frac{m_1 + m_2}{2(m_1 + 2m_2)}} \eta_3 - \sqrt{\frac{m_2}{2(m_1 + 2m_2)}} \eta_4.$$

In summary, we can say that with aid of group theory, one can obtain some normal modes, without the need of numerical diagonalizations, i.e without the aid of modern computers. The reason for such a success lies, of course, in the great degree of symmetry of such systems, which is geometric and physical, in the sense that the masses of the particles involved are equal, leading to simple expression for the center of mass. The other physical assumption involves the selection of the spring constants appearing in the expression of the Hamiltonian in conjunction with the fact that the coordinates of the heavy nucleus in the center were in quarantine.

In Example 3 of Chapter 6, we did not have such a success. The symmetry was low and the heavy artillery resulting from knowledge on group theory did not fire well.

7.4 Problems

7.3.1 Find the canonical (normal) coordinates and the corresponding frequencies of the system given in problem 6.3.5.

7.3.2 Do the same for problem 6.3.6.

7.3.3 Do the same for problem 6.3.7.

7.3.4 In the case of the molecule NH_3 including, in addition to the interactions between the hydrogen atoms discussed in the text, the interaction of the heavy nucleus and the rest as well, we have the additional terms

$$V = \frac{k'}{2}\left(\frac{1}{2}(x_4 - x_2) + \frac{1}{2}\sqrt{3}(y_4 - y_2)\right)^2$$

$$+ \frac{k'}{2}\left(\frac{1}{2}(x_4 - x_3) - \frac{1}{2}\sqrt{3}(y_4 - y_3)\right)^2 + \frac{k'}{2}(x_4 - x_1)^2.$$

Setting $\kappa = k'/k$ and $q_7 = x_4 = r\eta_7$, $q_8 = y_4 = r\eta_8$, we find the total matrix:

$$\begin{pmatrix}
k+\frac{1}{2} & 0 & -\frac{1}{4} & -\frac{\sqrt{3}}{4} & -\frac{1}{4} & \frac{\sqrt{3}}{4} & -kr & 0 \\
0 & \frac{3}{2} & -\frac{\sqrt{3}}{4} & -\frac{3}{4} & \frac{\sqrt{3}}{4} & -\frac{3}{4} & 0 & 0 \\
-\frac{1}{4} & -\frac{\sqrt{3}}{4} & \frac{\kappa}{4}+\frac{5}{4} & \frac{\sqrt{3}\kappa}{4}+\frac{\sqrt{3}}{4} & -1 & 0 & -\frac{\kappa r}{4} & -\frac{1}{4}\sqrt{3}\kappa r \\
-\frac{\sqrt{3}}{4} & -\frac{3}{4} & \frac{\sqrt{3}\kappa}{4}+\frac{\sqrt{3}}{4} & \frac{3\kappa}{4}+\frac{3}{4} & 0 & 0 & -\frac{1}{4}\sqrt{3}\kappa r & -\frac{3\kappa r}{4} \\
-\frac{1}{4} & \frac{\sqrt{3}}{4} & -1 & 0 & \frac{\kappa}{4}+\frac{5}{4} & -\frac{\sqrt{3}\kappa}{4}-\frac{\sqrt{3}}{4} & -\frac{\kappa r}{4} & \frac{1}{4}\sqrt{3}\kappa r \\
\frac{\sqrt{3}}{4} & -\frac{3}{4} & 0 & 0 & -\frac{\sqrt{3}\kappa}{4}-\frac{\sqrt{3}}{4} & \frac{3\kappa}{4}+\frac{3}{4} & \frac{1}{4}\sqrt{3}\kappa r & -\frac{3\kappa r}{4} \\
-\kappa r & 0 & -\frac{\kappa r}{4} & -\frac{1}{4}\sqrt{3}\kappa r & -\frac{\kappa r}{4} & \frac{1}{4}\sqrt{3}\kappa r & \frac{3\kappa r^2}{2} & 0 \\
0 & 0 & -\frac{1}{4}\sqrt{3}\kappa r & -\frac{3\kappa r}{4} & \frac{1}{4}\sqrt{3}\kappa r & -\frac{3\kappa r}{4} & 0 & \frac{3\kappa r^2}{2}
\end{pmatrix}.$$

with eigenvalues

$$\left\{ \frac{3}{2}, 0, 0, 0, \frac{1}{2}(2k+3), \frac{1}{2}(3kr^2 + 2k), \right.$$

$$\frac{1}{8}\left(6kr^2 - \sqrt{(6kr^2 + 4k + 12)^2 + 4(-72kr^2 - 24k)} + 4k + 12\right),$$

$$\left. \frac{1}{8}\left(6kr^2 + \sqrt{(6kr^2 + 4k + 12)^2 + 4(-72kr^2 - 24k)} + 4k + 12\right)\right\}.$$

Note the presence of the three zero frequencies, as expected, since the spurious degrees of freedom still exist. For $\kappa = \frac{m_{\text{H}}}{m_{\text{N}}}$ (the ratio of the masses) and $r = \sqrt{\kappa}$, the eigenvalues become

$$\{1.5, 0., 0., 0., 1.571, 0.079, 0.043, 3.036\}.$$

We find the already known solution, Application 4 of Section 7.3, with the additional almost zero frequencies.

7.3.5 Consider the $C_2Cl_2H_2$ molecule assuming it is in the C_{2v} form (problem 2.5.7). Construct the analog of the matrix, given in Table 6.1 in the case of the present symmetry. Show that the resulting representation is reduced as follows:

$$\Gamma(X) = 6\Gamma^1(X) + 2\Gamma^3(X) + 2\Gamma^4(X). \qquad (7.23)$$

Proceed as far as you can with the study of the normal oscillation modes.

7.3.6 Do the same with problem 7.3.5, considering that the molecule is as above, but is characterized by symmetry C_{2h}. In particular, show that

$$\Gamma(X) = 5\Gamma^1(X) + \Gamma^2(X) + 4\Gamma^4(X). \qquad (7.24)$$

Note: Regarding the reduction, the matrix of Eq. (5.31) may be useful, with proper identification of the elements.

Chapter 8

Space Groups

By space group, we understand the symmetry group which is characteristic of a given periodic system, such as an ideal crystal. This consists of the set of transformations which carry one point of the system to another. Thus, the space groups should contain point groups, such as the ones we have considered so far, as well as translation transformations. This imposes conditions on how the elements of the crystal repeat themselves so that they generate the whole crystal. As a result, only a fraction of the point group symmetries are compatible with the required space symmetry.

8.1 Introductory notions

Strictly speaking, the structure components that characterize a given symmetry must be repeated so that the whole space of a crystal is covered by them. We all know that the whole floor can be covered by placing side by side tiles of the shape of a triangle, a parallelogram (squares, diamonds, etc.) or regular hexagons, see Fig. 8.1. This cannot be done by placing circles or regular pentagons. Similarly, the whole 3D space can be covered by proper placing of prisms with bases shaped as triangles, parallelograms or regular hexagons.

In special cases, e.g. crystal structure, the coordinates are selected to have special characteristics, and the unit vectors along the axes do not form an orthogonal basis. Let us, for example, consider the case of regular hexagons, see Fig. 8.2, with basis vectors $\vec{\alpha}_1, \vec{\alpha}_2$. To see the geometric intricacies involved, we specify the vector $\vec{r}$ and the angle θ of two lines starting from a corner of a hexagon and ending at a distance of (n, m) regular hexagons along the two dimensions. Then, we imagine forming a cylinder by rolling up a strip of the lattice of width $|\vec{r}|$, i.e. with the cylinder axis perpendicular to $\vec{r}$.

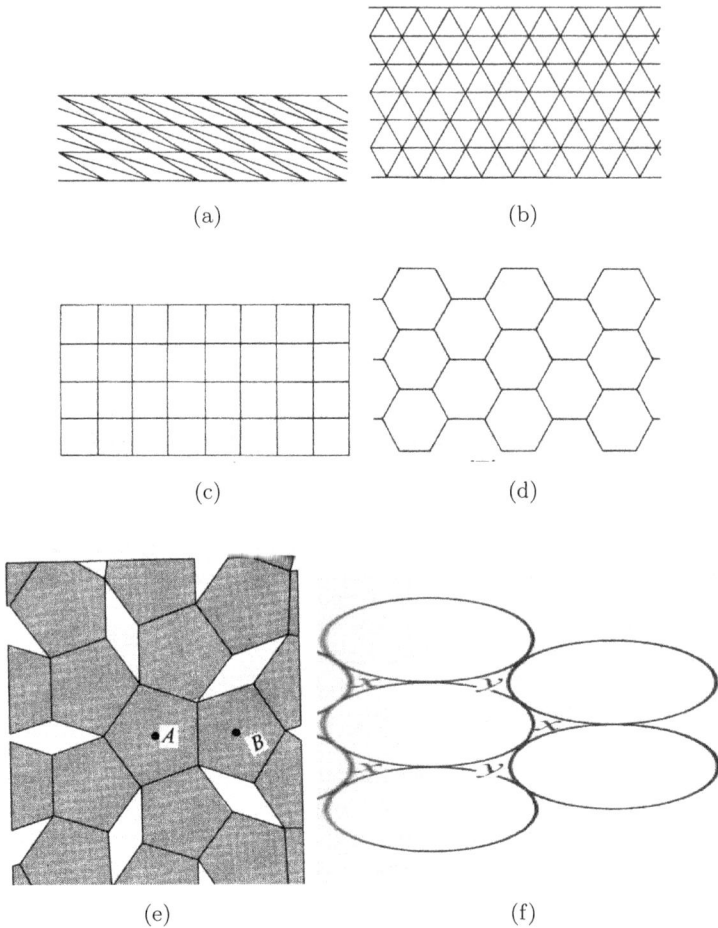

Fig. 8.1. Various shapes that can cover the whole plane: (a) scalene triangles combined to form a parallelogram (second-order axis), (b) equilateral triangles (third-order axis), (c) squares (fourth-order axis) and (d) hexagons (sixth-order axis). This cannot be accomplished by using: (e) regular pentagons and (f) circles or ellipses. Note that in case (a), we have $a \neq b$ and ϕ arbitrary, but in the other cases, $a = b$, while the angles are $2\pi/3$, $\pi/2$ and $\pi/3$ for (b), (c), and (d), respectively.

The vector $\mathbf{r}$ can be written as

$$\mathbf{r} = n\vec{\alpha}_1 + m\vec{\alpha}_2, \quad |\vec{\alpha}_1| = |\vec{\alpha}_2| = a. \tag{8.1}$$

From the properties of the regular hexagon, we easily find that the basis vectors $\vec{\alpha}_1, \vec{\alpha}_2$ form an angle $\theta_0 = \frac{\pi}{3}$. From Fig. 8.2, we find that the

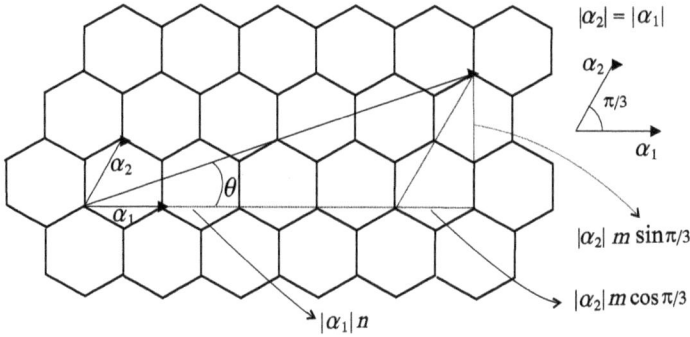

Fig. 8.2. A 2D hexagonal cell.

components of **r** along the direction $\vec{\alpha}_1$ and to that perpendicular to it are

$$r_1 = |\vec{\alpha}_1|\, n + |\vec{\alpha}_2|\, m \cos \frac{\pi}{3} = \left(n + \frac{1}{2}m\right) a,$$

$$r_2 = |\vec{\alpha}_2|\, m \sin \frac{\pi}{3} = \frac{\sqrt{3}}{2}\, m\, a.$$

From these, we obtain

$$\tan \theta = \frac{m\sqrt{3}}{2n + m}. \tag{8.2}$$

The length of **r** is obtained from the scalar product

$$|\vec{r}|^2 = (n\vec{\alpha}_1 + m\vec{\alpha}_2) \cdot (n\vec{\alpha}_1 + m\vec{\alpha}_2)$$
$$= a^2(n^2 + m^2 + 2nm \cos \theta_0) = a^2(n^2 + m^2 + nm).$$

If d is the diameter of the cylinder, $|\mathbf{r}| = \pi d$. Hence:

$$d = \frac{a}{\pi}\sqrt{n^2 + m^2 + nm}. \tag{8.3}$$

Remark. This arrangement is found in nano structures where the corners are occupied by carbon atoms. Thus, nano tubes are formed with very interesting properties.

8.2 The group of translations in space

A crystal, as opposed to amorphous materials and liquids, is character-ized by translational symmetry, i.e. by periodic repetition of its structural units.[1] The translation vector is not arbitrary, but it must be of the form

$$\mathbf{a} = n_1\mathbf{a}_1 + n_2\mathbf{a}_2 + n_3\mathbf{a}_3, \; n_1, n_2, n_3 = \text{integers}. \tag{8.4}$$

The constant vectors $\mathbf{a}_1, \mathbf{a}_2, \mathbf{a}_3$, however, need not be orthogonal. As a result, the two equivalent points $\mathbf{r}$ and $\mathbf{r}'$ of the crystal must be related via the relation

$$\mathbf{r}' = \mathbf{r} + \mathbf{a}. \tag{8.5}$$

The end of the vectors $\mathbf{a}_1, \mathbf{a}_2, \mathbf{a}_3$ form a cell, known as **Bravais lattice**; see Fig. 8.3 for the lattice in 2D and Fig. 8.4(a) for the same in 3D. The acceptable lattice should not contain in its inside any lattice positions. The choice of the basis vectors may not be unique, but the lattice volume must be the same. An important characteristic of a lattice is its metric given by

$$g_{ij} = \mathbf{a}_i \cdot \mathbf{a}_j. \tag{8.6}$$

Transformations of the form given by Eq. (8.5), constitute a group, called group of translations, see Example 2, Chapter 4, and it will be denoted by $T(\mathbf{a}) = e^{i\mathbf{a}\cdot\nabla/\hbar}$:

$$T(\mathbf{a}+\mathbf{b}) = T(\mathbf{a})T(\mathbf{b}), \quad T^{-1}(\mathbf{a}) = T(-\mathbf{a}), \tag{8.7}$$

which is Abelian.

8.3   The allowed crystal symmetries

In addition to translation, the lattice is characterized by a point symmetry associated with a group G_0. Suppose that in the space of the lattice vectors, we consider a set of transformations R. The demand that both $\mathbf{a}$ and $T\mathbf{a}$, $T \in R$ be lattice vectors restricts considerably the allowed groups G_0. The first condition is that G_0 must contain the element of inversion I. This is the result of the fact that, if there exists an element of translation symmetry $T(\mathbf{a})$, there must exist its inverse $T(-\mathbf{a})$. In other words, to every rotation, there must exist a rotational inversion, a rotation and reflection of the axis of symmetry. We already know, Example 3, Chapter 4, the form of

---

[1] To be exact, this holds only if the system is of infinite dimensions. From a mathematical point of view, this is not essential. The opposite sides of a finite sample, which consists of a very large number of structural elements, can be made to coincide, and one gets the topology of a ring.

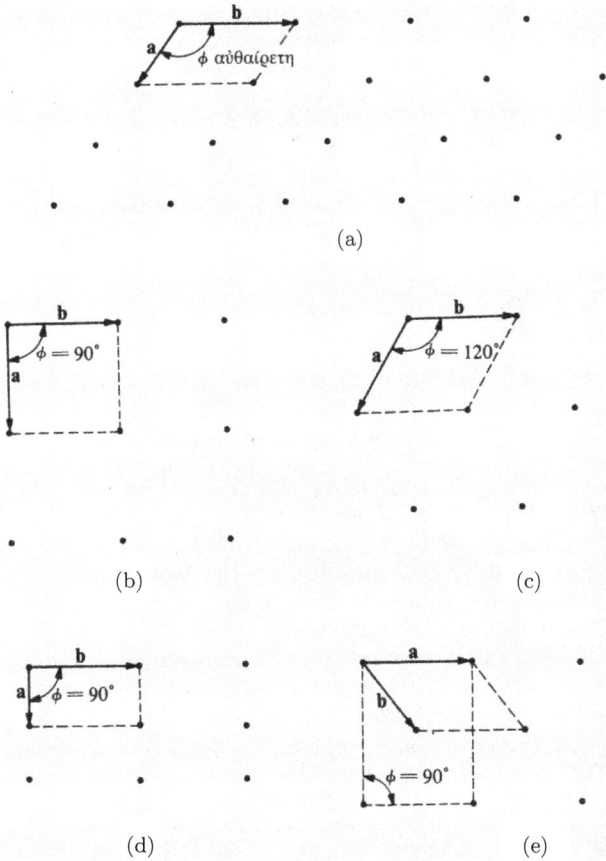

Fig. 8.3. Some 2D Bravais lattices.

rotation from which the reflection–rotation follows. We thus have matrices of the form

$$\rho = \begin{pmatrix} \cos\theta & \sin\theta & 0 \\ -\sin\theta & \cos\theta & 0 \\ 0 & 0 & 1 \end{pmatrix} \quad \text{(rotation)},$$

$$\rho = \begin{pmatrix} \cos\theta & \sin\theta & 0 \\ -\sin\theta & \cos\theta & 0 \\ 0 & 0 & -1 \end{pmatrix} \quad \text{(inversion)}. \tag{8.8}$$

If we now express it in the coordinate system of the axes a_1, a_2, a_3, we obtain a matrix of the form (ρ')

$$a_i' = Ta_i = \sum_j (\rho')_{ji} a_j, \quad (\rho')_{ji} = \text{integers}. \qquad (8.9)$$

The matrices (ρ) and $(\rho)'$, however, are similar, corresponding to two different bases. The similarity transformation S preserves, of course, the trace, i.e.

$$(\rho)' = S^{-1}(\rho)S \Rightarrow tr((\rho)') = tr((\rho)) \Rightarrow 2\cos\theta \pm 1 = \text{integer}. \qquad (8.10)$$

In other words,

$$\cos\theta = \pm 1, \pm\frac{1}{2}, 0 \Rightarrow \text{allowed axis order: } n = 1, 2, 3, 4, 6. \qquad (8.11)$$

If, however, there exists an axis C_n of order n, $n > 2$, it must contain symmetry planes, which pass through this axis. Thus, the symmetry must be of the type C_{nV}, $n > 2$, see Section 2.4.2. We have seen, however, that it must contain the inversion symmetry I. As a result, the allowed symmetries are:

$$S_2, C_{2h}, D_{2h}, D_{3h}, D_{4h}, D_{6h}, O_h, \qquad (8.12)$$

as listed in Section 2.4.2. This is the reason for the existence of only seven crystal systems, as indicated in Fig. 8.4.

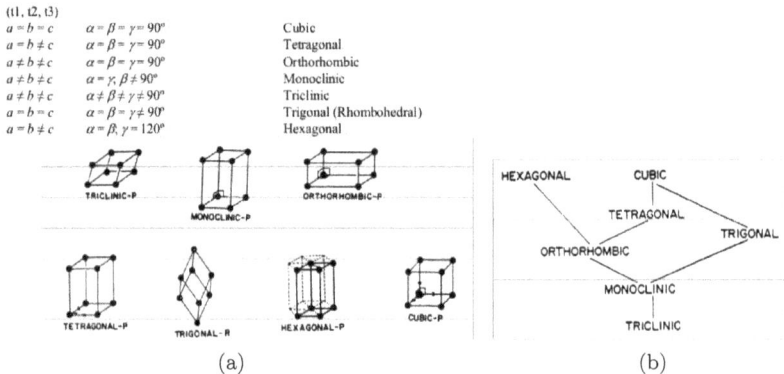

Fig. 8.4. The allowed crystal systems (a) and their hierarchy (b).

8.4 Space groups

We have seen that in the case of a crystal, we have two types of symmetry, the translational symmetry given by the transformations $T(\mathbf{a})$ and the point symmetry G_0. The total symmetry G containing both of them is

$$G = \{X \in G \Leftrightarrow X = T(\mathbf{a})R,\ R \in G_0\}, \tag{8.13}$$

i.e.

$$T(\mathbf{a})R\mathbf{r} \equiv R\mathbf{r} + \mathbf{a} \tag{8.14}$$

and

$$\begin{aligned} T(\mathbf{a})RT(\mathbf{a}')R'\mathbf{r} &= T(\mathbf{a})R(R'\mathbf{r} + \mathbf{a}') = RR'\mathbf{r} + R\mathbf{a}' + \mathbf{a} \\ &= T((R')^{-1}(\mathbf{a}' + R^{-1}\mathbf{a})RR'\mathbf{r}. \end{aligned} \tag{8.15}$$

The law of composition is

$$\begin{aligned} T(\mathbf{a})RT(\mathbf{a}')R' &= T((R')^{-1}(\mathbf{a}' + R^{-1}\mathbf{a})RR', \\ T(\mathbf{a}')R'T(\mathbf{a})R &= T((R)^{-1}(\mathbf{a} + (R')^{-1}\mathbf{a}')RR'. \end{aligned} \tag{8.16}$$

In other words, the commutation property is not valid even when the two elements of G_0 commute. The inverse transformation is given by

$$\begin{aligned} (T(\mathbf{a})R)^{-1}\mathbf{r} &= R^{-1}T^{-1}(\mathbf{a})\mathbf{r} = R^{-1}(\mathbf{r} - \mathbf{a}) = R^{-1}\mathbf{r} - R^{-1}\mathbf{a} \\ &= T(-\mathbf{a})R^{-1}r \Rightarrow (T(\mathbf{a})R)^{-1} = T(-\mathbf{a})R^{-1}. \end{aligned} \tag{8.17}$$

Furthermore, the operators $T(a)$ and R do not commute. Indeed,

$$RT(\mathbf{a})\mathbf{r} = R(\mathbf{r} + \mathbf{a}) = R\mathbf{r} + R\mathbf{a},\ T(\mathbf{a})R\mathbf{r} = R\mathbf{r} + \mathbf{a} \Rightarrow RT(\mathbf{a})\mathbf{r} \neq T(\mathbf{a})R\mathbf{r}.$$

Up to now, we have dealt with an "empty" crystal. In a real crystal, some aspects change as soon as the ions are placed in the lattice, leading to a change in symmetry. This is due not only to the symmetry of the Bravais lattice but also to the symmetry of the structural elements of the crystal as well. In the special case of diamond, this is shown in Fig. 8.5(a). There exist two interpenetrating cubic face-centered Bravais lattices, one at the beginning of the coordinates and the other along the main diagonal at a distance $d/4$ from the previous. Thus, the nearest neighbors of the first lattice are found in the second lattice. We see that the system has tetrahedral symmetry, $H_0 = T_d = \{T, TI\} \subset O_h$, followed by a translation by b, where

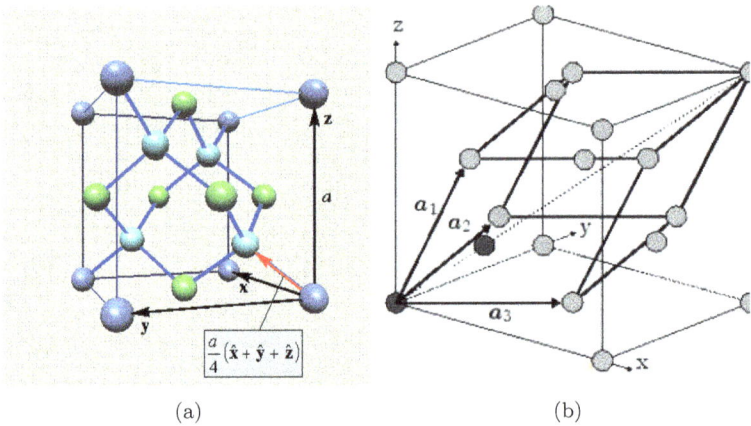

Fig. 8.5. (a) The lattice of diamond made up of four carbon atoms. It is composed of two interpenetrating cubic face-centered Bravais lattices along the diagonal of the cubic lattice at one-fourth of the length of the large diagonal, with basis vectors indicted by the two indicated points. In other words, it can be viewed as a face-centered cubic lattice of two points. (b) We show the basis vectors of the face-centered cubic lattice and the two atoms which determined the basis.

b is the rational part of a, here $1/4$. One could, in general, also consider transformations of the type

$$T(\mathbf{b})R, \ R \in H_0, \ H_0 \subset G_0 b, \ b = \text{rational part of } a.$$

Finally, it is possible to verify that the above elements do not constitute a group, since, among other things, the product of such transformations may lead to a translation $T(\mathbf{a})$. Thus, the group contains the elements

$$X = T(\mathbf{a})T(\mathbf{b})R = T(\mathbf{a}+\mathbf{b})R, \ R \in H_0, \ b = \text{integer part of } a. \quad (8.18)$$

Indeed,

$$T(\mathbf{a}+\mathbf{b})RT(\mathbf{a}'+\mathbf{b}')R' = T(\mathbf{a}+\mathbf{b})R(R'+\mathbf{a}'+\mathbf{b}')$$
$$= T(\mathbf{a}+\mathbf{b})(RR'+R(\mathbf{a}'+\mathbf{b}'))$$
$$= RR' + R(\mathbf{a}'+\mathbf{b}') + \mathbf{a} + \mathbf{b}$$
$$= T(\mathbf{a}+R\mathbf{a}'+\mathbf{b}+R\mathbf{b}')RR'. \quad (8.19)$$

Since, however, $R \in H_0$, the vector $R\mathbf{a}'$ is a lattice vector of the type $\mathbf{a}$ and $R\mathbf{b}'$ is a lattice vector of the type $\mathbf{b}$, as implied by Eq. (8.18), that is, the product is of the type XX'.

Beyond the accepted groups G_0, one may consider also their subgroups H_0. In other words, from the seven systems characterized by $G_0 = S_2, C_2h, D_2h, D_3h, D_4h, D_6h, O_h$, one obtains 32 subsystems belonging to the possible subgroups H_0 of G_0, as indicated in Table 8.1.

Table 8.1. The 32 crystal groups. We remind the reader that a bar above the order of the axis indicates that it is a rotation–reflection one, $\bar{n} = n + \bar{1}$.

System	Schoenflies symbol	Int. full Hermann-Mauguin	Int. short Hermann-Mauguin	Group order
Triclinic	C_1	1	1	1
	C_i or S_2	$\bar{1}$	$\bar{1}$	2
	C_{1h}	m	m	2
Monoclinic	C_2	2	2	2
	C_{2h}	$2/m$	$2/m$	4
	C_{2V}	$2mm$	$2mm$	4
Orthorombic	D_2	222	222	4
	D_{2h}	$2/m\,2/m\,2/m$	mmm	8
	C_4	4	4	4
	S_4	$\bar{4}$	$\bar{4}$	4
	S_{4h}	$4/m$	$4/m$	8
Square	D_{2d}	$\bar{4}2m$	$\bar{4}2m$	8
	C_{4V}	$4mm$	$4mm$	8
	D_d	422	42	8
	D_{4h}	$4/m\,2/m\,2/m$	$4/m\,mm$	16
	T	23	$23mm$	12
	T_h	$2/m\bar{3}$	$2/m\bar{3}$	24
Cubic	T_d	$\bar{4}3m$	$\bar{4}3m$	24
	O	432	432	24
	O_h	$4/m\,\bar{3}\,2/m$	$4/m\,\bar{3}\,2/m$	24
	C_3	3	3	3
	C_{3I} or S_6	$\bar{3}, \bar{3}$	6	
Triangular	C_{3V}	$3m$	$3m$	8
	D_3	32	32	6
	D_{3d}	$\bar{3}2$	$\bar{3}2$	12
	C_6	6	6	6
	C_{3h}	$\bar{6}$ or $3/m$	$\bar{6}$ or $3/m$	6
	C_{6h}	$6/m$	$6/m$	12
Hexagonal	D_{3h}	$\bar{6}m2$	$\bar{6}m2$	12
	C_{6V}	$6mm$	$6mm$	12
	D_6	622	62	12
	D_{6h}	$6/m\,2/m\,3/m$	$6/m\,mm2$	24

Before proceeding further, it is worth devoting some time to discussing some concepts and the notation used in crystal structure for the various systems.

8.4.1 *Crystal systems in 3D*

For our purposes, the most useful is the Schoenflies system, which has been discussed in Section 2.4.2. Another useful notation is that of Hermann-Mauguin, see Table 8.1. In the latter, the order of every unique axis is indicated. A unique axis is one which cannot result via a combination of other elements. Next, by m is indicated every unique plane of mirror symmetry (unique mirror plane). If any of these planes is perpendicular to any such axis, we put a line, "/", between the two (see Table 8.1). The inversion is not exhibited.

A group is called **conformal** or conformable when the space group, with the exception of the elements of translation, contains rotations around some axes, reflection, inversion and rotation–reflection axes. It does not include glide operators (glide planes) or helix operators (screw axes), which are sometimes used in crystal structures.

The characters of the irreducible representations of crystal groups are discussed in Section 8.5, and a system of their generators is explained in Section 8.6. A complete list of relevant tables has been included in Chapter 17.

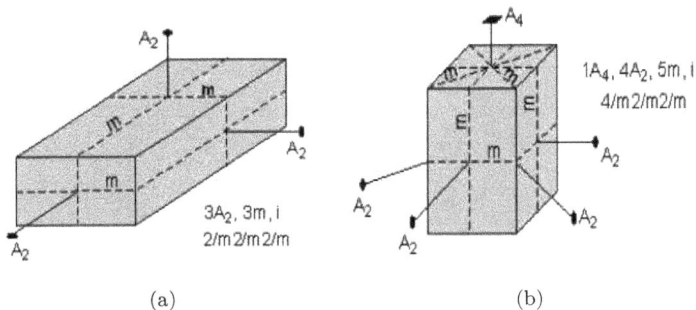

Fig. 8.6. (a) There exist three unique axes of second order and three unique mirror planes m, perpendicular to the axes. We thus write $2/m2/m2/m$. (b) There now exists a unique axis of fourth order and four axes of second order, from which two are perpendicular to faces and two are perpendicular to vertical edges. Of these, only two are unique, and the other two are a consequence of the existence of the fourth-order axis. In addition, there exist five mirror planes m, two perpendicular to the vertical faces, two pass through the vertical edges and one is horizontal. Of these, three are unique, while the others are a consequence of the presence of the fourth order axis. Every unique axis is perpendicular to a unique mirror plane. Thus, we write $4/m2/m2/m$.

8.4.2 *Crystal systems in two dimensions*

In recent years, the 2D crystal systems have become of great interest in various applications. The possible lattices are shown in Fig. 8.3. The allowed axes are of order $n = 2, 3, 4$ and 6. There exist five Bravais lattices in 2D characterized by two lattice vectors a_1, a_2 and the angle ϕ between them. There exist five lattice types and 13 conformal space groups summarized in Table 8.2. In addition, there exist four non-conformal ones (Dresselhaus *et al.*, 2002).

Table 8.2. The 13 conformal space groups in 2D. The integer n in the first column indicates that order of the axis perpendicular to the plane of the lattice, while m indicates the mirror plane. p indicates that the lattice is primitive, and c specifies that the lattice is centered.

Point group	Lattice	Notation full	Type	Notation abbreviated
$n = 1$	Elongated	$p1$	Conformal	$p1$
$n = 2$	$a_1 \neq a_2, \phi \neq \frac{\pi}{2}$	$p211$	Conformal	$p2$
m	orthogonal; p $a_1 \neq a_2, \phi = \frac{\pi}{2}$	p1m1	Conformal	pm
m	orthogonal; c $a_1 \neq a_2, \phi = \frac{\pi}{2}$	$c1m1$	Conformal	cm
mm	orthogonal; p $a_1 \neq a_2, \phi = \frac{\pi}{2}$	$p2mm$	Conformal	pmm
mm	orthogonal; c $a_1 \neq a_2, \phi = \frac{\pi}{2}$	$c2mm$	Conformal	cmm
$n = 4$	square; p $a_1 = a_2, \phi = \frac{\pi}{2}$	$p4$	Conformal	$p4$
$n = 4, mm$	square; p $a_1 = a_2, \phi = \frac{\pi}{2}$	$p4mm$	Conformal	$p4m$
$n = 3$	hexagon; p $a_1 = a_2, \phi = \frac{2\pi}{3}$	$p3$	Conformal	$p3$
$n = 3, m$	hexagon; p $a_1 = a_2, \phi = \frac{2\pi}{3}$	$p3m1$	Conformal	$p3m1$
$n = 3, m$	hexagon; p $a_1 = a_2, \phi = \frac{2\pi}{3}$	$p31m$	Conformal	$p31m$
$n = 6$	hexagon; p $a_1 = a_2, \phi = \frac{2\pi}{3}$	$p6$	Conformal	$p6$
$n = 6, mm$	hexagon; p $a_1 = a_2, \phi = \frac{2\pi}{3}$	$p6mm$	Conformal	$p6m$

8.5　Character tables of the irreducible representations of crystal groups

The character tables of the irreducible representations of crystal groups often do not include those that are the direct product of simple groups, such as the groups $D_{2h} = D_2 \otimes \{E, \sigma_h\}$. The reduction of D_2 is given in Table 17.4, and the reduction of $\{E, \sigma_h\}$ is trivial.

We should mention that in the tables of the irreducible representations of the crystal groups, a representative element of each class of the group is indicated. In the first column, the representation is indicated as follows: For the 1D ones, the symbol A (B) is used for the symmetric (antisymmetric) basis with respect to C_n, which is the rotation with the maximum order n, with the axis taken to be in the z direction. For this purpose, one could as well have chosen the behavior with respect to the mirror plane σ_h. The indices g (gerande) and u (ungerande) refer to the even and odd representations of the inversion operator I (see also Table 8.3). With the indices **E** and **F**, we denote the 2D and 3D representations, respectively. In addition to the characters, two more columns have been included. In the next to last, we have included the translation operators T_x, T_y, T_z along the axes x, y, z and the rotation ones R_x, R_y, R_z around the indicated axes of symmetry, respectively (the generators given are considered in a 3×3 representation). In the last column, we have included the projections α_{ij} of the polarization tensor or a suitable combination of them next to the

Table 8.3. The character table of D_{2h}. Note that $\sigma_x \Leftrightarrow \sigma_v(yz)$, $\sigma_y \Leftrightarrow \sigma_v(xz)$, $C_2\sigma_h \Leftrightarrow I$, $\sigma_h = C_2I$, $C_y\sigma_h = C_xI$, $C_x\sigma_h = C_yI$, that is, this group is isomorphic to $C_{2v} \otimes \{E, I\}$. This explains the symbolism of the table ($A \Leftrightarrow$ symmetric, $B \Leftrightarrow$ antisymmetric with respect to both $\sigma_h = C_2I$ and C_2, $g \Leftrightarrow$ eigenvalue $+1$ with respect to I, $u \Leftrightarrow$ eigenvalue -1 with respect to I.

$$D_{2h} \Leftrightarrow M_2 = \begin{pmatrix} 1 & 1 & 1 & 1 \\ 1 & 1 & -1 & -1 \\ 1 & -1 & 1 & -1 \\ 1 & -1 & -1 & 1 \end{pmatrix}, \{E, \sigma_h\} \Leftrightarrow M_1 = \begin{pmatrix} 1 & 1 \\ 1 & -1 \end{pmatrix} \Rightarrow D_{2h} = M_1 \otimes M_2:$$

	E	C_2	$C_v(yz)$	$C_v(xz)$	I	C_2I	$C_v(yz)I$	$C_v(xz)I$
A_{1g}	1	1	1	1	1	1	1	1
A_{2g}	1	1	-1	-1	1	1	-1	-1
B_{1g}	1	-1	1	-1	1	-1	1	-1
B_{2g}	1	-1	-1	1	1	-1	-1	1
A_{1u}	1	1	1	1	-1	-1	-1	-1
A_{2u}	1	1	-1	-1	-1	-1	1	1
B_{1u}	1	-1	1	-1	-1	1	-1	1
B_{2u}	1	-1	-1	1	-1	1	1	-1

corresponding operator in the previous column. The conventional notation has been retained as much as possible. The resulting tables are shown in 17.4–17.10 of the appendix, Chapter 17. More extensive tables can be found in appendix A, p. 479 of Dresselhaus *et al.* (2002).

We should, however, make a few comments here:

- In the case of an Abelian group, all irreducible representations are 1D. For the typical reader, we want to stress that in some cases, having to do with certain applications, two such representations have been grouped together under the symbol E (or $E_1, E_2, \ldots,$). Such examples are: the group C_4 in Table 17.6, the groups $C_{4h} = C_4 \otimes G_I$, $G_I = \{E, I\}$ and S_d in Table 17.7, the group C_6 in Table 17.8 and the group S_6 in Table 17.9.
- The irreducible representations for both groups T and T_h appear in the tables, for T in Table 17.9 and T_h in Table 17.10. The inclusion of T_h may seem redundant, since the two groups are related, $T_h = G_I \otimes T$. We thus know the representations of T_h can be obtained as a Kronecker product, see Section 4.5. Thus, for example, we find

$$
\begin{pmatrix} 1 & 1 \\ 1 & -1 \end{pmatrix} \otimes \begin{pmatrix} 1 & 1 & 1 \\ 2 & -1 & 2 \\ 3 & 0 & -1 \end{pmatrix} = \begin{pmatrix} 1 & 1 & 1 & 1 & 1 & 1 \\ 2 & -1 & 2 & 2 & -1 & 2 \\ 3 & 0 & -1 & 3 & 0 & -1 \\ 1 & 1 & 1 & -1 & -1 & -1 \\ 2 & -1 & 2 & -2 & 1 & -2 \\ 3 & 0 & -1 & -3 & 0 & 1 \end{pmatrix} = T_h.
$$

This way the irreducible representations of T_h in Table 17.10 were obtained. This solution, however, is not unique, and other equivalent forms are found in the literature. For more information, see, for example, Dresselhaus *et al.* (2002). For similar reasons, we have included in the tables the irreducible representations of both O and O_h.

8.6 Generators of the crystal groups

We recall that any group can be generated by a minimal subset of its elements by the group operations. These constitute a set of **generators** of the group. Thus, the number of generators for each group is unique, but not necessarily the set of the specific elements. One can consider the elements of a faithful representation of the group. The dimension of the representation may be arbitrary, but in practice, a 3×3 representation is found to be convenient. A 2×2 representation can be extended as we discussed in Section 4.4. The reader is encouraged to find the order of each group, Table 8.1, and confirm that a set of generators for each group can be selected as given in Tables 17.1–17.3 of the appendix, Chapter 17.

8.7 Problems

8.2.1 Given a plane lattice with lattice vectors a_1 a_2, $a_1 \neq a_2$ $\phi \neq \pi/2$:

  (i) Show that the point and translation symmetries are, respectively, $G_0 = \{E, C_2\}$ and $\{T_{a_1}, T_{a_2}\}$.

  (ii) Find the space symmetry.

  (iii) Find sub-symmetries, if any.

8.2.2 The point symmetry of a shape in a plane are $G_0 = C_{2v} = \{E, C_2, \sigma_x, \sigma_y\}$, where C_2 is a rotation around an axis perpendicular to the plane indicated as x, y.

  (i) Find the multiplication table of the corresponding space group.

  (ii) Is this group simple?

8.2.3 Do the same for the plane square lattice.

8.2.4 Do the same for the centered square lattice.

8.2.5 Do the same for the regular hexagonal plane lattice $\alpha = \pi/3$.

8.4.1 In some of the character tables given in Section 8.5, the relevant groups can take the form $G = G_I \otimes G_0$ $G_I = \{E, I\}$. Find the symmetry G_0. Find the character table of G from those of G_I and G_0. Check the relevant matrices. Is the result modified if one considers the reduction with respect to $G = G_0 \otimes G_I$?

8.5.1 Some of the groups included in the tables of generators in Section 8.6 have already been studied in various parts of the text. In such cases, construct all the elements of the relevant groups (multiplication table, number and content of classes, irreducible representations, etc.) starting from the generators. Comment on the results.

Chapter 9

Irreducible Representations
of Space Groups

In this chapter, after a careful study of the representations of translation symmetry, we study the irreducible representations of space groups. This will lead us to the Brillouin zone, which is a uniquely defined primitive cell in reciprocal space, k-space. Then, we apply our results to some systems of current practical interest, including graphene.

9.1 Irreducible representations of space groups

Let us, for simplicity, begin with the case of one dimension, i.e.

$$T(b)f(x) = f(x+b) \Rightarrow T(b) = e^{i\frac{bp_x}{\hbar}} \tag{9.1}$$

(see Section 4.7).

Let us suppose that we are dealing with a periodic function:

$$\phi(x+\lambda) = \phi(x) \Leftrightarrow \phi(x) = e^{2\pi i\frac{x}{\lambda}}. \tag{9.2}$$

In this case,

$$T(b) = e^{2\pi i\frac{b}{\lambda}}. \tag{9.3}$$

Indeed,

$$T(b)\phi(x) = e^{2\pi i\frac{b}{\lambda}}e^{2\pi i\frac{x}{\lambda}} = e^{2\pi i\frac{x+b}{\lambda}} = \phi(x+b).$$

The operators $T(b)$ constitute an Abelian group, which is continuous so long as the parameter b is continuous. It can become discrete if b takes

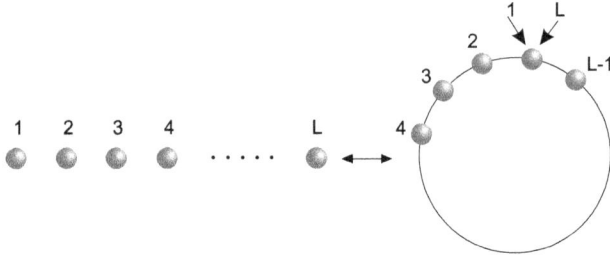

Fig. 9.1. Two topologically equivalent descriptions of a 1D crystal.

discrete values, e.g. $b = na_0$, $n = 0, 1, \ldots$. Then, the elements of the group are

$$T_n(b) = T(na_0) = \left(e^{2\pi i \frac{a_0}{\lambda}}\right)^n, \quad n = 0, 1, \ldots,$$

$$T(ma_0)T(na_0) = T((m+n)a_0).$$

The length a_0 can be chosen to specify the lattice points of the crystal. In reality, however, the crystal is not infinite, but it has a finite extent La_0. Consider now the mapping

$$na_0 \to T(na_0), T(na_0)^L = E \text{ if } \frac{a_0}{\lambda}L = 1 \Rightarrow a_0 = \frac{\lambda}{L}.$$

One, thus, sees that, topologically, the infinite 1D crystal is equivalent to the unit circle (Fig. 9.1). One representation of the group $\Gamma_n^{(m)}(L)$ is obtained by considering a basis characterized by a parameter $\xi = e^{2\pi i \frac{1}{L}}$ as follows:

$$|\xi_n\rangle = \begin{cases} 1, & n = 0, L \\ \xi^n, & n \neq 0, L \end{cases}, \quad n = 0, 1, \ldots, L, \quad \xi = e^{2\pi i \frac{1}{L}}.$$

In practice, one prefers a numbering beginning with 1 instead of 0, whence

$$|\xi_m\rangle = \begin{cases} \xi^L = 1, & m = 1 \\ \xi^{m-1}, & m \neq 1, \quad m = 2, 3, \ldots, L. \end{cases}$$

In this case, one considers as a generator the operator T_n:

$$\Gamma_n^{(m)}(L) = T_n|\xi^m\rangle = (\xi_m)^n \Rightarrow \Gamma_n^{(m)}(L)$$

$$= \begin{cases} 1 & m = 1 \text{ or } n = 1 \\ e^{2\pi i \frac{m-1}{L}(n-1)} & m, n = 2, 3, \ldots, L. \end{cases}$$

With the understanding that the values $n = 1$, $m = 1$ are written first, as it is a common practice in writing the irreducible representations and the elements of the group, we thus simply write

$$\Gamma_n^{(m)}(L) = e^{2\pi i \frac{m}{L} n}, \quad m, n = 1, 2, \ldots, L, \tag{9.4}$$

with m numbering the representations (lines of the matrix) and n the elements of the group (columns of the matrix). We often write

$$\Gamma_a^{(k)} = e^{ika}, \; k = 2\pi m L a_0, \; a = n a_0, \; m, n = 1, \ldots, L, \tag{9.5}$$

where k is the wave number and a the element of the translation.

In crude terms, the wave function describing the system, in the case of a periodic potential, is given by the Bloch relation:

$$\psi(x) = \Gamma_x^{(k)} u(x) = e^{ikx} u(x), \tag{9.6}$$

where $u(x)$ is a periodic function $u(x) = u(x+a)$. This function is obtained according to the rules of quantum mechanics with appropriate boundary conditions.

The above description corresponds to a 1D crystal with lattice length a, whose structure elements, e.g. L positive ions, are found in the positions na. Such systems actually exist, e.g. solid polymers and carbon nanotubes. In some instances, one finds solids with large anisotropy, in which case only one dimension can be considered relevant. In such cases, techniques of line groups, i.e. groups describing the symmetry of systems exhibiting translational periodicity in 1D, can be applied, leading to a simplification of such problems (Dresselhaus *et al.*, 2002). This book goes into a more complete treatment of the material discussed in this chapter, with many useful applications, in particular to solid-state physics.

As mentioned above, some 3D crystals can be highly anisotropic, as for example chain-type crystals which have line groups as subgroups of their space group. Whenever only one direction is relevant for some physical properties of a 3D system, one can expect to derive useful information by applying suitable line group approaches. The advantage of using line groups is their simplicity.

The generalization of Eq. (9.4) in the case of a 3D crystal is more or less obvious:

$$\Gamma_{n_1, n_2, n_3}^{(m_1, m_2, m_3)}(L_1, L_2, L_3) = e^{2\pi i \left(\frac{m_1}{L_1} n_1 + \frac{m_2}{L_2} n_2 + \frac{m_3}{L_3} n_3 \right)},$$

$$m_i, n_i = 1, 2, \ldots, L_i, \quad i = 1, 2, 3, \tag{9.7}$$

where (m_1, m_2, m_3) characterize the representation with the translational elements being

$$\mathbf{a} = n_1\mathbf{a}_1 + n_2\mathbf{a}_2 + n_3\mathbf{a}_3, \ \mathbf{a}_1, \mathbf{a}_2, \mathbf{a}_3 \text{ are lattice vectors,}$$

and $L_1\mathbf{a}_1, L_2\mathbf{a}_2, L_3\mathbf{a}_3$ the crystal dimensions. The space covered by these vectors defines a Bravais lattice.

One might try to generalize the above by writing the analog of Eq. (9.5) in 3D as

$$\Gamma_\mathbf{a}^{(\mathbf{k})} = e^{i\mathbf{k}\cdot\mathbf{a}}, \ \mathbf{k} = 2\pi\left(\frac{m_1}{L_1}\mathbf{a}_1, \frac{m_2}{L_2}\mathbf{a}_2, \frac{m_3}{L_3}\mathbf{a}_3\right), \quad \mathbf{a} = (n_1\mathbf{a}_1, n_2\mathbf{a}_2, n_3\mathbf{a}_3),$$

$$\mathbf{k}\cdot\mathbf{a} = 2\pi\left(\frac{m_1}{L_1}n_1 + \frac{m_2}{L_2}n_2 + \frac{m_3}{L_3}n_3\right). \tag{9.8}$$

This, however, is only true in the case of orthogonal lattice vectors. In the general case, we define the vectors $\mathbf{b}_1, \mathbf{b}_2, \mathbf{b}_3$, which are orthogonal to the vectors $\mathbf{a}_1, \mathbf{a}_2, \mathbf{a}_3$ such that $\mathbf{b}_i \cdot \mathbf{a}_i = 2\pi\delta_{ij}$. A convenient choice is

$$\mathbf{b}_1 = 2\pi\frac{\mathbf{a}_2 \otimes \mathbf{a}_3}{\mathbf{a}_1 \cdot (\mathbf{a}_2 \otimes \mathbf{a}_3)}, \quad \mathbf{b}_2 = 2\pi\frac{\mathbf{a}_3 \otimes \mathbf{a}_1}{\mathbf{a}_2 \cdot (\mathbf{a}_3 \otimes \mathbf{a}_1)}, \quad \mathbf{b}_3 = 2\pi\frac{\mathbf{a}_1 \otimes \mathbf{a}_2}{\mathbf{a}_3 \cdot (\mathbf{a}_1 \otimes \mathbf{a}_2)}. \tag{9.9}$$

Let us now define

$$\mathbf{k} = \left(\frac{m_1}{L_1}\mathbf{b}_1, \frac{m_2}{L_2}\mathbf{b}_2, \frac{m_3}{L_3}\mathbf{b}_3\right), \tag{9.10}$$

Then, provided that $\mathbf{a}$ is a lattice vector, one obtains

$$\Gamma_\mathbf{a}^{(\mathbf{k})} = e^{i\mathbf{k}\cdot\mathbf{a}}, \ \mathbf{k} = \left(\frac{m_1}{L_1}\mathbf{b}_1, \frac{m_2}{L_2}\mathbf{b}_2, \frac{m_3}{L_3}\mathbf{b}_3\right), \quad \mathbf{a} = (n_1\mathbf{a}_1, n_2\mathbf{a}_2, n_3\mathbf{a}_3),$$

$$\mathbf{k}\cdot\mathbf{a} = 2\pi\left(\frac{m_1}{L_1}n_1 + \frac{m_2}{L_2}n_2 + \frac{m_3}{L_3}n_3\right). \tag{9.11}$$

We must stress that the vector $\mathbf{k}$ must be chosen to satisfy the last relation of Eq. (9.11). The relations (9.9) are a convenient choice valid in 3D.

In 2D, a convenient choice is

$$\mathbf{a}_1 = (a_{1x}, a_{1y}), \quad \mathbf{a}_2 = (a_{2x}, a_{2y}) \Rightarrow$$

$$\mathbf{b}_1 = \frac{2\pi}{a_{1x}a_{2y} - a_{1y}a_{2x}}(a_{2y}, -a_{2x}), \quad \mathbf{b}_2 = \frac{2\pi}{a_{1x}a_{2y} - a_{1y}a_{2x}}(-a_{1y}, a_{1x}). \tag{9.12}$$

For given lattice symmetries, some lattice vectors are given in Table 9.1.

Table 9.1. The basic lattice vectors $\mathbf{a}_1$, $\mathbf{a}_2$ and $\mathbf{b}_1$, $\mathbf{b}_2$ in the inverse lattice space in 2D.

Lattice	$\mathbf{a}_1$	$\mathbf{a}_2$	$\mathbf{b}_1$	$\mathbf{b}_2$
Elongated; p	$a_1(1,0)$	$a_2(\cos\theta, \sin\theta)$	$\frac{2\pi}{a_1}(1, -\cot\theta)$	$\frac{2\pi}{a_2}(0, \frac{1}{\sin\theta})$
Orthogonal; p	$a_1(1,0)$	$a_2(0,1)$	$\frac{2\pi}{a_1}(1,0)$	$\frac{2\pi}{a_2}(0,1)$
Orthogonal; c	$\frac{1}{2}(a_1,a_2))$	$\frac{1}{2}(-a_1,a_2))$	$2\pi(\frac{1}{a_1}, \frac{1}{a_2})$	$2\pi(-\frac{1}{a_1}, \frac{1}{a_2})$
Square; p	$a(1,0))$	$a(0,1))$	$\frac{2\pi}{a}(1,0)$	$\frac{2\pi}{a}(0,1)$
Hexagon; p	$a(0,-1))$	$a(\frac{\sqrt{3}}{2}, \frac{1}{2})$	$\frac{2\pi}{a}(\frac{1}{\sqrt{3}}, -1)$	$\frac{2\pi}{a}(\frac{2}{\sqrt{3}}, 0)$

The vectors $\mathbf{b}_i$, no matter how they are defined, provided that they are orthogonal to the lattice vectors $\mathbf{a}_i$, constitute the **inverse lattice**. Two such vectors $\mathbf{k}$ and $\mathbf{k}'$ are **equivalent** if they differ by a vector $\mathbf{k}_0$ such that

$$\mathbf{k}' - \mathbf{k} = \mathbf{k}_0, \quad \mathbf{k}_0 = \ell_1\mathbf{b}_1 + \ell_2\mathbf{b}_2 + \ell_3\mathbf{b}_3, \quad \ell_1, \ell_2, \ell_3 \text{ are integers.} \qquad (9.13)$$

This can be seen by observing that

$$e^{i\mathbf{k}'\cdot\mathbf{a}} = e^{i(\mathbf{k}+\mathbf{k}_0)\cdot\mathbf{a}} = e^{i\mathbf{k}\cdot\mathbf{a}}e^{i\sum_i n_i\mathbf{a}_i\cdot\sum_j \ell_j\mathbf{b}_j} = e^{i\mathbf{k}\cdot\mathbf{a}}e^{2\pi i\sum_i n_i\ell_i} = e^{i\mathbf{k}\cdot\mathbf{a}}.$$

This means that in this case, there exists an invertible transformation S such that

$$\Gamma_a^{(\mathbf{k}')} = S^{-1}\Gamma_a^{(\mathbf{k})}S \Leftrightarrow S = e^{\pi i\left(\frac{\ell_1}{L_1}n_1 + \frac{\ell_2}{L_2}n_2 + \frac{\ell_3}{L_3}n_3\right)}. \qquad (9.14)$$

In such a case, the representations $\Gamma_a^{(\mathbf{k}')}$ and $\Gamma_a^{(\mathbf{k})}$ are equivalent. There exists a region of the vector $\mathbf{k}$ which determines completely the irreducible representations of the translation group. Thus, it is possible and convenient to consider a simply connected region in the space of the inverse lattice containing the origin. Any such choice must satisfy the following conditions:

(i) The region does not contain two equivalent vectors.
(ii) Every vector of the inverse lattice has an equivalent vector in the region.

In such a case, the region is a reduced Brillouin zone.
    We now note the following:

- The transformation given by Eq. (9.9) determines in the inverse lattice space a group G_0', isomorphic to G_0 of the Bravais lattice. Indeed, if R is any element of G_0, the following are true:

$$|a_i'\rangle = R|a_i\rangle, \ |b_i'\rangle = R|b_i\rangle \Rightarrow \langle a_i'|b_i'\rangle = \langle a_i|S^+S|b_i\rangle = \langle a_i|b_i\rangle = 2\pi\delta_{ij}.$$

- If **a** is a vector of the Bravais lattice, $R\mathbf{a}$ is a vector of the same lattice. Then, however, one finds
- If **k** is a vector of the inverse lattice, $R^{-1}\mathbf{k}$ is also a vector of the inverse lattice, i.e. it can be cast in the form of Eq. (9.10). Indeed,

$$\mathbf{k}' = R^{-1}\mathbf{k} = \sum_{i,j} \frac{m_i}{L_i}(\rho)^{-1}_{ji}\mathbf{b}_j = \sum_{i,j} \frac{m_i}{L_i}(\rho')_{ji}\mathbf{b}_j.$$

$(\rho')_{ji}$ are integers (recall the discussion leading to Eq. (8.10)). Thus,

$$\sum_i \frac{m_i}{L_i}(\rho')_{ji} = \text{rational} \Rightarrow \frac{m_j'}{L_j} = \text{rational} \Rightarrow \mathbf{k}' = \sum_j \frac{m_j'}{L_j}\mathbf{b}_j',$$

i.e. it is of the desired form, Eq. (9.10).

We can therefore summarize by saying that, as R runs through the elements of G_0, the operator R^{-1} goes over the elements of the group G_0', which is isomorphic to G_0. The two lattices give an equivalent description of the properties of the crystal.

9.2   The group of vector k of the reduced Brillouin zone

Let us suppose that we have found an irreducible representation of the space symmetry group, say, $\Gamma^i(X), X \in G$. Clearly, this is reducible if we limit ourselves to the translation subgroup $T(\mathbf{a})$. Furthermore, since this is Abelian, a basis can be found such that the representation $\Gamma^i(X)$ is diagonal whenever X is an element of translation, $X = T(\mathbf{a})$. This basis consists of the orthogonal vectors, which define the representation $\Gamma_{\mathbf{a}}^{(k)}$ of the previous section. Let us denote these vectors as $|e(\mathbf{k})\rangle$, where **k** is a vector in the Brillouin zone.

According to what we have mentioned above, $|e(\mathbf{k})\rangle$ are eigenvectors of the translation operators, and more specifically,

$$T(\mathbf{a})|e(\mathbf{k})\rangle = e^{i\mathbf{k}\cdot\mathbf{a}}|e(\mathbf{k})\rangle$$

(see Eq. (9.11)). Let us now consider the element $X \in G$, i.e.

$$X = T(\mathbf{a})T(\mathbf{b})R, \ R \in G_0. \tag{9.15}$$

Acting with this operator on the vector $|e(\mathbf{k})\rangle$, we get

$$|e(\mathbf{k})\rangle \Rightarrow |\mathbf{q}'\rangle = T(\mathbf{a})T(\mathbf{b})R|e(\mathbf{k})\rangle. \tag{9.16}$$

As a result, for a lattice vector $\mathbf{a}'$, we get

$$T(\mathbf{a}')|\mathbf{q}'\rangle = T(\mathbf{a}')T(\mathbf{a})T(\mathbf{b})R|e(\mathbf{k})\rangle = T(\mathbf{a})T(\mathbf{b})T(\mathbf{a}')R|e(\mathbf{k})\rangle.$$

But $T(\mathbf{a}')R = RT(R^{-1}\mathbf{a}')$ (see Section 4.7), hence

$$T(\mathbf{a}')|\mathbf{q}'\rangle = T(\mathbf{a})T(\mathbf{b})RT(R^{-1}\mathbf{a}')|e(\mathbf{k})\rangle = T(\mathbf{a})T(\mathbf{b})Re^{i\mathbf{k}\cdot R^{-1}\mathbf{a}'}|e(\mathbf{k})\rangle$$

$$= e^{i(R\mathbf{k})\cdot\mathbf{a}'}T(\mathbf{a})T(\mathbf{b})|e(\mathbf{k})\rangle = e^{i(R\mathbf{k})\cdot\mathbf{a}'}|\mathbf{q}'\rangle.$$

In other words, $|\mathbf{q}'\rangle$ is an eigenvector of the translation operator with eigenvalue $e^{i(R\mathbf{k})\cdot\mathbf{a}'}$. Therefore, it is nothing but a unit vector $|e(R\mathbf{k})\rangle$. We thus have

$$e(\mathbf{k}) \overset{T(\mathbf{a})T(\mathbf{b})R}{\longrightarrow} e(R\mathbf{k}).$$

This means that, if $e(\mathbf{k})$ is considered as a basis vector, one must also include the basis vector $e(R\mathbf{k})$, $R \in G_0$, as well.

Assume now that as R is moving in G_0, there exist at most ℓ elements $R_1, R_2, \ldots, R_\ell$, which yield the non-equivalent vectors

$$\mathbf{k}_i = R_i\mathbf{k}, \quad i = 1, 2, \ell.$$

The set of such vectors is called the **star of the vector k** of order ℓ. Consider now the set X of the elements of G, which leave the vector $\mathbf{k}$ invariant in the sense that

$$T(\mathbf{a})T(\mathbf{b})|e(\mathbf{k})\rangle = |e(\mathbf{k})\rangle. \tag{9.17}$$

This set constitutes a group $H(\mathbf{k})$, which is a subgroup of G. The group $H(\mathbf{k})$ contains all the translation operators. In case non-proper translation transformations do not exist, i.e. if $\mathbf{b} = 0$, this group will also contain some of the elements of G_0 and their product. This group is called the **group of the vector k**. Whenever the end of the vector $\mathbf{k}$ is found on the surface, the group $H(\mathbf{k})$ will contain as elements the transformations changing $\mathbf{k}$ to another vector equivalent to it.

We show that the group $H(\mathbf{k})$ resolves the group G in cosets, i.e. its left and right cosets satisfying

$$G = \{g_1H(\mathbf{k}), g_2H(\mathbf{k}), \ldots, g_\ell H(\mathbf{k}), g_1 = E, g_1H(\mathbf{k}) = H(\mathbf{k})\}. \tag{9.18}$$

It is clear that the elements $g_2, g_3, \ldots, g_\ell$ do not leave the vector $\mathbf{k}$ invariant, since the elements $g_i H(\mathbf{k})$, $i \neq 1$ would then be in $H(\mathbf{k})$. We now show that, if $g_i \neq g_j$, we must have $g_i \mathbf{k} \neq g_j \mathbf{k}$. This is a consequence of the fact that the relation $g_i \mathbf{k} = g_j \mathbf{k}$, $g_i \neq g_j$, leads to a contradiction. Indeed, in this case,

$$g_i \mathbf{k} = g_j \mathbf{k} \Rightarrow g_j^{-1} g_i \mathbf{k} = \mathbf{k} \Rightarrow g_j^{-1} g_i H(\mathbf{k}) = H(\mathbf{k}) \Rightarrow g_i H(\mathbf{k}) = g_j H(\mathbf{k}).$$

This, however, is absurd since two different cosets cannot have common elements (see Theorem 2 of Chapter 3). Thus, the relation given by Eq. (9.18) holds. This will be used later for the construction of the irreducible representations of G.

The star of $\mathbf{k}$ contains the elements

$$\mathbf{k}_i = g_i \mathbf{k}, \quad i = 1, 2, \ldots, \ell, \quad g_1 = E.$$

Let us now consider one of the vectors of the above star, e.g. $\mathbf{k}_\alpha$. Then, $H(\mathbf{k}_\alpha)$ will consist of the elements

$$Y \mathbf{k}_\alpha = \mathbf{k}_\alpha, \quad Y \in G,$$

$$Y g_\alpha \mathbf{k} = g_\alpha \mathbf{k} \Rightarrow g_\alpha^{-1} Y g_\alpha \mathbf{k} = \mathbf{k} \Rightarrow g_\alpha^{-1} Y g_\alpha \in H(\mathbf{k}) \Rightarrow H(\mathbf{k}_\alpha) = g_\alpha^{-1} H(\mathbf{k}) g_\alpha.$$

This means that the groups $H(\mathbf{k})$ and $H(\mathbf{k}_\alpha)$ are isomorphic. In other words, any element of the star can generate all the rest.

Example 1: Let us consider a crystal of the triclinic system with symmetry $S_2 = \{E, I\}$.

In this case, we have three different lattice lengths $\mathbf{a}_1, \mathbf{a}_2, \mathbf{a}_3$, while the inverse lattice is specified by the lengths given by Eq. (9.9) and the wave vector via the relation (9.11). We have a star of two elements $\mathbf{k}_1 = E\mathbf{k} = \mathbf{k}$, $\mathbf{k}_2 = I\mathbf{k} = -\mathbf{k}$. Hence, $H(\mathbf{k}) = \{T_\mathbf{a}\}$, and the group G contains the elements $\{T_\mathbf{a}\}, \{I T_\mathbf{a}\}$.

Example 2: We consider a crystal with a square lattice. The symmetry of the crystal is exhibited in Fig. 9.2(a'). In this case,

$$a_1 = a_2 = a, a_3 = b, \quad \mathbf{b}_1 = \frac{2\pi}{a} \hat{e}_1, \quad \mathbf{b}_2 = \frac{2\pi}{a} \hat{e}_2, \quad \mathbf{b}_3 = \frac{2\pi}{b} \hat{e}_3,$$

$$\mathbf{k} = \frac{2\pi}{a} \left(\frac{m_1}{L_1} \hat{e}_1 + \frac{m_2}{L_2} \hat{e}_2 \right) + \frac{2\pi}{b} \frac{m_3}{L_3} \hat{e}_3.$$

To find the stars, it is adequate to begin with vectors found in the first one-eighth of the square, e.g. those with ending points A, B, C, as shown

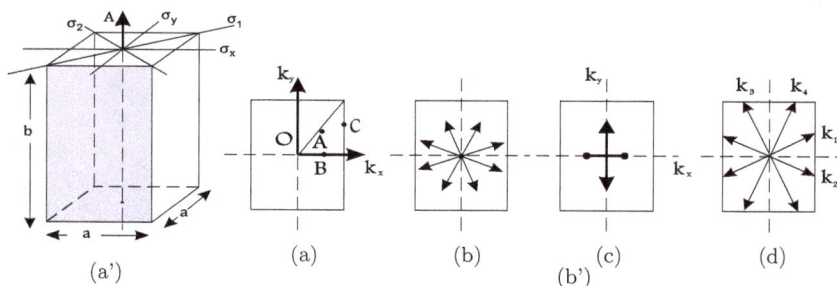

Fig. 9.2. (a') The elements of symmetry of the square lattice, a fourth-order axis and four symmetry planes containing the symmetry axis (group D_4). (b') The possible stars corresponding to the vectors with tops at the points A, B, C, as defined in panel (a), i.e. $(\vec{OA})$ in (b), $(\vec{OB})$ in (c) and $(\vec{OC})$ in (d).

in Fig. 9.2(b')(a). Starting with $(\vec{OA})$, we find the star of 9.2(b')(b), and similarly, from $(\vec{OB})$, we find that of 9.2(b')(c). Finally, beginning with $(\vec{OC})$, we find the star of 9.2(b')(d). We note that in this case, C lies on the Brillouin zone, and as a result, the star contains four vectors (those vectors that are not numbered are equivalent to the rest).

In the case of Fig. 9.2(b')(b), the only element of the group which leaves the vector $(\vec{OA})$ invariant is the identity element. Thus, $H(\mathbf{k}) = \{\mathbf{T}(\mathbf{a})\}$, that is, we have to deal with the translation of the lattice.

In the case of Fig. 9.2(b')(c), the vector $(\vec{OB})$ remains invariant under the two elements E, σ_x, and as a result, $H(\mathbf{k}) = \{\mathbf{T}(\mathbf{a}), \mathbf{T}(\mathbf{a})\sigma_x\}$. In the case of the star of Fig. 9.2(b')(d), only the identity element leaves the vector $(\vec{OC})$ invariant, that is, in this case, $H(\mathbf{k}) = \{\mathbf{T}(\mathbf{a})\}$.

9.3 Construction of the irreducible representations of the space group

In this section, we are going to construct the representations of space group symmetry,[1] supposing that we have already constructed the representations of the translation and point group symmetries separately.

First of all, since the elements of the translation group commute with each other, they can be separately diagonalized with a suitable choice of the basis. Indeed, in the basis $|e(\mathbf{k})\rangle$ we studied in the previous section,

---

[1]For further study and more realistic applications, the reader may refer to the more advanced book by Chen *et al.* (2002).

the representation of the group $T(\mathbf{a})$ is $\Gamma_{\mathbf{a}}^{(\mathbf{k})}$. Suppose now that $\sigma(\mathbf{k})$ is the space covered by this basis. Then, in this basis, any translation can be represented as follows:

$$\Gamma(T_{\mathbf{d}}) = e^{i\mathbf{k}\cdot\mathbf{d}}(\epsilon), \quad (\epsilon)_{ij} = \delta_{ij}, \quad i,j = 1,2,\ldots,n_d. \tag{9.19}$$

The representations $H(\mathbf{k})$ for which the relation (9.19) holds are called **regular**. We consider now all the elements of $H(\mathbf{k})$, which acting on an element of $\sigma(\mathbf{k})$, give another element belonging to $\sigma(\mathbf{k})$, that is, $\sigma(\mathbf{k})$ is an invariant subspace with respect to the elements of $H(\mathbf{k})$. Thus, the elements of $H(\mathbf{k})$ are represented in the space $\sigma(\mathbf{k})$. Consider now one of the remaining vectors $\mathbf{k}_i$ of the star of $\mathbf{k}$. The corresponding space $\sigma(\mathbf{k}_i)$ is found with the aid of g_i, as we have seen in the previous section. Equivalent representations $H(\mathbf{k}_i)$ are realized in every subspace $\sigma(\mathbf{k}_i)$. We show that, so long as the representation $\Gamma_{\mathbf{a}}^{(\mathbf{k})}$ is irreducible, the irreducible representations of $H(\mathbf{k}_i)$ are realized in the subspace $\sigma(\mathbf{k}_i)$.

Let us suppose that this is true, that is, there exists a subspace $\sigma(\mathbf{k}')$ which remains invariant under the action of the elements of $H(\mathbf{k}_i)$. Then, from this, we construct the subspaces

$$\sigma(\mathbf{k}_i') = g_i\sigma(\mathbf{k}'), \quad i = 1,2,\ldots,\ell \quad \text{and} \quad \sigma' = \sigma(\mathbf{k}_1') \oplus \sigma(\mathbf{k}_2') \oplus \cdots \oplus \sigma(\mathbf{k}_\ell').$$

Then, the space σ' is invariant with respect to all the elements X of G. Indeed, every element $X \in G$ must be of the form

$$X = g_i h, \ h \in H(\mathbf{k}).$$

In this case, however,

$$g_j h g_i H(\mathbf{k}) = g_j h g_i \sigma'(\mathbf{k}) = g_j g_i g_i^{-1} h g_i \sigma'(\mathbf{k}) = g_j g_i g_h' \sigma'(\mathbf{k}) = g_\ell h' \sigma'(\mathbf{k})$$
$$\Rightarrow g_j h \sigma'(\mathbf{k}_i) = g_\ell \sigma'(\mathbf{k})) \Rightarrow g_j h \sigma'(\mathbf{k}_i) = \sigma'(\mathbf{k}_\ell).$$

This means that, under the action of the element $X \in G$, the subspace $\sigma'(\mathbf{k}_i)$ is transformed to another one $\sigma'(\mathbf{k}_\ell)$ similar to it, that is,

$$X\sigma' \subset \sigma'.$$

The space σ' is, however, a subspace of the space in which $\Gamma_{\mathbf{a}}^{(\mathbf{k})}$ is assumed to be irreducible. This means that σ' cannot be a proper subspace of $\sigma(\mathbf{k})$.

In other words,

$$\sigma = \sigma(\mathbf{k}_1) \oplus \sigma(\mathbf{k}_2) \oplus \cdots \oplus \sigma(\mathbf{k}_\ell). \tag{9.20}$$

Proceeding the same way, we find

$$q_j \sigma(\mathbf{k}_i) = \begin{cases} \sigma(\mathbf{k}_\ell) & q_j g_i = g_\ell, q_j \notin H(\mathbf{k}) \\ \sigma(\mathbf{k}_i) & q_j \in H(\mathbf{k}). \end{cases}$$

Obviously,

$$q_j = g_j \notin H(\mathbf{k}) \Rightarrow g_\ell \neq g_i \Rightarrow \sigma(\mathbf{k}_i) \text{ different from } \sigma(\mathbf{k}_\ell).$$

Necessarily, therefore, only the elements of $H(\mathbf{k})$ are encountered in the spaces $\sigma(\mathbf{k}_i)$. Consequently, the representations of G are specified by the star of the vector and one irreducible representation $\Gamma^{(\alpha)}$ of the group $H(\mathbf{k})$. The resulting representation $\Gamma^{(\mathbf{k},\alpha)}$ has dimension ℓn_α and takes the form

$$\Gamma^{(\mathbf{k},\alpha)} = \Gamma^{(\mathbf{k})}(g_i) \otimes \Gamma^{(\alpha)}(h), \, h \in H(\mathbf{k}), X = g_i h. \tag{9.21}$$

For its construction, we distinguish two cases:

(i) The group does not have improper translation transformations.

In this case, it is adequate to construct the representations $T_a \, R \in G_0(\mathbf{k}) \subset G_0$, where

$$G_0(\mathbf{k}) \equiv \{R \in G_0, R\mathbf{k} = \mathbf{k}\} \tag{9.22}$$

and, by definition of subspace $\sigma(\mathbf{k})$,

$$\Gamma^{(\alpha)}(T(\mathbf{a})) = e^{i\mathbf{k}\cdot\mathbf{a}}(\epsilon), \, (\epsilon) \text{ diagonal } n_\alpha \times n_\alpha \text{ matrix.} \tag{9.23}$$

Let us now consider the element $X = T_{\mathbf{a}} h$, $h \in G_0(\mathbf{k})$. Then, since all the elements of $\sigma(\mathbf{k})$ are eigenstates of the translation operator $T_{\mathbf{a}}$, which is represented by a matrix multiple of the identity, the representation $\Gamma(X)$ is irreducible if and only if $\Gamma^{(\alpha)}(h)$ is irreducible, i.e.

$$\Gamma^{(\mathbf{k},\alpha)} = e^{i\mathbf{k}\cdot\mathbf{a}}\Gamma^{(\alpha)}(h) \, h \in G_0(\mathbf{k}). \tag{9.24}$$

(ii) The group contains improper translation transformations.

In this case, one encounters a difficulty stemming from the fact that the operators of the form $T_{\mathbf{b}} R$ do not constitute a group, in spite of the fact that the product of two such corresponds to a translation $T_{\mathbf{a}}$. We now distinguish two cases:

(a) The vector $\mathbf{k}$ is found inside the Brillouin zone, as, for example, in Figs. 9.2(b')b and 9.2(b')c.

In this case, the irreducible representations can be constructed as before via the group $G_0(\mathbf{k})$, since one can find a one-to-one correspondence between this group and $H(\mathbf{k})$.

Indeed, let us consider an element $h = T(\mathbf{a})T(\mathbf{b})R$ of $H(\mathbf{k})$ such that $R\mathbf{k} = \mathbf{k}, R \in G_0(\mathbf{k})$, which is characterized by the representation $\Gamma(h)$. In this case, the matrix

$$\Gamma'(R) = \Gamma(h)e^{-i\mathbf{k}\cdot(\mathbf{a}+\mathbf{b})}$$

is a representation of $G_0(\mathbf{k})$. Indeed,

$$\Gamma'(R_1)\Gamma'(R_2) = \Gamma(h_1)\Gamma(h_2)e^{-i\mathbf{k}\cdot(\mathbf{a_1}+\mathbf{b_1}+\mathbf{a_2}+\mathbf{b_2})}. \tag{9.25}$$

But

$$\begin{aligned}\Gamma(h_1)\Gamma(h_2) &= \Gamma(T(\mathbf{a_1}))\Gamma(T(\mathbf{b_1}))R_1\Gamma(T(\mathbf{a_2}))\Gamma(T(\mathbf{b_2}))R_2 \\ &= \Gamma(T(\mathbf{a_1}))\Gamma(T(\mathbf{b_1}))R_1T(R_1\mathbf{a_2})(R_1\mathbf{b_2})R_1R_2 \\ &\Rightarrow \Gamma(h_1)\Gamma(h_2) = \Gamma'(R_1R_2)e^{i\mathbf{k}\cdot(\mathbf{a_1}+\mathbf{b_1}+\mathbf{a_2}+\mathbf{b_2})}. \tag{9.26}\end{aligned}$$

and putting expression (9.26) into (9.25), we obtain

$$\Gamma'(R_1)\Gamma'(R_2) = \Gamma'(R_1R_2)e^{i\mathbf{k}\cdot(R_1\mathbf{a_2}+R_1\mathbf{b_2}-\mathbf{a_2}-\mathbf{b_2})}. \tag{9.27}$$

Since, however, the transformation R_1 is orthogonal, one finds that

$$\mathbf{k}\cdot(R_1\mathbf{a_2} + R_1\mathbf{b_2} - \mathbf{a_2} - \mathbf{b_2}) = (R^{-1}\mathbf{k} - \mathbf{k})\cdot(\mathbf{a_2} + \mathbf{b_2}).$$

Furthermore, since $\mathbf{k}$ lies inside the Brillouin zone and $R_1 \in G_0$, it follows that

$$R_1\mathbf{k} = \mathbf{k}, \quad R_1^{-1}\mathbf{k} = \mathbf{k}.$$

Hence,

$$\Gamma'(R_1)\Gamma'(R_2) = \Gamma'(R_1R_2), \quad R_1 \in G_0(\mathbf{k}), \quad R_2 \in G_0(\mathbf{k}). \tag{9.28}$$

With the same procedure, it is found that, if $\Gamma'(R)$ is a representation of the group $G_0(\mathbf{k})$, one can construct a representation of $H(\mathbf{k})$ as follows:

$$\Gamma(h) = e^{-i\mathbf{k}\cdot(\mathbf{a}+\mathbf{b})}, \tag{9.29}$$

(b) The vector $\mathbf{k}$ is found on the Brillouin zone, as, for example, shown in Fig. 9.2(b')d.

This means that the vector $\mathbf{k}$ has its endpoint on the Brillouin zone. As a result, $H(\mathbf{k})$ must necessarily contain transformations of the type

$$\mathbf{k} \overset{H(\mathbf{k})}{\longrightarrow} \mathbf{k} + \mathbf{k}_0, \tag{9.30}$$

where $\mathbf{k}_0$ is a vector of the inverse lattice, see Eq. (9.13). Then, Eq. (9.28) is no longer valid. We now distinguish two possibilities:

Case 1: Suppose there exists a rational number λ such that

$$\mathbf{k} = \lambda\mathbf{k}_0, \quad \lambda = \text{ rational.}$$

Since the translation operators are diagonal, see Eq. (9.19), there exist integers n_1, n_2, n_3 such that

$$\Gamma(T(\mathbf{k}_{0i})^{n_i+1}) = E, \quad i = 1,2,3, \tag{9.31}$$

where $(\mathbf{k}_{01}, \mathbf{k}_{02}, \mathbf{k}_{03})$ is a basis in the lattice space, e.g. $(\mathbf{b}_1, \mathbf{b}_2, \mathbf{b}_3)$ of Eq. (9.9). We now consider the group τ of the elements $T_{\ell_1,\ell_2,\ell_3}(\tau_1, \tau_2, \tau_3)$:

$$T_{\ell_1,\ell_2,\ell_3}(\tau_1, \tau_2, \tau_3) = \tau_1^{\ell_1}\tau_2^{\ell_2}\tau_3^{\ell_3}, \ell_i = 1,2,\ldots,n_i+1, \quad \tau_i^{n_i+1} = e, \quad i = 1,2,3,$$

where e is the identity element of the group.

This group is isomorphic to the translation group $T(\mathbf{a})$ with the correspondence

$$(T(\mathbf{k}_{0i}))^{n_i+\ell+1} \Leftrightarrow \tau_i, \ell = \text{integer,}$$
$$H'(\mathbf{k}) = T_{\ell_1,\ell_2,\ell_3}(\tau_1, \tau_2, \tau_3)T(\mathbf{a})R. \tag{9.32}$$

We now observe that, provided Eq. (9.31) is valid, the regular representations of $H'(\mathbf{k})$, i.e. those for which Eq. (9.19) holds, are isomorphic to those of $H(\mathbf{k})$. It is thus sufficient to construct the representations of $H'(\mathbf{k})$ under the assumption, of course, that the matrices τ_i, $i = 1,2,3$, will be chosen so that they lead to expressions of the form

$$\Gamma(T(\mathbf{a})) = e^{i\mathbf{k}\cdot\mathbf{a}}(\epsilon).$$

Case 2: $\mathbf{k} = \mathbf{k}' + \mathbf{k}''$, $\mathbf{k}' = \lambda\mathbf{k}_0$, $\lambda = $ rational,

where $\mathbf{k}''$ is found inside the Brillouin zone. In this case, we find the smallest integers n_1, n_2, n_3 for which we have

$$\left(e^{i\lambda\mathbf{k}_0.a_j}\right)^{n_j+1} = 1,$$

and we proceed as in case 1, that is, we construct operators τ_i, $i = 1,2,3$, and consider the group $H'(\mathbf{k}')$, which is isomorphic to $H(\mathbf{k})$. One can show that

$$\Gamma'\left(\tau_1^{\ell_1}\tau_2^{\ell_2}\tau_T^{\ell_3}(\mathbf{b})h\right) = \Gamma(h)e^{i\mathbf{k}''\cdot(\mathbf{a}+\mathbf{b})},$$

where Γ' is a representation of $H(\mathbf{k}')$ with the property that Eq. (9.19) is valid for the operators of the translation group.

Example 3: We consider a crystal in the triclinic system.
The relevant star has been discussed in Example 1. Since the star system contains only one star with only two vectors, i.e. $|\mathbf{k}_1\rangle = |\mathbf{k}\rangle$, $|\mathbf{k}_2\rangle = I|\mathbf{k}\rangle = -|\mathbf{k}\rangle$, there exists only one 2D representation. The translation element is a multiple of the identity, and thus, we have

$$\Gamma_{\mathbf{a}}^{(\mathbf{k})} = \begin{pmatrix} e^{i k_1 \cdot a} & 0 \\ 0 & e^{i k_2 \cdot a} \end{pmatrix}.$$

In addition, the group $\{E, I\}$ has two elements, and $H(\mathbf{k})$ contains only the identity element. Thus,

$$\Gamma^{(\mathbf{k})}(h) = \begin{pmatrix} 1 & 0 \\ 0 & 1 \end{pmatrix}, \quad \Gamma^{(\mathbf{k})}(g_1) = \begin{pmatrix} -1 & 0 \\ 0 & -1 \end{pmatrix},$$

$$\Gamma_{\mathbf{a}h}^{(\mathbf{k})} = \begin{pmatrix} e^{i k_1 \cdot a} & 0 \\ 0 & e^{i k_2 \cdot a} \end{pmatrix}, \quad \Gamma_{\mathbf{a}I}^{(\mathbf{k})} = \begin{pmatrix} -e^{i k_1 \cdot a} & 0 \\ 0 & -e^{i k_2 \cdot a} \end{pmatrix}.$$

Example 4: We construct the irreducible representations of the group discussed in Example 2 of the previous section.
     We examine each star separately:

(i) The star of $\vec{OA}$. The only operator which leaves the vector $\mathbf{k}$ invariant is the identity element. Thus, the representations are given by the eight vectors $\mathbf{k}_i$ of the star. Since the operator $\Gamma^{(k)}(T(\mathbf{a}))$ is diagonal, we have

$$\left(\Gamma^{(k)}(T(\mathbf{a}))\right)_{ij} = e^{i\mathbf{k}_i \cdot \mathbf{a}}\delta_{ij}, \quad i, j = 1, 2, \dots, 8.$$

The elements of G have the property of shifting the vectors of the star

$$g_i|k_j\rangle = g_i g_j |k\rangle = g_\ell |k\rangle, \quad g_\ell = g_i g_j,$$

hence

$$(\Gamma(g_i))_{jm} = \begin{cases} 1 & h_m = h_i h_j \\ 0 & \text{otherwise.} \end{cases}$$

This coincides with the regular representation $\Gamma^{\text{reg}}(g_i)$ of the group D_4:

$$\Gamma^{(k)}\left(T(\mathbf{a})g_i\right) = \Gamma^{(k)}\left(T(\mathbf{a})\right)\Gamma^{\text{reg}}(g_i).$$

This is not reducible even though, as we know, $\Gamma^{\text{reg}}(g_i)$ is, due to the presence of translations.

ii) The star of $\vec{OB}$. Now, the operators which leave the vector $\mathbf{k}$ invariant are $\{E, \sigma_y\}$. These constitute the group $H(\mathbf{k})$, isomorphic to C_2, with the representation

$$\Gamma^{(\alpha)}(E) = \begin{pmatrix} 1 & 0 \\ 0 & 1 \end{pmatrix}, \quad \Gamma^{(\alpha)}(\sigma_y) = \begin{pmatrix} 0 & 1 \\ 1 & 0 \end{pmatrix}.$$

The representation of the operators of translation are

$$\Gamma^{(k)}(T(\mathbf{a})) = \text{diag}\left\{e^{i\mathbf{k}_1 \cdot \mathbf{a}}, e^{i\mathbf{k}_2 \cdot \mathbf{a}}, e^{i\mathbf{k}_3 \cdot \mathbf{a}}, e^{i\mathbf{k}_4 \cdot \mathbf{a}}\right\}, \qquad (9.33)$$

where $\mathbf{k}_i$ are the four star vectors. The left cosets of $H(\mathbf{k})$ are

$$C_4 H(\mathbf{k}) = \{C_4, \sigma_2\}, C_4^2 H(\mathbf{k}) = \{C_4^2, \sigma_x\}, C_4^3 H(\mathbf{k}) = \{C_4^3, \sigma_1\},$$

$$\sigma_x H(\mathbf{k}) = \{C_4^2, \sigma_x\}, \sigma_1 H(\mathbf{k}) = \{C_4^3, \sigma_1\}, \sigma_2 H(\mathbf{k}) = \{C_4, \sigma_2\}.$$

We make the selection

$$g_1 = E \text{ (or } \sigma_y), \quad g_2 = C_4 \text{ (or } \sigma_2), \quad g_3 = C_4^2 \text{ (or } \sigma_x), \quad g_4 = C_4^3 \text{ (or } \sigma_1).$$

Hence,

$$g_i|\mathbf{k}_j\rangle = g_i g_j|\mathbf{k}\rangle = g_\ell|\mathbf{k}\rangle = |\mathbf{k}_\ell\rangle \Leftrightarrow g_\ell = g_i g_j.$$

We summarize this action in the following table:

	g_1	g_2	g_3	g_4
k_1	k_1	k_2	k_3	k_4
k_2	k_2	k_3	k_4	k_1
k_3	k_3	k_4	k_1	k_2
k_4	k_4	k_1	k_2	k_3.

The above table must be interpreted as follows: The column corresponding to g_i contains the result $g_j|\mathbf{k}\rangle$, where $|\mathbf{k}_j\rangle$ numbers the rows.

For example,

$$g_2|\mathbf{k}_1\rangle = |\mathbf{k}_2\rangle \Leftrightarrow 1 \text{ in position 1 of column 2,}$$

$$g_2|\mathbf{k}_2\rangle = |\mathbf{k}_3\rangle \Leftrightarrow 1 \text{ in position 3 of column 2, etc.}$$

This way, we find

$$\Gamma^{(\mathbf{k})}(g_1) = \begin{pmatrix} 1 & 0 & 0 & 0 \\ 0 & 1 & 0 & 0 \\ 0 & 0 & 1 & 0 \\ 0 & 0 & 0 & 1 \end{pmatrix}, \quad \Gamma^{(\mathbf{k})}(g_2) = \begin{pmatrix} 0 & 0 & 0 & 1 \\ 1 & 0 & 0 & 0 \\ 0 & 1 & 0 & 0 \\ 0 & 0 & 1 & 0 \end{pmatrix},$$

$$\Gamma^{(\mathbf{k})}(g_3) = \begin{pmatrix} 0 & 0 & 1 & 0 \\ 0 & 0 & 0 & 1 \\ 1 & 0 & 0 & 0 \\ 0 & 1 & 0 & 0 \end{pmatrix}, \quad \Gamma^{(\mathbf{k})}(g_4) = \begin{pmatrix} 0 & 1 & 0 & 0 \\ 0 & 0 & 1 & 0 \\ 0 & 0 & 0 & 1 \\ 1 & 0 & 0 & 0 \end{pmatrix}.$$

The resulting representations are of the form

$$\Gamma^{(\mathbf{k},\alpha)}(X) = \Gamma^{(k)}(g_i) \otimes \Gamma^{(\alpha)}(h) \Leftrightarrow X = g_i h$$

and, more explicitly,

$$\Gamma^{(\mathbf{k},\alpha)}(E) = \Gamma^{(k)}(E) \otimes \Gamma^{(\alpha)}(E), \Gamma^{(\mathbf{k},\alpha)}(C_4) = \Gamma^{(k)}(C_4) \otimes \Gamma^{(\alpha)}(E),$$

$$\Gamma^{(\mathbf{k},\alpha)}(C_4^2) = \Gamma^{(k)}(C_4^2) \otimes \Gamma^{(\alpha)}(E), \Gamma^{(\mathbf{k},\alpha)}(C_4^3) = \Gamma^{(k)}(C_4^3) \otimes \Gamma^{(\alpha)}(E),$$

$$\Gamma^{(\mathbf{k},\alpha)}(\sigma_x) = \Gamma^{(k)}(C_4) \otimes \Gamma^{(\alpha)}(\sigma_y), \Gamma^{(\mathbf{k},\alpha)}(\sigma_y) = \Gamma^{(k)}(E) \otimes \Gamma^{(\alpha)}(\sigma_y),$$

$$\Gamma^{(\mathbf{k},\alpha)}(\sigma_1) = \Gamma^{(k)}(C_4^2) \otimes \Gamma^{(\alpha)}(\sigma_y), \Gamma^{(\mathbf{k},\alpha)}(\sigma_2) = \Gamma^{(k)}(C_4) \otimes \Gamma^{(\alpha)}(\sigma_y).$$

The translation operator is diagonal. Therefore,

$$\Gamma^{(\mathbf{k},\alpha)}(T(\mathbf{a})) = \Gamma^{(k)}(T(\mathbf{a}) \otimes \Gamma^{(\alpha)}(E),$$

where $\Gamma^{(k)}(T(\mathbf{a}))$ is given by Eq. (9.33).
  Thus, finally,

$$\Gamma^{(\mathbf{k},\alpha)}(T(\mathbf{a})g_i h) = \Gamma^{(\mathbf{k},\alpha)}(T(\mathbf{a}))\Gamma^{(\mathbf{k},\alpha)}(g_i h).$$

This 8×8 representation is irreducible due to the presence of translation, even though the representations $\Gamma^{(k,\alpha)}(g_i h)$ are reducible.

(iii) The star of $\vec{OC}$. In this case, the only operator that leaves $\mathbf{k}$ invariant is the identity element. As a result, the group $H(\mathbf{k})$ contains only one element. From Fig. 9.2(b')(d), one gets

$$\mathbf{k}_1 = \mathbf{k}\,(g_1 = E), \quad \mathbf{k}_2 = \sigma_x\mathbf{k}, \quad \mathbf{k}_3 = C_4\mathbf{k}, \quad \mathbf{k}_4 = \sigma_1\mathbf{k}.$$

This action is summarized in the following table:

	$g_1 = E$	$g_2 = \sigma_x$	$g_3 = C_4$	$g_4 = \sigma_1$
k_1	k_1	k_2	k_3	k_4
k_2	k_2	k_3	k_4	k_1
k_3	k_3	$-k_4$	k_1	$-k_2$
k_4	k_4	$-k_3$	$-k_2$	$k_1.$

Note that

$$\sigma_2|\mathbf{k}\rangle = -|\mathbf{k}_4\rangle, \quad C_4^3|\mathbf{k}\rangle = -|\mathbf{k}_3\rangle, \quad \sigma_y|\mathbf{k}\rangle = -|\mathbf{k}_2\rangle.$$

The matrix $\Gamma^{(\mathbf{k})}(T(\mathbf{a}))$ is given by Eq. (9.33) and

$$\Gamma^{(\mathbf{k})}(E) = \begin{pmatrix} 1 & 0 & 0 & 0 \\ 0 & 1 & 0 & 0 \\ 0 & 0 & 1 & 0 \\ 0 & 0 & 0 & 1 \end{pmatrix}, \quad \Gamma^{(k)}(\sigma_x) = \begin{pmatrix} 0 & 1 & 0 & 0 \\ 1 & 0 & 0 & 0 \\ 0 & 0 & 0 & -1 \\ 0 & 0 & -1 & 0 \end{pmatrix},$$

$$\Gamma^{(\mathbf{k})}(C_4) = \begin{pmatrix} 0 & 0 & 1 & 0 \\ 0 & 0 & 0 & -1 \\ 1 & 0 & 0 & 0 \\ 0 & 1 & 0 & 0 \end{pmatrix}, \quad \Gamma^{(k)}(\sigma_1) = \begin{pmatrix} 0 & 0 & 0 & 1 \\ 0 & 0 & 1 & 0 \\ 0 & 1 & 0 & 0 \\ 1 & 0 & 0 & 0 \end{pmatrix}.$$

From the above, making use of the multiplication table of D_4, one obtains the remaining matrices:

$$\Gamma^{(\mathbf{k})}(\sigma_2) = \Gamma^{(k)}(\sigma_x)\Gamma^{(\mathbf{k})}(C_4) = \begin{pmatrix} 0 & 0 & 0 & -1 \\ 0 & 0 & 1 & 0 \\ 0 & -1 & 0 & 0 \\ -1 & 0 & 0 & 0 \end{pmatrix},$$

$$\Gamma^{(\mathbf{k})}(C_4^3) = \Gamma^{(k)}(\sigma_x)\Gamma^{(\mathbf{k})}(\sigma_1) = \begin{pmatrix} 0 & 0 & 1 & 0 \\ 0 & 0 & 0 & 1 \\ -1 & 0 & 0 & 0 \\ 0 & -1 & 0 & 0 \end{pmatrix},$$

$$\Gamma^{(\mathbf{k})}(\sigma_y) = \Gamma^{(k)}(C_4)\Gamma^{(k)}(\sigma_1) = \begin{pmatrix} 0 & 1 & 0 & 0 \\ -1 & 0 & 0 & 0 \\ 0 & 0 & 0 & 1 \\ 0 & 0 & 1 & 0 \end{pmatrix},$$

$$\Gamma^{(\mathbf{k})}(C_4^2) = \Gamma^{(k)}(C_4)\Gamma^{(k)}(C_4) = \begin{pmatrix} 1 & 0 & 0 & 0 \\ 0 & -1 & 0 & 0 \\ 0 & 0 & 1 & 0 \\ 0 & 0 & 0 & -1 \end{pmatrix}.$$

From these, one finally obtains the required 4×4 irreducible matrices:

$$\Gamma^{(\mathbf{k})}(T(\mathbf{a})X) = \Gamma^{(\mathbf{k})}(T(\mathbf{a}))\Gamma^{(\mathbf{k})}(X), \quad X \in D_4.$$

For more information regarding the construction of the irreducible representations of space groups, the reader is referred to more specialized books, e.g. Dresselhaus *et al.* (2002) Chapters 9 and 10. Furthermore, in Appendix C of the same reference, selected tables and figures for 3D space groups in real space and in reciprocal space are presented. For the benefit of the reader, we mention that the real space tables and figures given in the first part of the appendix (Section C.1) pertain mainly to crystallographic information and are used for illustrative purposes in various chapters of this book. The tables which pertain to reciprocal space appear in the second part of the appendix (Section C.2) and are mainly the tables for the group of wave vectors for various high-symmetry points in the Brillouin zone for various cubic space groups and other space groups selected for illustrative purposes.

9.4   Graphene — A remarkable example of a 2D crystal

Graphene can be considered a thin sheet of graphite. It possesses important properties, and its discovery in 2004 by two Russian scientists, Andre Geim and Konstantin Novoselov, of the University of Manchester, was almost immediately recognized as very important and led to them being awarded the Nobel Prize in 2010.

Graphene has the structure of a honeycomb. In order to reproduce this honeycomb structure, we need only repeat the two symmetrically different carbon atoms A and B, as shown in Fig. 9.3(a). In this form, one cannot find the symmetries of the structure, since it does not correspond to a Bravais lattice. On the other hand, if one considers the two neighboring atoms A and B as a unit (shaded), one sees the structure of Fig. 9.3(b)

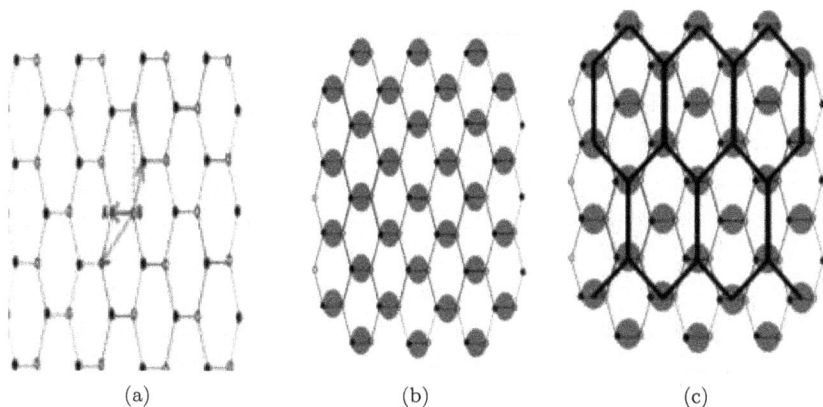

Fig. 9.3. The structure of graphene composed of regular hexagons is shown. At the corners of the hexagons there exist two different carbon atoms A and B as shown in panel (a). The two neighboring atoms A and B are considered a unit and are shaded as shown in panel (b). The shaded units thus constitute a Bravais lattice, as shown in panel (c).

leading to the lattice of Fig. 9.3(c). The lattice symmetry is well known, see the review article by Castro-Neto *et al.* (2019). In Fig. 9.4, one can see the elementary Bravais lattice, the interpenetrating triangular sub-lattices with lattice vectors $\mathbf{a}_1$ and $\mathbf{a}_2$ and the three closest neighbors specified by the unit vectors δ_i, $i = 1, 2, 3$. On the right, the inverse Brillouin lattice is shown, with lattice vectors $\mathbf{b}_1$ and $\mathbf{b}_2$.

The vectors $\mathbf{a}_1$ and $\mathbf{a}_2$ can be written as

$$\mathbf{a}_1 = \frac{a}{2}(3, \sqrt{3}), \quad \mathbf{a}_2 = \frac{a}{2}(3, -\sqrt{3}).$$

The corresponding vectors $\mathbf{b}_1$ and $\mathbf{b}_2$ can be obtained with the methods described in Section 9.1. The result is

$$\mathbf{b}_1 = \frac{2\pi}{3a}(1, \sqrt{3}), \quad \mathbf{b}_2 = \frac{2\pi}{3a}(1, -\sqrt{3})$$

such that

$$\mathbf{a}_i \cdot \mathbf{b}_j = 2\pi \delta_{i,j}.$$

The angle between $\mathbf{a}_1$ and $\mathbf{a}_2$ is $\frac{\pi}{4}$, while between $\mathbf{b}_1$ and $\mathbf{b}_2$ is $\frac{3\pi}{4}$.

The point symmetry of this crystal is C_{6v}. It consists of one symmetry axis of order 6 and six symmetry planes containing this axis and (a) passing through opposite corners of the hexagon or (b) passing through the middle

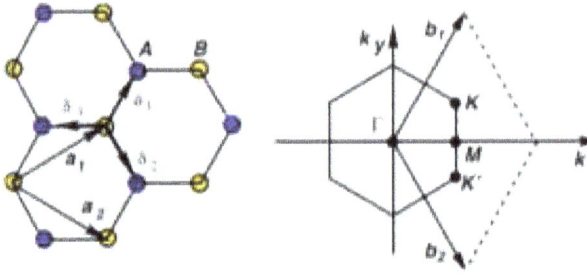

Fig. 9.4. The structure of graphene as given by Castro-Neto *et al.* (2019). On the left, one can see the elementary Bravais lattice, the interpenetrating triangular sub-lattices with lattice vectors $\mathbf{a}_1$ and $\mathbf{a}_2$ and the three closest neighbors specified by the unit vectors δ_i, $i = 1, 2, 3$. On the right, one can see the Brillouin zone and the Dirac points K and K'.

points of the opposite sides of the hexagon. Thus, the system possesses 12 elements and is commonly denoted as $6mm$ in the Hermann–Mauguin notation. It is characterized by four 1D irreducible representations indicated as A_1, A_2, B_1 and B_2 and two 2D ones indicated as E_1 and E_2, see Table 17.9. All its elements can be generated from two generators, see Table 17.3.

9.4.1 *The irreducible representations of the space group connected with point symmetry C_{6v}*

The group C_{6v} is characterized by two generators, one rotation A and one reflection σ_1. In 2 dimensions the 12 elements are represented as:

$$E = A^6 = \begin{pmatrix} 1 & 0 \\ 0 & 1 \end{pmatrix}, \quad A = \begin{pmatrix} \frac{1}{2} & \frac{\sqrt{3}}{2} \\ -\frac{\sqrt{3}}{2} & \frac{1}{2} \end{pmatrix}, \quad A^2 = \begin{pmatrix} -\frac{1}{2} & \frac{\sqrt{3}}{2} \\ -\frac{\sqrt{3}}{2} & -\frac{1}{2} \end{pmatrix},$$

$$A^3 = \begin{pmatrix} -1 & 0 \\ 0 & -1 \end{pmatrix}, \quad A^4 = \begin{pmatrix} -\frac{1}{2} & -\frac{\sqrt{3}}{2} \\ \frac{\sqrt{3}}{2} & -\frac{1}{2} \end{pmatrix}, \quad A^5 = \begin{pmatrix} \frac{1}{2} & -\frac{\sqrt{3}}{2} \\ \frac{\sqrt{3}}{2} & \frac{1}{2} \end{pmatrix},$$

$$\sigma_1 = \begin{pmatrix} 1 & 0 \\ 0 & -1 \end{pmatrix}, \quad \sigma_2 = \begin{pmatrix} \frac{1}{2} & -\frac{\sqrt{3}}{2} \\ -\frac{\sqrt{3}}{2} & -\frac{1}{2} \end{pmatrix}, \quad \sigma_3 = \begin{pmatrix} -\frac{1}{2} & -\frac{\sqrt{3}}{2} \\ -\frac{\sqrt{3}}{2} & \frac{1}{2} \end{pmatrix},$$

$$\sigma_4 = \begin{pmatrix} -1 & 0 \\ 0 & 1 \end{pmatrix}, \quad \sigma_5 = \begin{pmatrix} -\frac{1}{2} & \frac{\sqrt{3}}{2} \\ \frac{\sqrt{3}}{2} & \frac{1}{2} \end{pmatrix}, \quad \sigma_6 = \begin{pmatrix} \frac{1}{2} & \frac{\sqrt{3}}{2} \\ \frac{\sqrt{3}}{2} & -\frac{1}{2} \end{pmatrix}.$$

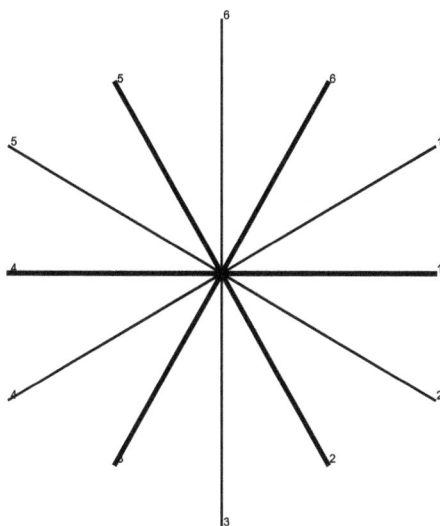

Fig. 9.5. The star of OK (fine line) and OM (thick line). The star of OK' is the same with that of OK, simply rotated by the corresponding angle, $\pi/3$. The index shows the initial position in the sense $1, 2, \ldots, 6$.

Of special interest are the vectors **k** associated with the points K, K' and M, characterized by high symmetry. Proceeding as above, Section 9.3, we find the stars of OK, OK' and OM.

- The star of OK. Acting on OK with group elements, we find the vectors shown in Fig. 9.5 (fine lines). These elements remain invariable under the action of the rotation elements of the group $H = \{A^i, i = 1, \ldots, 6\}$. The left cosets are

$$\sigma_1 H = \sigma_2 H = \sigma_3 H = \sigma_4 H = \sigma_5 H = \sigma_6 H = \{\sigma_1, \sigma_2, \sigma_3, \sigma_4, \sigma_5, \sigma_6\}$$

$$g_1 = E \text{ or } \sigma_6, \quad g_2 = A \text{ or } \sigma_1, \quad g_3 = A^2 \text{ or } \sigma_2, \quad g_4 = A^3 \text{ or } \sigma_3,$$

$$g_5 = A^4 \text{ or } \sigma_4, \quad g_6 = A^5 \text{ or } \sigma_5.$$

Hence, we find

$$\Gamma^{\mathbf{k}_K}(g_1) = \begin{pmatrix} 1 & 1 & 0 & 0 & 0 & 0 \\ 0 & 1 & 1 & 0 & 0 & 0 \\ 0 & 0 & 1 & 1 & 0 & 0 \\ 0 & 0 & 0 & 1 & 1 & 0 \\ 0 & 0 & 0 & 0 & 1 & 1 \\ 1 & 0 & 0 & 0 & 0 & 1 \end{pmatrix}, \quad \Gamma^{\mathbf{k}_K}(g_2) = \begin{pmatrix} 0 & 0 & 0 & 0 & 0 & 1 \\ 1 & 0 & 0 & 0 & 0 & 0 \\ 0 & 1 & 0 & 0 & 0 & 0 \\ 0 & 0 & 1 & 0 & 0 & 0 \\ 0 & 0 & 0 & 1 & 0 & 0 \\ 0 & 0 & 0 & 0 & 1 & 0 \end{pmatrix},$$

$$\Gamma^{\mathbf{k}_K}(g_3) = \begin{pmatrix} 0 & 0 & 0 & 0 & 1 & 0 \\ 0 & 0 & 0 & 0 & 0 & 1 \\ 1 & 0 & 0 & 0 & 0 & 0 \\ 0 & 1 & 0 & 0 & 0 & 0 \\ 0 & 0 & 1 & 0 & 0 & 0 \\ 0 & 0 & 0 & 1 & 0 & 0 \end{pmatrix}, \quad \Gamma^{\mathbf{k}_K}(g_4) = \begin{pmatrix} 0 & 0 & 0 & 1 & 0 & 0 \\ 0 & 0 & 0 & 0 & 1 & 0 \\ 0 & 0 & 0 & 0 & 0 & 1 \\ 1 & 0 & 0 & 0 & 0 & 0 \\ 0 & 1 & 0 & 0 & 0 & 0 \\ 0 & 0 & 1 & 0 & 0 & 0 \end{pmatrix},$$

$$\Gamma^{\mathbf{k}_K}(g_5) = \begin{pmatrix} 0 & 0 & 1 & 0 & 0 & 0 \\ 0 & 0 & 0 & 1 & 0 & 0 \\ 0 & 0 & 0 & 0 & 1 & 0 \\ 0 & 0 & 0 & 0 & 0 & 1 \\ 1 & 0 & 0 & 0 & 0 & 0 \\ 0 & 1 & 0 & 0 & 0 & 0 \end{pmatrix}, \quad \Gamma^{\mathbf{k}_K}(g_6) = \begin{pmatrix} 0 & 1 & 0 & 0 & 0 & 0 \\ 0 & 0 & 1 & 0 & 0 & 0 \\ 0 & 0 & 0 & 1 & 0 & 0 \\ 0 & 0 & 0 & 0 & 1 & 0 \\ 0 & 0 & 0 & 0 & 0 & 1 \\ 1 & 0 & 0 & 0 & 0 & 0 \end{pmatrix}.$$

The space representation is

$$\Gamma^{k,a}(q_i)$$

$$= \Gamma^{\mathbf{k}_K}(g_i) \otimes \mathrm{diag}\left(e^{i(\mathbf{k}_K)_1 a}, e^{i(\mathbf{k}_K)_2 a}, e^{i(\mathbf{k}_K)_3 a}, e^{i(\mathbf{k}_K)_4 a} e^{i(\mathbf{k}_K)_5 a}, e^{i(\mathbf{k}_K)_6 a}\right),$$

where $(\mathbf{k}_K)_i$, $i = 1, 2, \ldots, 6$, are the vectors of the star of OK.
- The star of OM. Proceeding similarly, we find the star of Fig. 9.5 (thick lines). Now,

$$g_1 = E \text{ or } \sigma_1, \quad g_2 = A \text{ or } \sigma_2, \quad g_3 = A^2 \text{ or } \sigma_3, \quad g_4 = A^3 \text{ or } \sigma_4,$$

$$g_5 = A^4 \text{ or } \sigma_5, \quad g_6 = A^5 \text{ or } \sigma_5.$$

Furthermore, $\Gamma^{\mathbf{k}_M}(g_i) = \Gamma^{\mathbf{k}_K}(g_i)$, and the translation element has the same form, except that now, $(\mathbf{k}_M)_i$, $i = 1, 2, \ldots, 6$ are the vectors of the star of OM.
- The star of OK'. It is similar to that of the OK case, except for the fact that the vectors of the star are now rotated by an angle $\pi/3$ in the clockwise direction. Now,

$$g_1 = E \text{ or } \sigma_2, \quad g_2 = A \text{ or } \sigma_3, \quad g_3 = A^2 \text{ or } \sigma_4, \quad g_4 = A^3 \text{ or } \sigma_5,$$

$$g_5 = A^4 \text{ or } \sigma_6, \quad g_6 = A^5 \text{ or } \sigma_1.$$

9.5 Problems

9.1.1 Construct the inverse of a square lattice of length a, and find the reduced Brillouin zone.

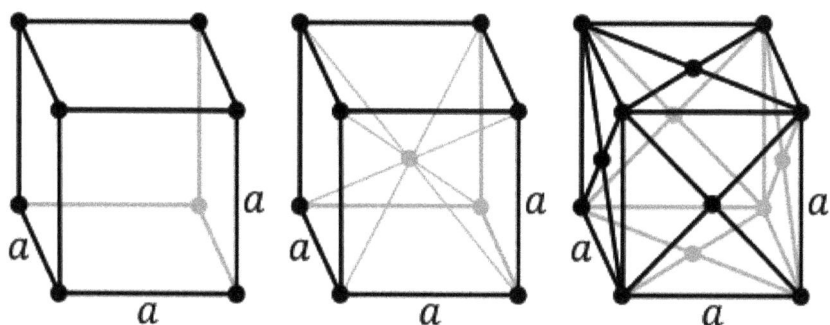

Fig. 9.6. From left to right: simple, space-centered and face-centered cubic lattices.

9.1.2 Do the same for a hexagonal lattice for regular hexagon with a side length of a.

9.1.3 Do the same for a cubic lattice (Fig. 9.6).

9.1.4 Do the same for a space-centered cubic lattice (Fig. 9.6).

9.1.4 Do the same for a face-centered cubic lattice (Fig. 9.6).

9.2.1 Show that, if $\mathbf{k}$ is found inside the reduced Brillouin zone, the same is true for $R\mathbf{k}, R \in G_0$.

Chapter 10

Normal Modes in Crystals

In Chapter 7, we studied the normal modes in the case of molecules. Analogous phenomena can occur in solid crystals. In this case, of course, we have to deal with a large number of constituents of the order of Avogadro's number, which interact in many ways, e.g. via interactions of the Van der Waals type, electrostatic or covalent bond, and depend on the relative distance of the interacting constituents. Clearly, in this case, the problem is more complex than in the case of molecules, so some approximations are called for. In particular, one considers only the interactions between first neighbors or at most between the next to first neighbors. This is assumed even in the case of long-range forces, such as the electrostatic interaction, due to some sort of screening. Often, a harmonic interaction is employed, which we use in this chapter.

10.1 Introduction

Suppose we have a large number of particles of the order of Avogadro's number interacting in a number of ways, e.g. via Van der Waals, electrostatic or covalent bond, which depend on the relative distance of the interacting constituents:

$$V = \sum_{i \neq j} V(\mathbf{x}_i - \mathbf{x}_j). \tag{10.1}$$

Under the influence of interactions of this type, once the system is moved away from its equilibrium position, one will note the appearance of oscillations. Of special interest are those extending over the entire crystal, which, in a similar fashion as in the case of molecules, are called **normal modes**.

Finding the normal modes of a solid of given symmetry is a problem of considerable difficulty, and as in the case of molecules we have discussed in Chapter 7, two approximations are adopted:

(i) Only the forces between the first neighbors or at most between the next to first neighbors are considered important. This assumption is made even in the case of long-range forces, such as the electrostatic force, since interactions between distant particles are weakened due to screening effects.

(ii) The interaction is approximated to be harmonic:

$$V(\mathbf{r}_i - \mathbf{r}_j) \approx \frac{1}{2} C (\mathbf{r}_i - \mathbf{r}_j)^2. \tag{10.2}$$

This is a good approximation, provided the system is not moving too far from equilibrium.

10.2   Normal modes for 1D crystal with one kind of particle

Let us consider a 1D crystal with the same molecules along the chain with one molecule in position u_n and closest neighbors at positions u_{n-1}, u_{n+1}, as shown in Eq. (10.3). The distance between neighbors at equilibrium is equal to the lattice length a:

$$\cdots \quad \bullet \quad \bullet \quad \bullet \quad \cdots \tag{10.3}$$
$$u_{n-1} \ u_n \ u_{n+1}$$

(i) Classical approximation:

If we ignore the interactions of the rest of the molecules, the equation of motion becomes

$$m \frac{d^2 u_n}{dt^2} = C(u_{n-1} - u_n) - C(u_n - u_{n+1}). \tag{10.4}$$

From our experience with ordinary differential equations, we seek a solution of the form

$$u_n = \sum_k Q_k e^{i(kna - \omega t)}. \tag{10.5}$$

Putting this expression into the differential equation, we find

$$\frac{m}{C} \sum_k \frac{d^2 Q_k}{dt^2} e^{i(kna-\omega t)}$$

$$= \sum_k Q_k e^{i(k(n-1)a-\omega t)} + \sum_k Q_k e^{i(k(n+1)a-\omega t)} - 2 \sum_k Q_k e^{i(kna-\omega t)}$$

$$= \sum_k Q_k \left(e^{-ika} + e^{ika} - 2 \right) e^{i(kna-\omega t)}$$

$$= \sum_k Q_k \left(2\cos ka - 2 \right) e^{i(kna-\omega t)}. \qquad (10.6)$$

From this expression, we get

$$\omega^2 = \frac{2C}{m}(1 - \cos ka) \Rightarrow \omega(k) = \sqrt{\frac{2C}{m}(1 - \cos ka)} \Rightarrow \omega(k) = 2\omega_0 \left| \sin \frac{ka}{2} \right|,$$
$$(10.7)$$

with $\omega_0 = \sqrt{\frac{C}{m}}$.

The last expression of Eq. 10.7 is the **dispersion equation**. Now, we see that k must be found in the Bravais lattice, that is in the interval $-\frac{\pi}{a} \leq k \leq \frac{\pi}{a}$, and since the function is even, $0 \leq k \leq \frac{\pi}{a}$. This is exhibited in Fig. 10.1.

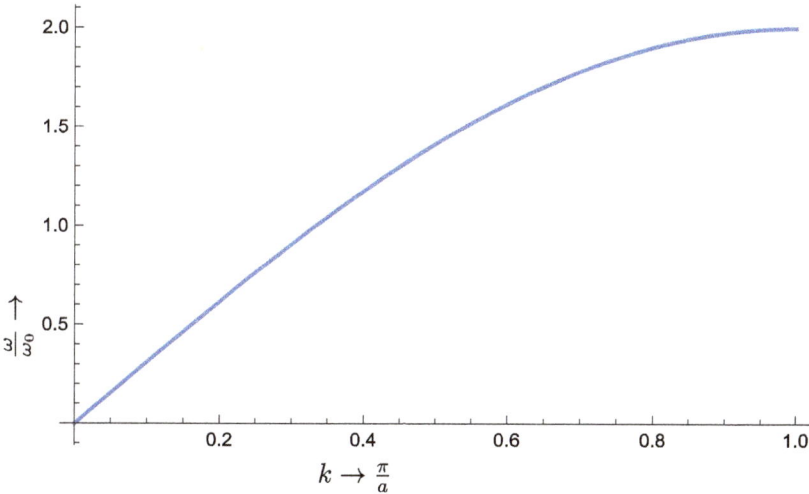

Fig. 10.1. The dispersion relation $\omega = \omega(k)$ in the case of 1D crystal with the same molecules and lattice length a.

(ii) Quantum treatment:

In quantum mechanics, we recognize the above as a typical problem of a quantum mechanical harmonic oscillator, which can be easily generalized in more than 1D. In this case, we consider the Hamiltonian function (operator) for the system. It can be written as

$$\mathcal{H} = \sum_i \frac{\mathbf{p}_i^2}{2m} + \frac{1}{2} m\omega_0^2 \sum_{i \neq j} (\mathbf{x}_i - \mathbf{x}_j)^2, \tag{10.8}$$

where i, j are indices numbering the structure elements (molecules).[1] The canonical coordinates are defined as

$$\mathbf{Q_k} = \frac{1}{\sqrt{N}} \sum_\ell e^{ika\ell} \mathbf{q}_\ell, \quad \mathbf{\Pi_k} = \frac{1}{\sqrt{N}} \sum_\ell e^{-ika\ell} \mathbf{p}_\ell, \tag{10.9}$$

where $\mathbf{k}$ is the wave number.

The commutation relations are

$$[\mathbf{x}_\ell, \mathbf{p}_m] = \hbar \delta_{\ell,m}, \ [\mathbf{x}_\ell, \mathbf{x}_m] = 0, \ [\mathbf{p}_\ell, \mathbf{p}_m] = 0$$

$$[\mathbf{Q_k}, \mathbf{\Pi_{k'}}] = \frac{1}{N} \sum_{\ell,m} e^{ika\ell} e^{-ik'am} [\mathbf{q}_\ell, \mathbf{p}_m] = i\hbar \frac{1}{N} \sum_\ell e^{i(k-k')\ell a} = i\hbar \delta_{\mathbf{k,k'}}$$

$$[\mathbf{Q_k}, Q_{\mathbf{k'}}] = 0, \ [\mathbf{\Pi_k}, \mathbf{\Pi_{k'}}] = 0. \tag{10.10}$$

One can now convince oneself that

$$\mathbf{x}_\ell \mathbf{x}_{\ell+m} = \sum_{\mathbf{k,k'}} \mathbf{Q_k Q_{k'}} e^{iak\ell} e^{iak'(\ell+m)} = \sum_{\mathbf{k,k'}} \mathbf{Q_k Q_{k'}} e^{ia\ell(\mathbf{k+k'})} e^{iamk'}$$

$$= \sum_{\mathbf{k,k'}} \mathbf{Q_k Q_{k'}} \delta_{\mathbf{k,-k'}} e^{iamk'} = \sum_{\mathbf{k}} \mathbf{Q_k Q_{-k}} e^{-iamk}, \tag{10.11}$$

$$\sum_\ell \mathbf{p}_\ell^2 = \sum_{\mathbf{k}} \mathbf{\Pi_k \Pi_{-k}}. \tag{10.12}$$

---

[1]In the present case, the vector formulation is adequate. It is, however, understood that in reality one needs dN coordinates, where d is the space dimension, for example, $2N$ coordinates in 2D, see Section 7.1.

It can easily be found that

$$\frac{1}{2}m\omega_0^2 \sum_{i \neq j} (\mathbf{x}_i - \mathbf{x}_j)^2 = \frac{1}{2}m\omega_0^2 \sum_{\mathbf{k}} \mathbf{Q_k Q_{-k}} \left(2 - e^{ika} - e^{-ika}\right)$$

$$= \frac{1}{2}m\omega_0^2 \sum_{\mathbf{k}} \mathbf{Q_k Q_{-k}} 4 \left(\sin \frac{ka}{2}\right)^2 = \frac{1}{2}m \sum_{\mathbf{k}} \omega_k^2 \mathbf{Q_k Q_{-k}},$$

$$(10.13)$$

where

$$\omega_k = 2\omega_0 \left(\sin \frac{ka}{2}\right), \tag{10.14}$$

just as we found above.

10.3 Normal modes in 1D crystal with two kinds of molecules

The problem now is a bit harder. Let us suppose that the arrangement is as in Fig. 10.2.

$$\frac{M_1}{C} \frac{d^2 u_s}{dt^2} = (v_{s-1} - u_s) - (u_s - v_s), \tag{10.15}$$

$$\frac{M_2}{C} \frac{d^2 v_s}{dt^2} = (u_s - v_s) - (v_s - u_{s+1}). \tag{10.16}$$

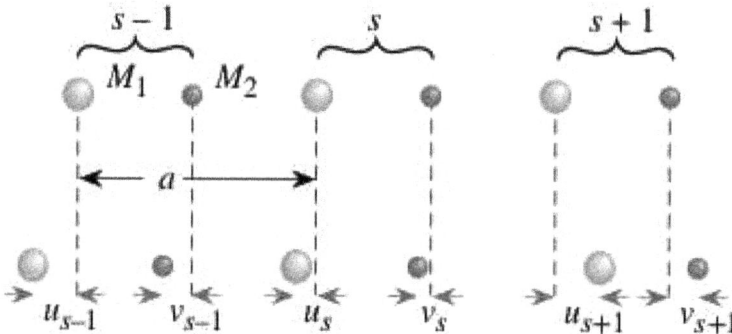

Fig. 10.2. The nearest neighbors of molecules of mass M_1 in positions u_i, $i = s - 1$, $s, s+1$ and M_2 in positions v_i, $i = s - 1, s, s+1$, placed in alternate positions.

Proceeding as above, we find the solution to the system of two equations searching for such of the type

$$u_s = u_0 e^{-i(ksa - \omega t)}, \ v_s = v_0 e^{-i(ksa - \omega t)}.$$

This way, we are led to the relations

$$-\frac{M_1}{C}\omega^2 u_0 = -2u_0 + v_0 \left(1 + e^{-ika}\right),$$

$$-\frac{M_2}{C}\omega^2 v_0 = \left(1 + e^{ika}\right) + u_0 \left(1 + e^{ika}\right).$$

(10.17)

This system has a non-zero solution, provided that

$$\det \begin{pmatrix} \frac{M_1}{C}\omega^2 - 2 & 1 + e^{-ika} \\ 1 + e^{ika} & \frac{M_1}{C}\omega^2 - 2 \end{pmatrix}$$

$$= \frac{M_1}{C}\frac{M_2}{C}\omega^4 - 2\left(\frac{M_1}{C} + \frac{M_2}{C}\right)\omega^2 + 4 - 2(1 + \cos ka) = 0. \quad (10.18)$$

The last equation has two solutions:

$$\omega_{\pm}^2 = \frac{C}{M_1} + \frac{C}{M_2} \pm \sqrt{\left(\frac{C}{M_1} + \frac{C}{M_2}\right)^2 - 4\frac{C^2}{M_1 M_2}\left(\sin\frac{ka}{2}\right)^2}. \quad (10.19)$$

From the above equations, we find the two dispersion relations:

$$\omega_o(k) = \omega_+ = \sqrt{\frac{C}{M_1} + \frac{C}{M_2} + \sqrt{\left(\frac{C}{M_1} + \frac{C}{M_2}\right)^2 - 4\frac{C^2}{M_1 M_2}\left(\sin\frac{ka}{2}\right)^2}},$$

(optical dispersion) (10.20)

$$\omega_a(k) = \omega_- = \sqrt{\frac{C}{M_1} + \frac{C}{M_2} - \sqrt{\left(\frac{C}{M_1} + \frac{C}{M_2}\right)^2 - 4\frac{C^2}{M_1 M_2}\left(\sin\frac{ka}{2}\right)^2}},$$

(acoustic dispersion) (10.21)

These two types of dispersion are exhibited in Fig. 10.3. In the optical type, the dispersion function exhibits a maximum at $k = 0$, which is $\sqrt{2\left(\frac{C}{M_1} + \frac{C}{M_2}\right)}$ and a minimum at $k = \pi/a$, which is $\sqrt{2C/\min(M_1, M_2)}$. In the acoustic mode, the frequency becomes zero at $k = 0$ and gets to a maximum at $k = \pi/a$, which is $\sqrt{2C/\max(M_1, M_2)}$. This behavior, in the name and in similarity with that of Fig. 10.1, is understood by

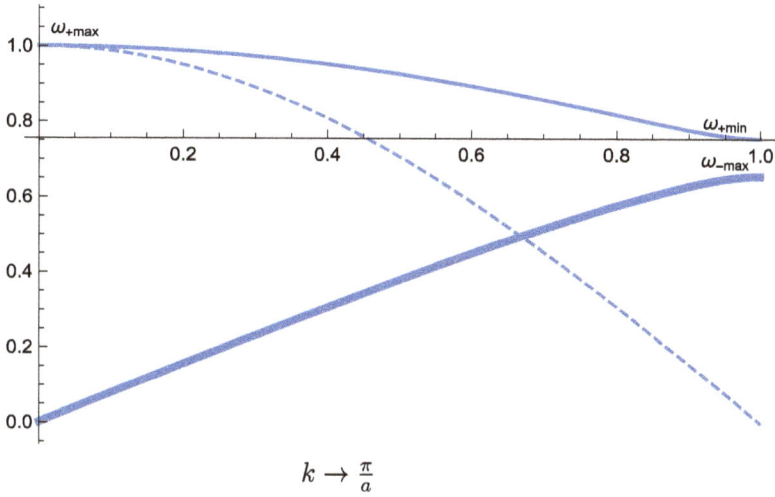

Fig. 10.3. The dispersion relation $\omega = \omega(k)$ for $0 \leq k \leq \pi/a$ in the case of 1D crystal with two different molecules placed in alternate positions along a lattice of length a. The fine solid line indicates the optical oscillation type with a maximum value $\omega_{+\,\text{max}} = \sqrt{2(C/M_1 + C/M_1)}$ at $k = 0$ and a minimum $\omega_{+\,\text{min}} = \sqrt{2(C/(\min(M_1 M_2))}$ at $k = \pi/a$. The thick solid line indicates the acoustic oscillation type, which starts at zero for $k = 0$ and attains a maximum value $\omega_{-\,\text{max}} = \sqrt{2(C/(\max(M_1 M_2))}$ at $k = \pi/a$. The dashed line corresponds to the case of $M_1 \approx M_2$.

observing that

$$\omega_a(k) \approx 2\omega_0 \left| \sin \frac{ka}{2} \right|, \quad \omega_0 = \sqrt{\frac{C}{2(M_1 + M_2)}}. \tag{10.22}$$

10.4 Normal oscillation modes of graphene

We now examine the normal modes of one layer of graphene, which is clearly a 2D problem. The lattice symmetry is already known, see Section 9.4.

For the study of this system, we need six coordinates, which we select according to the positions of the end points of the relevant vectors as

$$\delta_3 \leftrightarrow (x_1, 0), \ a_1 \leftrightarrow (x_2, 0), \ \delta_1 \leftrightarrow (x_3, y_3), \ \delta_2 \leftrightarrow (x_4, y_4).$$

In such a case, the potential function in the harmonic approximation takes the form

$$\frac{V}{k} = \frac{1}{2}\left(\left(\frac{1}{2}(x_2 - x_3) - \frac{\sqrt{3}y_3}{2} \right)^2 + \left(\frac{1}{2}(x_2 - x_4) - \frac{\sqrt{3}y_4}{2} \right)^2 + (x_1 - x_2)^2 \right). \tag{10.23}$$

Ordering now the six points as

$$x_i \to q_i, \quad i = 1, \ldots, 4, \quad y_3 \to q_5, \quad y_4 \to q_6,$$

we find

$$\frac{V}{k} = \frac{1}{2}\left((q_1 - q_2)^2 + \left(\frac{1}{2}(q_2 - q_3) - \frac{\sqrt{3}q_5}{2}\right)^2 + \left(\frac{1}{2}(q_2 - q_4) - \frac{\sqrt{3}q_6}{2}\right)^2\right).$$

$$(10.24)$$

10.4.1   The classical equation of motion approach: A prelude

We find it useful to consider a procedure analogous to the classical approximation of Section 10.2.

With the potential of Eq. (10.24), the equation of motion of particle A found at the end of the lattice point a_1 is

$$\frac{m}{C}\frac{d^2 q_2}{dt^2} = -\frac{\partial V}{\partial q_2} \Rightarrow \frac{m}{C}\frac{d^2 q_2}{dt^2} = q_1 - \frac{3q_2}{2} + \frac{q_3}{4} + \frac{q_4}{4} + \frac{\sqrt{3}q_5}{4} + \frac{\sqrt{3}q_6}{4}.$$

$$(10.25)$$

Now, setting

$$q_1 = (n_x - 1)a, \quad q_2 = n_x a, \quad q_3 = (n_x + 1)a, \quad q_4 = (n_x + 1)a,$$

$$q_5 = (n_y + 1), \quad q_6 = (n_y - 1)$$

and seeking solutions of the differential equation of the form of Eq. (10.5), i.e.

$$u_n = \sum_{k_x, k_y} Q_{k_x, k_y} e^{i(k_x n_x a + k_y n_y a - \omega t)}, \qquad (10.26)$$

we find

$$\frac{m}{C}\frac{d^2 q_2}{dt^2} = \sum_{k_x, k_y} Q_{k_x, k_y} e^{i(k_x n_x a + k_y n_y a - \omega t)}$$

$$= \sum_{k_x, k_y} A^{(0)}(k_x, ky) Q_{k_x, k_y} e^{i(k_x n_x a + k_y n_y a - \omega t)}, \qquad (10.27)$$

$$A^{(0)}(k_x, k_y) = \sum_j \phi_j \mathbf{R}_j^{(0)}$$

$$\times \left(e^{-ik_x a} - \frac{3}{2} + \frac{1}{4}e^{ik_x a} + \frac{1}{4}e^{ik_x a} + \frac{\sqrt{3}}{4}e^{k_y a} + \frac{\sqrt{3}}{4}e^{-k_y a} \right),$$

$$(10.28)$$

$$\phi = (e^{-ik_x a}, 1, e^{ik_x a}, e^{ik_x a}, e^{ik_y a}, e^{-ik_y a}),$$

$$(10.29)$$

$$\mathbf{R}^{(0)} = (1, -3/2, 1/4, 1/4, \sqrt{3}/4, \sqrt{3}/4),$$

$$(10.30)$$

$$A^{(0)}(k_x, k_y) = A_R^{(0)}(k_x, k_y) + iA_I^{(0)}(k_x, k_y),$$

$$(10.31)$$

$$A_R^{(0)}(k_x, k_y)$$
$$= (\cos k_x a, -3/2, (1/4)\cos k_x a, (1/4)\cos k_x a, (\sqrt{3}/4)\cos k_y a, (\sqrt{3}/4)\cos k_y a),$$
$$A_I^{(0)}(k_x, k_y)$$
$$= (-\sin k_x a, 0, (1/4)\sin k_x a, (1/4)\sin k_x a, (\sqrt{3}/4)\sin k_y a, -(\sqrt{3}/4)\sin k_y a).$$

$$(10.32)$$

This way, we obtain

$$\omega^2 = \omega_0^2 \sqrt{\left(A_R^{(0)}(k_x, k_y)\right)^2 + \left(A_I^{(0)}(k_x, k_y)\right)^2},$$

$$(10.33)$$

$$\omega = \omega_0 \left(\left(A_R^{(0)}(k_x, k_y)\right)^2 + \left(A_R^{(0)}(k_x, k_y)\right)^2\right)^{1/4}.$$

$$(10.34)$$

The dispersion relation is exhibited in Fig. 10.4. The 2D plot corresponds to the data $\mathbf{k}_1 \leftrightarrow \mathbf{b}_1 = \left(2\pi/(3a), 2\pi/(\sqrt{3}a)\right)$ of Fig. 9.4.

10.4.2   *The method of diagonalizing the potential*

Proceeding as in Section 7.3, we find that the matrix v corresponding to the potential of Eq. (10.24) is

$$v = \begin{pmatrix} \frac{1}{2} & -\frac{1}{2} & 0 & 0 & 0 & 0 \\ -\frac{1}{2} & \frac{3}{4} & -\frac{1}{8} & -\frac{1}{8} & -\frac{\sqrt{3}}{8} & -\frac{\sqrt{3}}{8} \\ 0 & -\frac{1}{8} & \frac{1}{8} & 0 & \frac{\sqrt{3}}{8} & 0 \\ 0 & -\frac{1}{8} & 0 & \frac{1}{8} & 0 & \frac{\sqrt{3}}{8} \\ 0 & -\frac{\sqrt{3}}{8} & \frac{\sqrt{3}}{8} & 0 & \frac{3}{8} & 0 \\ 0 & -\frac{\sqrt{3}}{8} & 0 & \frac{\sqrt{3}}{8} & 0 & \frac{3}{8} \end{pmatrix}.$$

$$(10.35)$$

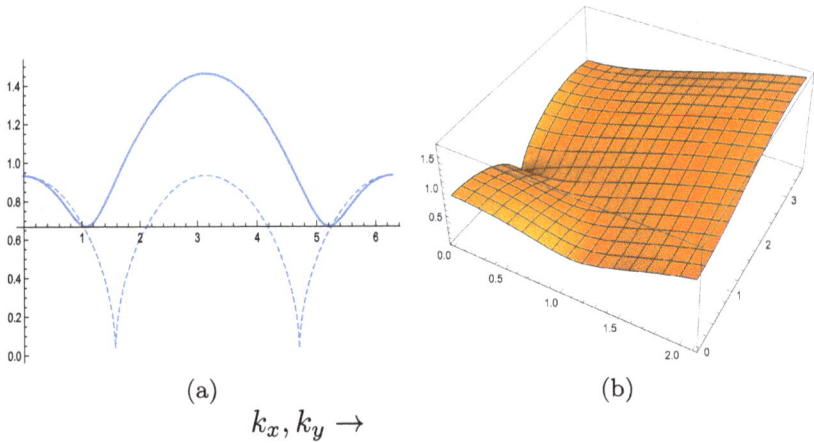

$$k_x, k_y \to$$

Fig. 10.4. The dispersion relation $\omega = \omega(k_x, k_y)$ in the case of graphene: (a) for $0 \le k_x \le 2\pi/a$, $k_y = 0$ (solid curve) and $k_x = 0$, $\le k_y \le 2\pi/a$ (dashed line); and (b) The dispersion relation in 2D with boundaries related to $\mathbf{k}_1 - (2\pi/3a, 2\pi/\sqrt{3}a)$ (see text). All curves correspond to the data of Eq. (10.34).

Its eigenvalues are $(5/4, 1/2, 1/2, 0, 0, 0)$, and the corresponding eigenvectors are given by the columns of the matrix

$$
S =
\begin{pmatrix}
\mathbf{R}^{(1)} & \mathbf{R}^{(2)} & \mathbf{R}^{(3)} & \mathbf{R}^{(4)} & \mathbf{R}^{(5)} & \mathbf{R}^{(6)} \\
\hline
\frac{2}{\sqrt{15}} & -\frac{1}{\sqrt{5}} & -\sqrt{\frac{2}{15}} & \frac{\sqrt{\frac{3}{13}}}{2} & \sqrt{\frac{6}{65}} & \frac{1}{2} \\
-\sqrt{\frac{3}{5}} & 0 & 0 & \frac{\sqrt{\frac{3}{13}}}{2} & \sqrt{\frac{6}{65}} & \frac{1}{2} \\
\frac{1}{2\sqrt{15}} & 0 & \frac{\sqrt{\frac{5}{6}}}{2} & \frac{\sqrt{\frac{3}{13}}}{2} & -\frac{9\sqrt{\frac{3}{130}}}{2} & \frac{1}{2} \\
\frac{1}{2\sqrt{15}} & \frac{1}{\sqrt{5}} & -\frac{1}{2\sqrt{30}} & -\frac{3\sqrt{\frac{3}{13}}}{2} & \frac{\sqrt{\frac{3}{130}}}{2} & \frac{1}{2} \\
\frac{1}{2\sqrt{5}} & 0 & \frac{\sqrt{\frac{5}{2}}}{2} & 0 & \frac{\sqrt{\frac{13}{10}}}{2} & 0 \\
\frac{1}{2\sqrt{5}} & \sqrt{\frac{3}{5}} & -\frac{1}{2\sqrt{10}} & \frac{2}{\sqrt{13}} & \frac{3}{2\sqrt{130}} & 0
\end{pmatrix}.
\qquad (10.36)
$$

We note that, due to the fact that the masses of the particles are the same, the kinetic energy remains diagonal. As a result, the equation of motion is given by the relation

$$
\frac{d}{dt}\left(\frac{\partial \mathcal{L}}{\partial \dot{\mathbf{R}}^{(\alpha)}}\right) = \frac{\partial \mathcal{L}}{\partial \mathbf{R}^{(\alpha)}}, \quad \mathcal{L} = T - V, \qquad (10.37)
$$

and as a result, it leads to the relations

$$(\omega^\alpha)^2 = \omega_0^2 \lambda^\alpha \sqrt{\left(A_R^{(\alpha)}(k_x, k_y)\right)^2 + \left(A_I^{(\alpha)}(k_x, k_y)\right)^2}, \tag{10.38}$$

$$\omega^\alpha = \omega_0 \sqrt{\lambda^\alpha} \left(\left(A_R^{(\alpha)}(k_x, k_y)\right)^2 + \left(A_I^{(\alpha)}(k_x, k_y)\right)^2\right)^{1/4}, \tag{10.39}$$

$$A^{(\alpha)}(k_x, k_y) = A_R^{(\alpha)}(k_x, k_y) + i A_I^{(\alpha)}(k_x, k_y), \tag{10.40}$$

$$A^{(\alpha)}(k_x, k_y) = \sum_j \phi_j R_j^{(\alpha)} \left(e^{ik_x a} + \frac{\sqrt{3}}{4} e^{k_y a} + \frac{\sqrt{3}}{4} e^{-k_y a}\right). \tag{10.41}$$

We thus see that we have three dispersion relations corresponding to the eigenvalues $\lambda^{(1)} = 5/4$, $\lambda^{(2)} = 1/2$, $\lambda^{(3)} = 1/2$. The eigenvector $\mathbf{R}^{(1)}$ is uniquely determined, while $\mathbf{R}^{(2)}$ and $\mathbf{R}^{(3)}$ constitute a convenient set of two orthogonal vectors corresponding to the degenerate value $1/2$. The remaining coordinates are spurious. In the case of the vector $\mathbf{R}^{(6)}$, the spurious status is obvious (motion of the center of mass). Regarding the status of the rest, it is not at all obvious what it is due to, since their selection is arbitrary due to the degeneracy of the relevant eigenvalue.

The three dispersion relations are exhibited in Fig. 10.5.

It is worth noting that the vector

$$\mathbf{R}^{(0)} = (1, -3/2, 1/4, 1/4, \sqrt{3}/4, \sqrt{3}/4), \tag{10.42}$$

after normalization, coincides with the vector $\mathbf{R}^{(1)}$. This way, one can understand the similarity between Figs. 10.5(a) and 10.5(b). The diagonalization supplies, of course, all the dispersion relations.

At this point, we make a detour.

Remark 1: At this point, it is worth noting that the diagonalization method is required not only in more than 1D. Let us, for example, consider the 1D example of Section 10.2, restricted to the potential between the nearest neighbors. In the harmonic approximation, one gets

$$V(x_1, x_2) = \frac{1}{2}(x_1 - x_2)^2 + \frac{1}{2}(x_2 - x_3)^2 \Rightarrow v = \begin{pmatrix} 1 & -1 & 0 \\ -1 & 2 & -1 \\ 0 & -1 & 1 \end{pmatrix}. \tag{10.43}$$

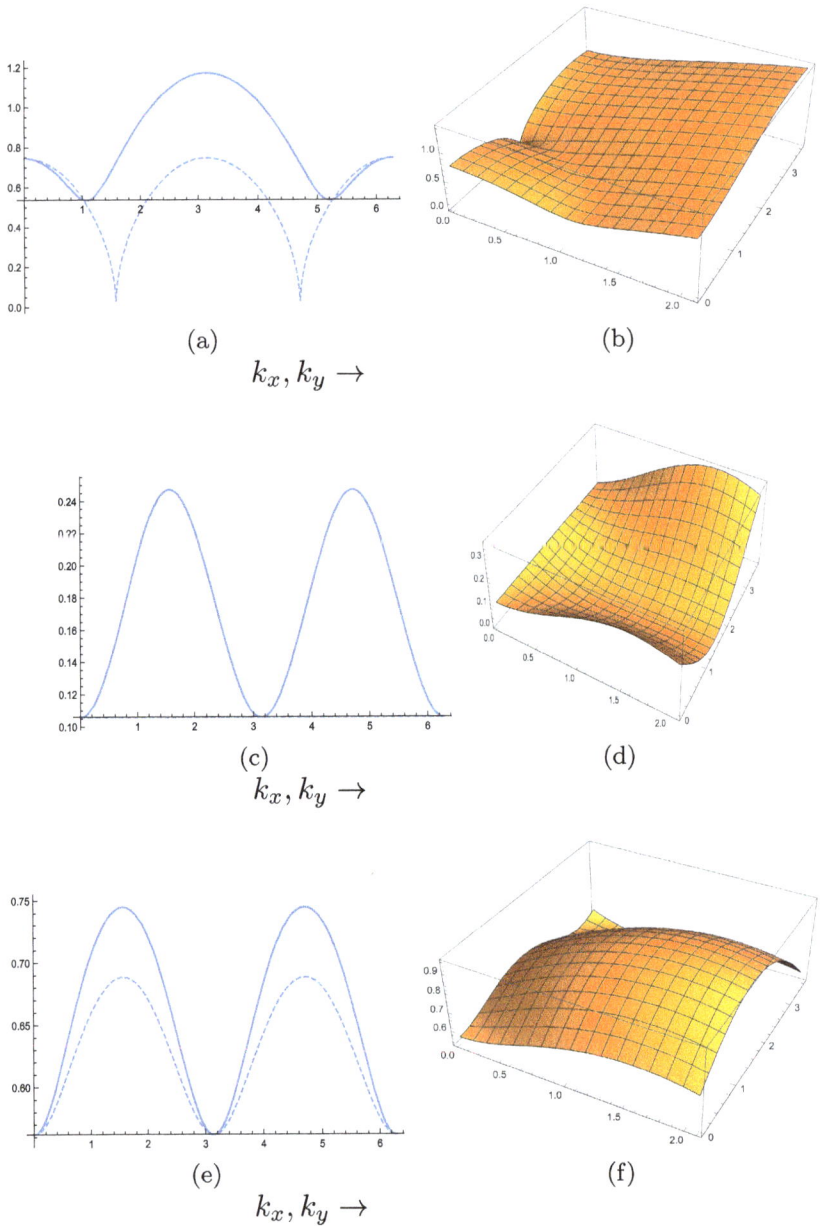

(a)

$k_x, k_y \rightarrow$

(b)

(c)

$k_x, k_y \rightarrow$

(d)

(e)

$k_x, k_y \rightarrow$

(f)

Fig. 10.5. The dispersion relation $\omega = \omega(k_x, k_y)$ in the case of graphene: (a) and (b) in the case of $\mathbf{R}^{(1)}$ corresponding to $\lambda^{(1)} = 5/4$; (c) and (d) in the case of $\mathbf{R}^{(2)}$; (e) and (f) in the case of $\mathbf{R}^{(3)}$. The rest of the symbols are as in Fig. 10.4.

The eigenvalues of this matrix are $(3,1,0)$, and the corresponding eigenvectors are the columns of the matrix

$$S = \begin{pmatrix} \mathbf{R}^{(1)} & \mathbf{R}^{(2)} & \mathbf{R}^{(3)} \\ -\frac{1}{\sqrt{6}} & -\frac{1}{\sqrt{2}} & \frac{1}{\sqrt{3}} \\ \frac{2}{\sqrt{6}} & 0 & \frac{1}{\sqrt{3}} \\ -\frac{1}{\sqrt{6}} & \frac{1}{\sqrt{2}} & \frac{1}{\sqrt{3}} \end{pmatrix}. \tag{10.44}$$

In the current case,

$$\phi = (e^{-ika}, 1, e^{ika}), \tag{10.45}$$

and hence,

$$A_R^{(1)}(kx) = \frac{2}{\sqrt{6}}(1 - \cos kx), \quad A_I^{(1)}(kx) = 0, \tag{10.46}$$

$$\omega^{(1)} = \omega_0 \sqrt{3} \left(\left(\frac{2}{\sqrt{6}}(1 - \cos kx) \right)^2 \right)^{1/4},$$

$$= \omega_0 2^{1/4} 3^{1/4} \sqrt{1 - \cos kx} = \omega_0 \left(\frac{3}{2} \right)^{1/4} 2 \left| \sin \frac{kx}{2} \right|, \tag{10.47}$$

$$A_R^{(2)}(kx) = 0, \quad A_I^{(1)}(kx)\sqrt{2}\sin kx), \tag{10.48}$$

$$\omega^{(2)} = \omega_0 \sqrt{2} \sin kx. \tag{10.49}$$

The obtained results are exhibited in Fig. 10.6.

It should be noted that the eigenvector corresponding to the frequency $\omega^{(2)}$ contains two components not corresponding to the nearest neighbors. With the diagonalization of the Hamiltonian matrix, however, this admixture became effective through its interaction with the nearest neighbor. The eigenvector $\mathbf{R}^{(3)}$ corresponds to the motion of the center of mass, and it is linked to a spurious oscillation with zero frequency.

Returning to the consideration of graphene, it is constructive to say a few words about the symmetry of system A and its three neighbors. It is clear that the only symmetry group appearing is G containing two elements, $G = \{E, \sigma_x\}$, where σ_x is a reflection with respect to the axis x,

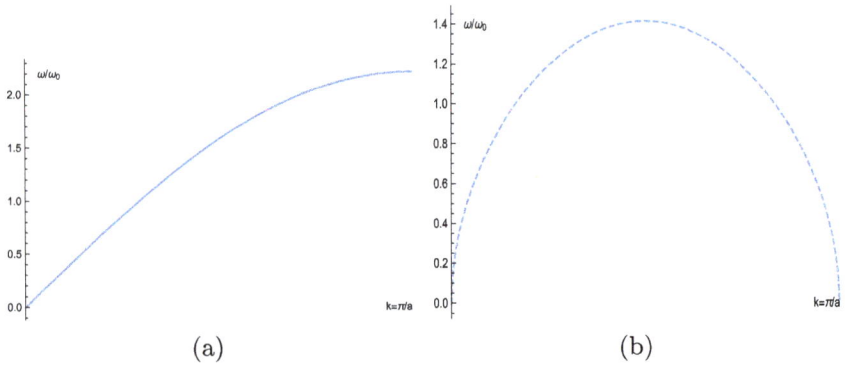

Fig. 10.6. As in Fig. 10.1: (a) for the frequency $w^{(1)}$ and (b) for the frequency $w^{(2)}$.

The 6D representation of G can take the form

$$\Gamma(E) = \begin{pmatrix} 1 & 0 & 0 & 0 & 0 & 0 \\ 0 & 1 & 0 & 0 & 0 & 0 \\ 0 & 0 & 1 & 0 & 0 & 0 \\ 0 & 0 & 0 & 1 & 0 & 0 \\ 0 & 0 & 0 & 0 & 1 & 0 \\ 0 & 0 & 0 & 0 & 0 & 1 \end{pmatrix}, \quad \Gamma(\sigma_x) = \begin{pmatrix} 1 & 0 & 0 & 0 & 0 & 0 \\ 0 & 1 & 0 & 0 & 0 & 0 \\ 0 & 0 & 0 & 1 & 0 & 0 \\ 0 & 0 & 1 & 0 & 0 & 0 \\ 0 & 0 & 0 & 0 & 0 & 1 \\ 0 & 0 & 0 & 0 & 1 & 0 \end{pmatrix}.$$

$$(10.50)$$

The irreducible representations of this group are 1D:

$$\Gamma^{(1)}(E) = \Gamma^{(1)}(\sigma_x) = 1, \ \Gamma^{(2)}(E) = 1, \ \Gamma^{(2)}(\sigma_x) = -1.$$

The 6D representation of the group is, of course, reducible, and the following relation holds:

$$\Gamma(X) = \alpha_1 \Gamma^1(X) \oplus \alpha_2 \Gamma^2(X).$$

The integers α_i can be found via the character relations, (see Eq. (5.24)) as follows:

$$\alpha_1 = \frac{1}{2}\left(\chi(E)\chi^1(E) + \chi(\sigma_x)\chi^1(\sigma_x)\right) = \frac{1}{2}(6 \times 1 + 2 \times 1) = 4,$$

$$\alpha_2 = \frac{1}{2}\left(\chi(E)\chi^2(E) + \chi(\sigma_x)\chi^2(\sigma_x)\right) = \frac{1}{2}(6 \times 1 + 2 \times (-1)1) = 2$$

$$\Gamma(X) = 4\Gamma^1(X) \oplus 2\Gamma^2(X).$$

$$(10.51)$$

As a result, according to Wigner's theorem, see Section 6.2, the Hamiltonian matrix is reduced into two sub-matrices with dimensions 4×4 and 2×2.

For matrices with relatively small dimensions, the benefit stemming from employing the symmetry is small, and it may not be worth the trouble of invoking it. This is the reason why one prefers the straightforward method of matrix diagonalization. In the case of large dimensions, however, the use of symmetry toward the solution becomes a must. In any case, it has the additional advantage of giving some physical meaning to the obtained results.

PART B

Special topics and applications

Chapter 11

The Symmetric Group S_n

The symmetric group S_n, see Section 2.2, is defined for every $n = 1, 2, 3, \ldots$.
It is perhaps the most important discrete group both from a mathematical point of view as well as because of its applications. We recall Cayley's Theorem 2, see Section 3.1, which states that every discrete group of order n is isomorphic to a subgroup of S_n. An elementary exposition has been made in Section 2.3. We will see in this chapter that S_n can be used in the construction of many body wave functions, composed of indistinguishable particles, which must have a specific symmetry under the exchange of any two particles, e.g. symmetric or antisymmetric. As we shall see in this chapter, this is not a trivial problem, especially when the wave functions are defined not only in ordinary space but in composite function spaces, e.g. ordinary and spin space. Furthermore, the symmetric group is very useful in labeling the representations of the classical continuous groups, see for example, Vergados (2017).

11.1 The role of the symmetric group S_n in quantum mechanics

Let us suppose that a system of n indistinguishable particles is characterized by the variables $y_1, y_2, \ldots, y_n$, with $y_i = (\mathbf{r}_i, \mathbf{s}_i, \text{etc.})$. Let us further suppose that the Hamiltonian describing the system is invariant under transformations of the symmetry S_n. Then, according to Wigner's theorem, see Section 6.2, the resulting eigenfunctions transform according to the irreducible representations of S_n. Furthermore, different representations do not mix, unless they happen to be degenerate. From the set of solutions, the physically acceptable ones are the functions Ψ_F, which are antisymmetric under the exchange of any two particles, or Ψ_B, which are symmetric under the exchange of any two particles. In the first instance, the particles are called Fermions and are characterized by half-integral spin. In the second case, the particles are characterized by integral spin and are called Bosons. Thus,

$$(i,j)\Psi_F(y_1, y_2, \ldots, y_i, \ldots, y_j, \ldots, y_n) = -\Psi_F(y_1, y_2, \ldots, y_i, \ldots, y_j, \ldots, y_n)$$
(11.1)

or

$$(i,j)\Psi_B(y_1, y_2, \ldots, y_i, \ldots, y_j, \ldots, y_n) = \Psi_B(y_1, y_2, \ldots, y_i, \ldots, y_j, \ldots, y_n),$$
(11.2)

where (i,j) is the transposition permutation of S_n and the y_i's correspond to one kind of variable, e.g. ordinary space (orbital) $\mathbf{r}_i$ or spin $\mathbf{s}_i$. In the case of only one variable, the study of the representations of S_n would be of little importance. In practice, however, one needs to study the case of the wave function depending on both orbital and spin symmetries. Then, the wave function is a product of an orbital and a spin part. In that case, we have to consider all possible permutations p of S_n, writing

$$\Psi_F = \frac{1}{dim([f])} \sum_{p \in S_n} (-1)^{\epsilon(p)} \Psi_{[f]}(\mathbf{r}_1, \mathbf{r}_2, \ldots, \mathbf{r}_n)\chi_{[\tilde{f}]}(\mathbf{s}_1, \mathbf{s}_2, \ldots, \mathbf{s}_n),$$

$$\epsilon(p) = \begin{cases} 0 & p = \text{even} \\ 1 & p = \text{odd}, \end{cases}$$
(11.3)

$$\Psi_B = \frac{1}{dim([f])} \sum_{p \in S_n} \Psi_{[f]}(\mathbf{r}_1, \mathbf{r}_2, \ldots, \mathbf{r}_n)\chi_{[f]}(\mathbf{s}_1, \mathbf{s}_2, \ldots, \mathbf{s}_n).$$
(11.4)

In other words, in this case, one needs not only the symmetric S or antisymmetric A representations of S_n, but many representations $[f]$.[1]

Example 1: For illustration purposes, we are going to consider the rather trivial case of S_2. A suitable basis is $|1\rangle = \psi_\alpha(1)\psi_\beta(2)$, $|2\rangle = \psi_\alpha(2)\psi_\beta(1)$.

---

[1]We will see later that in the case of Fermions, the function $\Psi_{[f]}(\mathbf{r}_1, \mathbf{r}_2, \ldots, \mathbf{r}_n)$ is connected with that of spin $\chi_{[\tilde{f}]}(\mathbf{s}_1, \mathbf{s}_2, \ldots, \mathbf{s}_n)$, see Section 11.2.

In this basis, we encounter the representation

$$T(E) = \begin{pmatrix} 1 & 0 \\ 0 & 1 \end{pmatrix}, \quad T((12)) = \begin{pmatrix} 0 & 1 \\ 1 & 0 \end{pmatrix}.$$

This basis is reduced by diagonalizing $T((12))$:

$$\Gamma(E) = \begin{pmatrix} 1 & 0 \\ 0 & 1 \end{pmatrix}, \quad \Gamma((12)) = \begin{pmatrix} 1 & 0 \\ 0 & -1 \end{pmatrix} \Rightarrow$$

$$\Psi_B = \frac{1}{\sqrt{2}}(E + (12))(\psi_\alpha(1)\psi_\beta(2)), \quad \Psi_F = \frac{1}{\sqrt{2}}(E - (12))\psi_\alpha(1)\psi_\beta(2).$$

or

$$\Psi_B = \frac{1}{\sqrt{2}}(\psi_\alpha(1)\psi_\beta(2) + \psi_\alpha(2)\psi_\beta(1)),$$

$$\Psi_F = \frac{1}{\sqrt{2}}(\psi_\alpha(1)\psi_\beta(2) - \psi_\alpha(2)\psi_\beta(1)).$$

Let us now consider the case of particles with spin. Then, a basis can be

$$|1\rangle = \psi_\alpha(1)\psi_\beta(2)\chi_\gamma(1)\chi_\delta(2), |2\rangle = \psi_\alpha(2)\psi_\beta(1)\chi_\gamma(1)\chi_\delta(2),$$

$$|3\rangle = \psi_\alpha(1)\psi_\beta(2)\chi_\gamma(2)\chi_\delta(1), |4\rangle = \psi_\alpha(2)\psi_\beta(1)\chi_\gamma(2)\chi_\delta(1) \Rightarrow$$

$$T(12) = \begin{pmatrix} 0 & 1 & 1 & 0 \\ 1 & 0 & 0 & 1 \\ 1 & 0 & 0 & 1 \\ 0 & 1 & 1 & 0 \end{pmatrix} \Rightarrow S = \begin{pmatrix} \frac{1}{2} & \frac{1}{2} & \frac{1}{2} & \frac{1}{2} \\ \frac{1}{2} & -\frac{1}{2} & \frac{1}{2} & -\frac{1}{2} \\ \frac{1}{2} & -\frac{1}{2} & -\frac{1}{2} & \frac{1}{2} \\ \frac{1}{2} & \frac{1}{2} & -\frac{1}{2} & -\frac{1}{2} \end{pmatrix},$$

where S is a matrix diagonalizing $T(12)$. The resulting eigenfunctions can be written as

$$\Psi^1(B) = \frac{1}{2}(|1\rangle + |2\rangle + |3\rangle + |4\rangle), \quad \Psi^2(B) = \frac{1}{2}(|1\rangle - |2\rangle - |3\rangle + |4\rangle),$$

$$\Psi^1(F) = \frac{1}{2}(|1\rangle + |2\rangle - |3\rangle - |4\rangle), \quad \Psi^2(F) = \frac{1}{2}(|1\rangle - |2\rangle + |3\rangle - |4\rangle).$$

More explicitly,

$$\Psi^1(F) = \frac{1}{\sqrt{2}}(E + (12))\psi_\alpha(1)\psi_\beta(2) \times \frac{1}{\sqrt{2}}(E - (12))\chi_\gamma(1)\psi_\delta(2), \quad (11.5)$$

$$\Psi^2(F) = \frac{1}{\sqrt{2}}(E - (12))\psi_\alpha(1)\psi_\beta(2) \times \frac{1}{\sqrt{2}}(E + (12))\chi_\gamma(1)\psi_\delta(2), \quad (11.6)$$

$$\Psi^1(B) = \frac{1}{\sqrt{2}}(E + (12))\psi_\alpha(1)\psi_\beta(2) \times \frac{1}{\sqrt{2}}(E + (12))\chi_\gamma(1)\psi_\delta(2), \quad (11.7)$$

$$\Psi^2(B) = \frac{1}{\sqrt{2}}(E - (12))\psi_\alpha(1)\psi_\beta(2) \times \frac{1}{\sqrt{2}}(E - (12))\chi_\gamma(1)\psi_\delta(2), \quad (11.8)$$

or

$$\Psi^1(F) = \frac{1}{\sqrt{2}}(\psi_\alpha(1)\psi_\beta(2) + \psi_\alpha(2)\psi_\beta(1)) \times \frac{1}{\sqrt{2}}(\chi_\gamma(1)\psi_\delta(2) - \chi_\gamma(2)\psi_\delta(1)),$$
$$(11.9)$$

$$\Psi^2(F) = \frac{1}{\sqrt{2}}(\psi_\alpha(1)\psi_\beta(2) - \psi_\alpha(2)\psi_\beta(1)) \times \frac{1}{\sqrt{2}}(\chi_\gamma(1)\psi_\delta(2) + \chi_\gamma(2)\psi_\delta(1)),$$
$$(11.10)$$

$$\Psi^1(B) = \frac{1}{\sqrt{2}}(\psi_\alpha(1)\psi_\beta(2) + \psi_\alpha(2)\psi_\beta(1)) \times \frac{1}{\sqrt{2}}(\chi_\gamma(1)\psi_\delta(2) + \chi_\gamma(2)\psi_\delta(1)),$$
$$(11.11)$$

$$\Psi^2(B) = \frac{1}{\sqrt{2}}(\psi_\alpha(1)\psi_\beta(2) - \psi_\alpha(2)\psi_\beta(1)) \times \frac{1}{\sqrt{2}}(\chi_\gamma(1)\psi_\delta(2) - \chi_\gamma(2)\psi_\delta(1)).$$
$$(11.12)$$

We observe that in the case of Bosons under T_{12}, the space part has the same symmetry as that of the spin, while in the case of Femions, the space part and the spin part have opposite symmetry, yielding the expected overall symmetry under T_{12}.

11.2   The irreducible representations of the symmetric group S_n

We have seen in Section 3.3 that the classes of S_n are determined by the natural numbers $\nu_1, \nu_2, \ldots, \nu_n$ characterizing the cyclic structure of the elements of S_n, i.e.

$$\nu_1 + 2\nu_2 + 3\nu_3 \cdots + n\nu_n = n \Leftrightarrow \{1^{\nu_1}, 2^{\nu_2}, 3^{\nu_3}, \ldots, n^{\nu_n}\}.$$

If we now define the quantities f_i:

$$\nu_1 + \nu_2 + \nu_3 \cdots + \nu_n = f_1, \ \nu_2 + \nu_3 \cdots + \nu_n = f_2,$$

$$\nu_3 \cdots + \nu_n = f_3, \ldots, \nu_n = f_n,$$

the non-negative integers f_i satisfy the relations

$$f_1 \geq f_2 \geq f_3 \cdots \geq f_n, \ f_1 + f_2 + f_3 \cdots + f_n = n. \tag{11.13}$$

The set of non-negative integers $f_1, f_2, f_3, \ldots, f_n$ satisfying the relation (11.13) is called a **partition** of n.

Such a partition can be represented by a **Young tableau** defined as follows:

- The Young tableau consists of n boxes arranged in at most n rows, each row beginning at the same vertical straight line.
- Each row contains at most n boxes.
- The length of each row no longer than the previous one.

A typical Young tableau $[f_1, f_2, f_3]$ is

In the above example, $f_1 = 9$, $f_2 = 5$, $f_3 = 3$ for S_9.

The allowed representations are classified by Young tableaux containing f_1 boxes in the first row, f_2 boxes in the second row and f_3 in the third row, etc., such that $f_1 \geq f_2 \geq f_3, \ldots, f_{n-1} \geq f_n$, with $f_1 + f_2 + \cdots + f_r = n$.

- A single row pattern represents a symmetric representation.
- A single column pattern represents an antisymmetric representation.
- In all other cases, we have mixed symmetry.

11.3 Construction of functions of a given S_n symmetry

In each box, we put one of n different indices, e.g. $1, 2, \ldots, n$. The Young tableau is called **standard**, if the indices cannot decrease in any row, e.g. from left to right, and must increase in any column, e.g. from up to down. Sometimes, this arrangement will be denoted by

$$[f_1, f_2, \ldots, f_r],$$

where r is the number of rows, f_1 is the number of boxes in the first row, f_2 in the second row, etc. We indicate as c the number of columns in a given tableau. In this counting, we do not consider rows and columns with one box.

For $n = 2$ objects, we have two possibilities:

$$\boxed{\begin{array}{|c|c|} \hline 1 & 2 \\ \hline \end{array}}, \qquad \boxed{\begin{array}{|c|} \hline 1 \\ \hline 2 \\ \hline \end{array}},$$

corresponding to [2] $(r = 1, c = 0)$ and $\begin{bmatrix} 1 \\ 1 \end{bmatrix} \equiv [1^2]$ $(r = 0, c = 1)$. Both are standard.

For $n = 3$ objects, the standard tableaux are:

$$\begin{array}{|c|c|c|} \hline 1 & 2 & 3 \\ \hline \end{array}, \quad \begin{array}{|c|c|} \hline 1 & 2 \\ \hline 3 \\ \cline{1-1} \end{array}, \quad \begin{array}{|c|c|} \hline 1 & 3 \\ \hline 2 \\ \cline{1-1} \end{array}, \quad \begin{array}{|c|} \hline 1 \\ \hline 2 \\ \hline 3 \\ \hline \end{array}. \tag{11.14}$$

In the first tableau [3], we have $r = 1, c = 0$, in the tableau [2,1], we have $r = 1, c = 1$ and in tableau [1,1,1], we have $r = 0, c = 1$.

We will see that each tableau designates a representation of the symmetric group, and each of the standard ones that go with it corresponds to a vector associated with this representation. We see that in the above example, the [2,1] representation of the symmetric group S_3 is 2D. The other two are 1D.

11.3.1 *The Young operators*

The k numbers in a given row indicate that we should consider all possible permutations of $S_k, k = 1, 2, \ldots, r$, involving the numbers present in that row. Then, sum them up with a + sign regardless of whether they are even or odd. The resulting operator is symmetric. The ℓ numbers in a given column indicate that we should consider all possible permutations of S_ℓ, $\ell = 1, 2, \ldots, c$. Then, construct an operator summing them up with the sign + or − dependent on whether they are even or odd permutations.

Let us begin by considering a problem in which two particles can occupy two levels, say α and β.

(i) Consider the symmetric case [2], $r = 1, c = 0$. Suppose that the two particles occupy both levels. The occupation is designated by assigning to each level a number indicating the particle occupying it. Only the permutations associated with the row need be considered, and they are the identity E and the permutation $p_{12} = (12)$. The corresponding Young operator is $T([2]) = E + (12)$. If the particles occupy different levels, the symmetric wave function can be obtained by applying the operator in any product of the two functions, e.g. particle 1 occupies the level α and 2 the level β, i.e. $\alpha(1)\beta(2)$. Thus,

$$\psi_{\alpha,\beta}^{[2]} = \frac{1}{\sqrt{2}} T\left([2]\right) |\alpha(1)\beta(2)\rangle = \frac{1}{\sqrt{2}} (\alpha(1)\beta(2) + \alpha(2)\beta(1)). \qquad (11.15)$$

The factor $\frac{1}{\sqrt{2}}$ was introduced to make $\psi_{\alpha,\beta}$ normalized, once the single-particle states have been normalized.

If the two particles can occupy the same level, the solution is trivial, and one does not need the power of Young tableaux. Clearly, the symmetric combinations are

$$\psi_{\alpha,\alpha}^{[2]} = \alpha(1)\alpha(2), \quad \psi_{\beta,\beta}^{[2]} = \beta(1)\beta(2).$$

(ii) Consider the anti-symmetric case [1,1], ($r = 0, c = 1$). Only the permutations associated with the column need to be considered, and they are the identity E and the permutation $p_{12} = (12)$ as before, but now the corresponding Young operator is $T\left([1^2]\right) = E - (12)$.

Clearly, in this case, the two particles must occupy different levels (Pauli principle):

$$\psi_{\alpha,\beta}^{[1^2]} = \frac{1}{\sqrt{2}} T\left([1^2]\right) |\alpha(1)\beta(2)\rangle = \frac{1}{\sqrt{2}} (\alpha(1)\beta(2) - \alpha(2)\beta(1)). \qquad (11.16)$$

Let us now continue by considering three particles which can occupy three levels, say α, β and γ. Then:

(a) Consider the symmetric case [3], (r=1,c=0).
Then, $T([3]) = E + (12) + (13) + (23) + (123) + (132)$ and

$$\psi_{\alpha,\beta,\gamma}^{[3]} = \frac{1}{\sqrt{6}} T\left([6]\right) |\alpha(1)\beta(2)\gamma(3)\rangle$$

$$= \frac{1}{\sqrt{6}} (\alpha(1)\beta(2)\gamma(3) + \alpha(2)\beta(1)\gamma(3) + \alpha(3)\beta(2)\gamma(1)$$

$$+ \alpha(1)\beta(3)\gamma(2) + \alpha(2)\beta(3)\gamma[1] + \alpha(3)\beta(1)\gamma[2]). \qquad (11.17)$$

If two particles can occupy the same level, say α, and the other is put in level γ, one finds

$$\psi_{\alpha,\gamma}^{[3]} = \frac{1}{\sqrt{3}} T\left([6]\right) |\alpha(1)\alpha(2)\gamma(3)\rangle$$

$$= \frac{1}{\sqrt{3}} (\alpha(1)\alpha(2)\gamma(3) + \alpha(2)\alpha(3)\gamma(1) + \alpha(1)\alpha(3)\gamma(2)). \qquad (11.18)$$

The other two possibilities can be found similarly.

(b) Consider the completely anti symmetric case $[1^3]$, $(r = 0, c = 1)$. Then, $T\left([1^3]\right) = E - (12) - (13) - (23) + (123) + (132)$ and

$$\psi_{\alpha,\beta,\gamma}^{[1^3]} = \frac{1}{\sqrt{6}} T\left([1^3]\right) |\alpha(1)\beta(2)\gamma(3)\rangle$$

$$= \frac{1}{\sqrt{6}} (\alpha(1)\beta(2)\gamma(3) - \alpha(2)\beta(1)\gamma(3) - \alpha(3)\beta(2)\gamma(1)$$

$$- \alpha(1)\beta(3)\gamma(2) + \alpha(2)\beta(3)\gamma[1] + \alpha(3)\beta(1)\gamma[2]). \quad (11.19)$$

(iii) Consider the mixed symmetry case $[2, 1]$, $(r = 1, c = 1)$. We distinguish two possibilities:

(a) The numbers 1,2 are in the row, while in the column, we have 1,3. In the row, we associate the operator $T([2])_S = E + (12)$, while in the column, wo have the operator $T([1^2])_A = E - (13)$. Thus,

$$T([2,1]_1) = T([1^2])_A T([2])_S = (E - (13))(E + (12))$$

$$= E + (12) - (13) - (132), \quad \psi_{\alpha,\beta,\gamma}^{[21]_1} = \frac{1}{2} T([21]_1)|\alpha(1)\beta(2)\gamma(3)\rangle$$

$$= \frac{1}{2} (\alpha(1)\beta(2)\gamma(3) + \alpha(2)\beta(1)\gamma(3)$$

$$- \alpha(3)\beta(2)\gamma(1) - \alpha(3)\beta(1)\gamma(2)). \quad (11.20)$$

(b) In the other possibility, the 2 and 3 are interchanged. Thus, we find

$$T([2,1]_2) = (E - (12))(E + (13)) = E - (12) + (13) - (123),$$

$$\psi_{\alpha,\beta,\gamma}^{[21]_2} = \frac{1}{2} T\left([21]_2\right) |\alpha(1)\beta(2)\gamma(3)\rangle$$

$$= \frac{1}{2} (\alpha(1)\beta(2)\gamma(3) - \alpha(2)\beta(1)\gamma(3) + \alpha(3)\beta(2)\gamma(1)$$

$$- \alpha(2)\beta(3)\gamma(1)). \quad (11.21)$$

In the above, we assumed that all single-particle wave functions are different. Obviously, if two of them are the same, one cannot construct an antisymmetrizer with respect to them. The reader is encouraged to apply the analogous to the above procedure in this case.

The above can be generalized. We construct the completely symmetric operators S_{ℓ_i} associated with each row ℓ_i of the Young tableau and the completely antisymmetric operators A_{r_i} corresponding to each column r_i of the Young tableau:

$$\hat{Y}_{S_{\ell_i}} = \sum_{p_{\ell_i}} p_{\ell_i}, \quad p_{\ell_i} = \text{permutation of the numbers } a_1, a_2, \ldots, a_{n(\ell_i)} \text{of line } \ell_i,$$

$$\hat{Y}_{A_{r_i}} = \sum_{p_{r_i}} (-1)^{p_{r_i}} p_{r_i}, \quad p_{r_i} = \text{permutation of } b_1, b_2, \ldots, b_{n(r_i)} \text{of column } r_i,$$

where $n(\ell_i)$ and $n(r_i)$ are the lengths of line ℓ_i and column r_i, respectively, and

$$(-1)^{p_{r_i}} = \begin{cases} 1 & \text{even permutation } p_{r_i}, \\ -1 & \text{odd permutation } p_{r_i}. \end{cases}$$

The corresponding basis vector is

$$\psi_f = \left(\sum_{r_i} \frac{1}{n(r_i)!} \hat{Y}_{A_{r_i}} \right) \left(\sum_{\ell_i} \frac{1}{n(\ell_i)!} \hat{Y}_{S_{\ell_i}} \right) \psi,$$

$$\psi = \psi_\alpha(q_1)\psi_\beta(q_2), \ldots, \psi_\omega(q_n), \tag{11.22}$$

where ψ is the product of the single-particle states of the space considered with the operators acting on the independent variables (coordinates) q_i of the functions.[2] Frequently, we will write $\psi_\gamma(i)$ instead of $\psi_\gamma(q_i)$.

The functions obtained this way constitute a set (Hammermesh, 1964), which characterizes the irreducible representations S_n.

We note that the resulting function is antisymmetric with respect to the indices of the operation performed last, i.e. of $\left(\sum_{r_i} \frac{1}{n(r_i)!} \hat{Y}_{A_{r_i}} \right)$. In relation to the effect of the operator which acted first, i.e. $\left(\sum_{\ell_i} \frac{1}{n(\ell_i)!} \hat{Y}_{S_{\ell_i}} \right)$, all we can say is that the function cannot be antisymmetric with respect to the indices involved in it.

The Young operators discussed above are very useful in the study of the symmetric group. In particular, they very useful in the construction of the irreducible representation s of S_n.

---

[2]There exists an equivalent description with the operator's action on the index of these functions. Then, the antisymmetrization is performed in rows and symmetrization on the columns.

11.3.2   An example — The Irreducible representations of S_3

We begin by writing down the Young operators of S_3. According to what we have seen above, we have the following Young tableaux:

$$\boxed{\begin{array}{|c|c|c|} \hline 1 & 2 & 3 \\ \hline \end{array}} \; , \quad \begin{array}{c} \boxed{\begin{array}{|c|c|} \hline 1 & 2 \\ \hline 3 \end{array}} \end{array} \; , \quad \begin{array}{c} \boxed{\begin{array}{|c|c|} \hline 1 & 3 \\ \hline 2 \end{array}} \end{array} \; , \quad \begin{array}{c} \boxed{\begin{array}{|c|} \hline 1 \\ \hline 2 \\ \hline 3 \end{array}} \end{array} . \tag{11.23}$$

The first has only one line. Thus,

$$\hat{Y}[3] = \hat{Y}_{S(123)} = (E + (12) + (13) + (23) + (123) + (132)).$$

The last one has only one column. Thus,

$$\hat{Y}[1^3] = \hat{Y}_{A(123)} = (E - (12) - (13) - (23) + (123) + (132)).$$

For the mixed 2D symmetry, we have

$$\begin{array}{|c|c|} \hline 1 & 2 \\ \hline 3 \end{array} \Leftrightarrow \hat{Y} = (E - (13))(E + (12)),$$

$$\hat{Y} = (E - (13))(E + (12)) = E + (12) - (13) - (13)(12)$$
$$= E + (12) - (13) - (132),$$

$$\begin{array}{|c|c|} \hline 1 & 3 \\ \hline 2 \end{array} \Leftrightarrow \hat{Y} = (E - (12))(E + (13)),$$

$$\hat{Y} = (E - (12))(E + (13)) = E - (12) + (13) - (12)(13)$$
$$= E - (12) + (13) - (123).$$

Example 2: We begin with the trivial consideration of the 1D representations of S_3.

These are the representations associated with $f = [3]$ and $f = [1^3]$. The corresponding wave functions, in a slightly different notation than in Example 1, are

$$\psi([3]) = \frac{1}{\sqrt{6}}\hat{Y}_{[3]} (\psi_\alpha(1)\psi_\beta(2)\psi_\gamma(3)),$$

$$\psi([1^3]) = \frac{1}{\sqrt{6}}\hat{Y}_{[1^3]} (\psi_\alpha(1)\psi_\beta(2)\psi_\gamma(3)). \tag{11.24}$$

The corresponding functions have been constructed in Eqs. (11.17) and (11.19).

Example 3: We find the 2D representation of S_3.
For the 2D one, we begin with the vector

$$\begin{array}{|c|c|}\hline 1 & 2 \\\hline 3 \\\cline{1-1}\end{array} \Leftrightarrow \hat{Y}_1 = (E - (13))(E + (12)).$$

We then define the operator

$$\hat{Y}_2 = (12)\hat{Y}_1 = (12)(E - (13))(E + (12))$$
$$= ((12) - (12)(13))(E + (12)) = ((12) - (123))(E + (12)).$$

We will show that the functions

$$\psi^{(1)} = \frac{1}{2}\hat{Y}_1\left(\psi_\alpha(1)\psi_\beta(2)\psi_\gamma(3)\right), \ \psi^{(2)} = \frac{1}{2}\hat{Y}_2\left(\psi_\alpha(1)\psi_\beta(2)\psi_\gamma(3)\right) \quad (11.25)$$

constitute a basis in the space of the 2D representation. Indeed,

$$(13)\hat{Y}_1 = -\hat{Y}_1, \ (12)\hat{Y}_1 = \hat{Y}_2,$$
$$(23)\hat{Y}_1 = (23)(E - (13))(E + (12)) = (23) - (23)(13))(E + (12))$$
$$= ((23) - (132))(E + (12)) \equiv \hat{Y}_3,$$
$$\hat{Y}_1 - \hat{Y}_2 - \hat{Y}_3 = (E - (13) - (12) + (123) - (23) + (132))(E + (12))$$
$$= A(E + (12))$$

But $A(E + (12))\left(\psi_\alpha(1)\psi_\beta(2)\psi_\gamma(3)\right) = 0 \Rightarrow \hat{Y}_1 = \hat{Y}_2 + \hat{Y}_3.$
Hence,

$$(23)\hat{Y}_1 = \hat{Y}_1 - \hat{Y}_2.$$

Similarly,

$$(123)\hat{Y}_1 = ((123) - (123)(13))(E + (12)) = ((123) - (12))(E + (12)).$$

However,

$$\hat{Y}_1 - \hat{Y}_3 = (E - (13) - (23) + (132))(E + (12))$$
$$= (A + (12) - (123))(E + (12))$$
$$= ((12) - (123))(E + (12)) \Rightarrow (123)\hat{Y}_1 = -\hat{Y}_2,$$
$$(132)\hat{Y}_1 = ((132) - (132)(13))(E + (12)) = ((132) - (23))(E + (12))$$
$$= -\hat{Y}_3 = \hat{Y}_2 - \hat{Y}_1.$$

As a result, writing $\psi = \psi_\alpha(1)\psi_\beta(2)\psi_\gamma(3)$, we obtain

$$(13)\psi^{(1)} = \frac{1}{2}(13)\hat{Y}_1\psi = -\psi^{(1)},$$

$$(13)\psi^{(2)} = \frac{1}{2}(13)(12)\hat{Y}_1\psi = \frac{1}{2}(132)\hat{Y}_1\psi = \frac{1}{2}(\hat{Y}_2 - \hat{Y}_1)\psi = \psi^{(2)} - \psi^{(1)} \Rightarrow$$

$$T(E) = \begin{pmatrix} 1 & 0 \\ 0 & 1 \end{pmatrix}, \quad T(13) = \begin{pmatrix} -1 & -1 \\ 0 & 1 \end{pmatrix}, \tag{11.26}$$

$$(12)\psi^{(1)} = \frac{1}{2}\hat{Y}_1\psi = \psi^{(2)}, \ (12)\psi^{(2)} = \frac{1}{2}\hat{Y}\psi = \psi^{(1)} \Rightarrow$$

$$T(12) = \begin{pmatrix} 0 & 1 \\ 1 & 0 \end{pmatrix}, \tag{11.27}$$

$$(23)\psi^{(1)} = \frac{1}{2}\hat{Y}_3\psi = \frac{1}{2}(\hat{Y}_1 - \hat{Y}_2)\psi = \psi^{(1)} - \psi^{(2)},$$

$$(23)\psi^{(2)} = \frac{1}{2}(23)(12)\hat{Y}_1\psi = \frac{1}{2}(123)\hat{Y}_1\psi = -\frac{1}{2}\hat{Y}_2\psi = -\psi^{(2)} \Rightarrow$$

$$T(23) = \begin{pmatrix} 1 & 0 \\ -1 & -1 \end{pmatrix}, \tag{11.28}$$

$$(123)\psi^{(1)} = \frac{1}{2}(123)\hat{Y}_1\psi = -\frac{1}{2}\hat{Y}_2\psi = -\psi^{(2)},$$

$$(123)\psi^{(2)} = \frac{1}{2}(123)(12)\hat{Y}_1\psi = \frac{1}{2}(23)\hat{Y}_1\psi = \frac{1}{2}(\hat{Y}_1 - \hat{Y}_2)\psi = \psi^{(1)} - \psi^{(2)} \Rightarrow$$

$$T(123) = \begin{pmatrix} 0 & 1 \\ -1 & -1 \end{pmatrix}, \tag{11.29}$$

$$(132)\psi^{(1)} = \frac{1}{2}(132)\hat{Y}_1\psi = \frac{1}{2}(\hat{Y}_2 - \hat{Y}_1)\psi = -\psi^{(1)} + \psi^{(2)},$$

$$(132)\psi^{(2)} = \frac{1}{2}(132)(12)\hat{Y}_1\psi = \frac{1}{2}(13)\hat{Y}_1\psi = -\frac{1}{2}\hat{Y}_1\psi = -\psi^{(1)} \Rightarrow$$

$$T(132) = \begin{pmatrix} -1 & -1 \\ 1 & 0 \end{pmatrix}. \tag{11.30}$$

The above basis, however, is not orthogonal. One can easily see that the matrix of overlaps, ovlp, involving the scalar products is

$$\text{ovlp} = \begin{pmatrix} \langle \psi^{(1)}|\psi^{(1)}\rangle & \langle \psi^{(1)}|\psi^{(2)}\rangle \\ \langle \psi^{(2)}|\psi^{(1)}\rangle & \langle \psi^{(2)}|\psi^{(2)}\rangle \end{pmatrix} = \begin{pmatrix} 1 & \frac{1}{2} \\ \frac{1}{2} & 1 \end{pmatrix}. \tag{11.31}$$

A convenient orthonormal basis can be obtained by taking the eigenvectors of the matrix ovlp, which are

$$|1\rangle = \frac{1}{\sqrt{3}}(\psi^{(1)} + \psi^{(2)}), \quad |2\rangle = (\psi^{(1)} - \psi^{(2)}),$$

$$S = \begin{pmatrix} \frac{1}{\sqrt{3}} & 1 \\ \frac{1}{\sqrt{3}} & -1 \end{pmatrix}, \quad S^{-1} = \begin{pmatrix} \frac{\sqrt{3}}{2} & \frac{\sqrt{3}}{2} \\ \frac{1}{2} & -\frac{1}{2} \end{pmatrix}.$$

Thus, one obtains the unitary representation

$$\Gamma(X) = S^{-1}T(X)S.$$

That is, explicitly

$$\Gamma(12) = \begin{pmatrix} 1 & 0 \\ 0 & -1 \end{pmatrix}, \quad \Gamma(13) = \begin{pmatrix} -\frac{1}{2} & -\frac{\sqrt{3}}{2} \\ -\frac{\sqrt{3}}{2} & \frac{1}{2} \end{pmatrix}, \quad \Gamma(23) = \begin{pmatrix} -\frac{1}{2} & \frac{\sqrt{3}}{2} \\ \frac{\sqrt{3}}{2} & \frac{1}{2} \end{pmatrix},$$

$$\Gamma(123) = \begin{pmatrix} -\frac{1}{2} & -\frac{\sqrt{3}}{2} \\ \frac{\sqrt{3}}{2} & -\frac{1}{2} \end{pmatrix}, \quad \Gamma(132) = \begin{pmatrix} -\frac{1}{2} & \frac{\sqrt{3}}{2} \\ -\frac{\sqrt{3}}{2} & -\frac{1}{2} \end{pmatrix}.$$

This is the standard form of the 2D S_3 representation obtained most simply, if one could guess the basis

$$|1\rangle = \frac{1}{\sqrt{6}}(\psi(1) + \psi(2) - 2\psi(3)), \quad |2\rangle = \frac{1}{\sqrt{2}}(\psi(1) - \psi(2)), \qquad (11.32)$$

as the reader can easily verify. This is usually referred to as the standard form. One can avoid the tedious path of the Young tableaux, but this requires imagination!

The above set of matrices can be brought into another frequently used form by an additional unitary transformation:

$$V = \begin{pmatrix} \frac{1}{2} & -\frac{\sqrt{3}}{2} \\ \frac{\sqrt{3}}{2} & \frac{1}{2} \end{pmatrix},$$

which diagonalizes the permutation (23). Thus, we find

$$\Gamma(E) = \begin{pmatrix} 1 & 0 \\ 0 & 1 \end{pmatrix}, \quad \Gamma(12) = \begin{pmatrix} -\frac{1}{2} & -\frac{\sqrt{3}}{2} \\ -\frac{\sqrt{3}}{2} & \frac{1}{2} \end{pmatrix}, \quad \Gamma(13) = \begin{pmatrix} -\frac{1}{2} & \frac{\sqrt{3}}{2} \\ \frac{\sqrt{3}}{2} & \frac{1}{2} \end{pmatrix},$$

$$\Gamma(23) = \begin{pmatrix} 1 & 0 \\ 0 & -1 \end{pmatrix} \Gamma(123) = \begin{pmatrix} -\frac{1}{2} & -\frac{\sqrt{3}}{2} \\ \frac{\sqrt{3}}{2} & -\frac{1}{2} \end{pmatrix}, \quad \Gamma(132) = \begin{pmatrix} -\frac{1}{2} & \frac{\sqrt{3}}{2} \\ -\frac{\sqrt{3}}{2} & -\frac{1}{2} \end{pmatrix}.$$

Remark: The above process could have been applied considering the

Young tableau $\begin{array}{|c|c|} \hline 1 & 3 \\ \hline 2 \\ \cline{1-1} \end{array}$.

11.4   Construction of symmetric and antisymmetric functions in two spaces

In this section, we are going to construct completely symmetric and anti-symmetric functions for single-particle states defined in two spaces, involving, for example, orbital and spin degrees of freedom. We will merely outline the procedure here, and we refer the reader to the literature for details, see, for example, Hammermesh (1964).

11.4.1   *Construction of antisymmetric functions*

Definition: The *adjoined (conjugate)* of a given Young tableau $[f]$ is the tableau $[\tilde{f}]$ obtained from that of $[f]$ by interchanging the role of rows and columns. For example,

Whenever $[f] = [\tilde{f}]$, the tableau is called self-adjoined as, for example, in

the case of the tableau $\begin{array}{|c|c|} \hline & \\ \hline \end{array}$.

Theorem: *The product of a space function with another one depending on spin is antisymmetric when it can be cast in the form*

$$[1^n] = [f] \times [\tilde{f}], \tag{11.33}$$

where $[f]$ *is an irreducible representation of S_n and $\times$ indicates the direct product.*

Before proceeding to the proof, we state the following lemma.

Lemma: *Consider one irreducible representation $[f]$ of S_n. Then, we have*

$$[f] \times [1^n] = [\tilde{f}]. \tag{11.34}$$

The proof is quite simple. Since $[1^n]$ is 1D, $[f] \times [1^n]$ is irreducible, say $[f']$. Then, we construct the functions $\Psi_{[f]}$ and $\Psi_{[1^n]}$. The first is antisymmetric with respect to the indices included in the columns of $[f]$, while the second is antisymmetric with respect to all indices. Therefore, the product gives a function which is symmetric with respect to the indices included in the columns of the tableau $[f']$. Thus, $[f']$ is nothing but the adjoined diagram of $[f]$, $[f'] = [\tilde{f}]$.

We now proceed with the proof of the theorem. From Eq. (11.34), one derives the relation of characters:

$$\chi^{[\tilde{f}]}(p) = \chi^{[f]}(p)\chi^{[1^n]}(p), \ p \in S_n. \tag{11.35}$$

However,

$$\chi^{[1^n]}(p) = (-1)^{\epsilon(p)} \Rightarrow$$

$$\chi^{[\tilde{f}]}(p) = \chi^{[f]}(p)(-1)^{\epsilon(p)}, \quad p \in S_n. \tag{11.36}$$

We now consider the product $[f] \times [f']$, which is reducible. Thus, it can be cast in the form:

$$[f] \times [f'] = \alpha_{[1^n]}[1^n] \oplus \cdots, \tag{11.37}$$

$$\alpha_{[1^n]} = \frac{1}{n!} \sum_{p \in S_n} \chi^{[f]}(p)\chi^{[f']}(p)$$

$$= \frac{1}{n!} \sum_{p \in S_n} \chi^{[\tilde{f}]}(p)(-1)^{\epsilon(p)}(-1)^{\epsilon(p)}\chi^{[f']}(p) \Rightarrow$$

$$\alpha_{[1^n]} = \frac{1}{n!} \sum_{p \in S_n} \chi^{[\tilde{f}]}(p)\chi^{[f']}(p) = \begin{cases} 1 & \text{if} [f'] = [\tilde{f}] \\ 0 & \text{if } [f'] \neq [\tilde{f}] \end{cases},$$

which constitutes a proof of the theorem, yielding the relation

$$\Psi_F^{[f]} = \frac{1}{\dim([f])} \sum_{p \in S_n} (-1)^{\epsilon(p)} \Psi_{[f]}(\mathbf{r}_1, \mathbf{r}_2, \ldots, \mathbf{r}_n) \chi_{[\tilde{f}]}(\mathbf{s}_1, \mathbf{s}_2, \ldots, \mathbf{s}_n),$$

$$\epsilon(p) = \begin{cases} 0 & p = \text{even} \\ 1 & p = \text{odd} \end{cases}. \tag{11.38}$$

11.4.2   *Construction of symmetric functions*

We proceed as above but now considering the 1D completely symmetric case $[n]$, $\chi^{[n]}(p) = 1$. Hence,

$$[f] \times [f'] = \alpha_{[n]}[n] \oplus \cdots \Rightarrow \alpha_{[n]} = \begin{cases} 1 & \text{if } [f'] = [f] \\ 0 & \text{if} [f'] \neq [f] \end{cases}, \tag{11.39}$$

$$\Psi_B^{[f]} = \frac{1}{dim([f])} \sum_{p \in S_n} \Psi_{[f]}(\mathbf{r}_1, \mathbf{r}_2, \ldots, \mathbf{r}_n) \chi_{[f]}(\mathbf{s}_1, \mathbf{s}_2, \ldots, \mathbf{s}_n). \tag{11.40}$$

As we have mentioned earlier, the above results hold whenever the interaction is spin independent. In the general case, the above functions only constitute a basis, meaning that the eigenfunctions of the Hamiltonian can be expressed as

$$\Psi_F = \sum_{[f]} \Psi_F^{[f]}, \quad \Psi_B = \sum_{[f]} \Psi_B^{[f]}. \tag{11.41}$$

11.5   Kronecker products and the emergence of tensors

A tensor is the generalization of the notion of vector. A vector is any quantity (v^1, v^2, v^3) that, under rotations, transforms exactly like the coordinates (x^1, x^2, x^3). Thus,

$$x'^i = \alpha^i_j x^j \text{(coordinates)} \Leftrightarrow v'^i = \alpha^i_j v^j \text{(vector)}. \tag{11.42}$$

In the above expressions, the repeated indices are assumed to be summed over. We have deliberately used upper indices to specify the components of the vector quantity. The transformation matrix is purposely written with one lower and one upper index, but it is understood that the lower index refers to the rows and the upper index to the columns.

The product of two vectors is transformed as

$$v'^{i_1} v'^{i_2} = \alpha^{i_1}_{j_1} \alpha^{i_2}_{j_2} v^{j_1} v^{j_2}.$$

Any quantity $F^{i_1 i_2}$ transforming like this, i.e.

$$F'^{\,i_1 i_2} = \alpha^{i_1}_{j_1} \alpha^{i_2}_{j_2} F^{j_1 j_2}, \tag{11.43}$$

is called a **second rank tensor**. In the above expression again, summation over repeated indices is understood.

The product of two vectors is not, of course, the only second rank tensor, but any tensor can be expressed as a linear combination of terms of the form of Eq. (11.43). A linear combination of tensors of a given rank is also a tensor of the same rank. Of particular interest are the combinations

$$S^{ij} = F^{ij} + F^{ji} (\text{symmetric}), \quad A^{ij} = F^{ij} - F^{ji} (\text{anti-symmetric}).$$

It is easy to see that this symmetry does not change under the above transformations, i.e.

$$\alpha^{i}_{k} \alpha^{j}_{m} S^{km} = (S^{ij})', \quad \alpha^{i}_{k} \alpha^{j}_{m} A^{km} = (A^{ij})'.$$

The above concept can be generalized to an arbitrary rank r:

$$F^{i_1, i_2, \ldots, i_r} = \alpha^{i_1}_{j_1} \alpha^{i_2}_{j_2}, \ldots, \alpha^{i_r}_{j_r} F^{j_1, j_2, \ldots, j_r}. \tag{11.44}$$

This transformation is represented by the Kronecker product:

$$T = \left(\underbrace{(\alpha) \otimes (\alpha) \otimes \quad \cdots \quad (\alpha) \otimes (\alpha)}_{r - \text{times}} \right).$$

Unfortunately, the representation T is not, in general, irreducible. In this case, we say that the associated tensor is reducible. In the reduction, the discrete symmetry group S_r will play an important role.

11.6 Construction of tensors of a given symmetry

The construction of completely symmetric tensors is easy. All one has to do is to apply all possible permutations on the r indices on any tensor of the form of Eq. (11.44), i.e.

$$F^{S(1,2,\ldots,r)} = SF^{i_1,i_2,\ldots,i_r}, \quad S = \sum_{\{j_1,j_2,\ldots,j_r\}} \begin{pmatrix} i_1 & i_2 & \cdots & i_r \\ j_1 & j_2 & \cdots & j_r \end{pmatrix}.$$

Similarly, for antisymmetric tensors,

$$F^{A(1,2,\ldots,r)} = AF^{i_1,i_2,\ldots,i_r}, \quad A = \sum_{\{j_1,j_2,\ldots,j_r\}} (-1)^{\pi(i,j)} \begin{pmatrix} i_1 & i_2 & \cdots & i_r \\ j_1 & j_2 & \cdots & j_r \end{pmatrix},$$

where $(-1)^{\pi(i,j)}$ is $+1$ if the permutation is even or -1 if it is odd.

In practice, we need to consider tensors of mixed symmetry. For this purpose, the Young tableau (diagram) is very useful.[3] As we have discussed in Section 11.2, it consists of r boxes arranged in at most r rows with each row containing at most r boxes. In each box, we put one of r different indices, e.g. $1, 2, \ldots, r$. It may be useful to remind the reader here that in a standard Young tableau, the indices cannot decrease in any row, e.g. from left to right, and must increase in any column, e.g. from up to down. Sometimes, this arrangement will be denoted by

$$[f_1, f_2, \ldots, f_s],$$

where s is the number of rows, f_1 is the number of boxes in the first row, f_2 in the second row, etc. Recall that for $r = 2$, we have two possibilities:

corresponding to $[2]$ and $\begin{bmatrix} 1 \\ 1 \end{bmatrix} \equiv [1^2]$. Both are standard. For three objects, the standard tableaux are

                                 (11.45)

There is only one way to fill in the first arrangement of boxes (single row) to form a standard tableau, and the same is true for the last one (single column). In the middle arrangement, however, the numbers can be placed

---

[3] For applications of the Young tableaux to physics, we refer the reader to the literature (Chen *et al.*, 2002).

in two different ways. For each Young tableau $[f]$, we can associate a Young operator $Y[f]$. For $r = 2$, we have the following:

- The symmetric case.

$$Y\left(\boxed{1\ 2}\right) = 1 + P_{12} \Rightarrow T\left(\boxed{i_1\ i_2}\right) = Y\left(\boxed{1\ 2}\right) F^{i_1,i_2} =$$

$$F^{i_1,i_2} + F^{i_2,i_1}.$$

Since $i_1 \le i_2$ for three objects, we have the following tensor possibilities: (i) those with identical indices $F^{i_1 i_1}$, $F^{i_2 i_2}$, $F^{i_3 i_3}$ which are already symmetric.

(ii) three tensors with different indices:

$$T\left(\boxed{i_1\ i_2}\right) = Y\left(\boxed{1\ 2}\right) F^{i_1 i_2} = F^{i_1 i_2} + F^{i_2 i_1},$$

$$T\left(\boxed{i_1\ i_3}\right) = Y\left(\boxed{1\ 3}\right) F^{i_1 i_3} = F^{i_1 i_3} + F^{i_3 i_1}$$

$$T\left(\boxed{i_2\ i_3}\right) = Y\left(\boxed{2\ 3}\right) F^{i_2 i_3} = F^{i_2 i_3} + F^{i_3 i_2}.$$

In general, for n objects, we have

$$N(n, [2]) = n + \binom{n}{2} = n + \frac{1}{2}n(n-1) = n(n+1).$$

- The antisymmetric case:

$$Y\left(\begin{array}{c}\boxed{1}\\\boxed{2}\end{array}\right) = 1 - P_{12}.$$

We normally write for $i_1 < i_2$,

$$Y\left(\begin{array}{c}\boxed{1}\\\boxed{2}\end{array}\right) = 1 - (12) \Rightarrow T\left(\begin{array}{c}\boxed{i_1}\\\boxed{i_2}\end{array}\right) = Y\left(\begin{array}{c}\boxed{1}\\\boxed{2}\end{array}\right) F^{i_1 i_2} = F^{i_1,i_2} - F^{i_2,i_1},$$

i.e. the usual antisymmetric tensor normally written as $x^{i_1} \otimes x^{i_2}$. For 3 objects, we have

$$Y\left(\begin{array}{c}\boxed{1}\\\boxed{2}\end{array}\right), \quad Y\left(\begin{array}{c}\boxed{1}\\\boxed{3}\end{array}\right), \quad Y\left(\begin{array}{c}\boxed{2}\\\boxed{3}\end{array}\right).$$

In general, for n objects, we have

$$N\left(n, [1^2]\right) = \binom{n}{2} = \frac{1}{2}n(n-1).$$

possibilities. Let us now consider the antisymmetric tensors of rank $r = 3$, corresponding to $[f] = [1^3]$, i.e. the last diagram of Eq. (11.14). For this, take three objects i_1, i_2, i_3. We have

$$Y \begin{pmatrix} \boxed{\begin{matrix} 1 \\ 2 \\ 3 \end{matrix}} \end{pmatrix} = 1 - (12) - (13) - (23) + (123) + (132) \Rightarrow$$

$$T \begin{pmatrix} \boxed{\begin{matrix} i_1 \\ i_2 \\ i_3 \end{matrix}} \end{pmatrix} - Y \begin{pmatrix} \boxed{\begin{matrix} 1 \\ 2 \\ 3 \end{matrix}} \end{pmatrix} F^{i_1 i_2 i_3} = F^{i_1 i_2 i_3} - F^{i_2 i_1 i_3} - F^{i_3 i_2 i_1}$$

$$- F^{i_1 i_3 i_2} + F^{i_2 i_3 i_1} + F^{i_3 i_1 i_2}.$$

There exist

$$N\left(n, [1^3]\right) = \binom{n}{3} = \frac{1}{6}n(n-1)(n-2)$$

possibilities.

For the symmetric case, we similarly find

$$Y\left(\boxed{1\,|\,2\,|\,3}\right) = 1 + (12) + (13) + (23) + (123) + (132) \Rightarrow$$

$$T\left(\boxed{i_1\,|\,i_2\,|\,i_3}\right) = Y\left(\boxed{1\,|\,2\,|\,3}\right) F^{i_1 i_2 i_3}$$

$$= F^{i_1 i_2 i_3} + F^{i_2 i_1 i_3} + F^{i_3 i_2 i_1} + F^{i_1 i_3 i_2} + F^{i_2 i_3 i_1} + F^{i_3 i_1 i_2}.$$

For $i_1 \leq i_2 \leq i_3$, there exist:

$$N\left(n, [3]\right) = \binom{n+2}{3} = \frac{(n+2)!}{3!(n-1)!} = \frac{1}{6}n(n+1)(n+2)$$

possibilities.

To proceed further, we need a recipe on how to deal with diagrams with more than one row and one column. Let us suppose that we have k rows and s columns. Then, construct the operator

$$Y_S = Y_{S_k} Y_{S_{k-1}} \cdots Y_{S_2}, Y_{S_1},$$

where Y_{S_i} is the symmetric Young operator corresponding to the row i. Next, we construct a similar operator corresponding to the columns

$$Y_A = Y_{A_s} Y_{A_{s-1}} \cdots Y_{A_2}, Y_{A_1},$$

where Y_{A_i} is the antisymmetric Young operator corresponding to the column i. The Young operator corresponding to the diagram $[f]$ is defined to be

$$Y[f] = Y_A Y_S.$$

For example, for $r = 3$, $k = 2$ and $s = 2$, i.e. corresponding to the symmetry $[2, 1]$, we have

$$Y\left(\begin{array}{|c|c|}\hline 1 & 2 \\\hline 3 \\\hline\end{array}\right) = (1 - P_{13})(1 + P_{12}), \ Y\left(\begin{array}{|c|c|}\hline 1 & 3 \\\hline 2 \\\hline\end{array}\right) = ((1 - P_{12})(1 + P_{13}) \Rightarrow$$

$$T\left(\begin{array}{|c|c|}\hline i_1 & i_2 \\\hline i_3 \\\hline\end{array}\right) = Y\left(\begin{array}{|c|c|}\hline 1 & 2 \\\hline 3 \\\hline\end{array}\right) F^{i_1 i_2 i_3} = (1 - P_{13})(1 + P_{12}) F^{i_1 i_2 i_3}$$

$$= (1 - P_{13})\left(F^{i_1 i_2 i_3} + F^{i_2 i_1 i_3}\right) = \left(F^{i_1 i_2 i_3} + F^{i_2 i_1 i_3}\right)$$
$$- \left(F^{i_3 i_2 i_1} + F^{i_2 i_3 i_1}\right).$$

Similarly,

$$T\left(\begin{array}{|c|c|}\hline i_1 & i_3 \\\hline i_2 \\\hline\end{array}\right) = Y\left(\begin{array}{|c|c|}\hline 1 & 3 \\\hline 2 \\\hline\end{array}\right) F^{i_1 i_2 i_3} = (1 - P_{12})(1 + P_{13}) F^{i_1 i_2 i_3}$$

$$= (1 - P_{12})\left(F^{i_1 i_2 i_3} + F^{i_3 i_2 i_1}\right) = \left(F^{i_1 i_2 i_3} + F^{i_3 i_2 i_1}\right)$$
$$- \left(F^{i_2 i_1 i_3} + F^{i_3 i_1 i_2}\right).$$

Thus for $n = 3$, we have the following tensor possibilities:

$$T\left(\begin{array}{|c|c|}\hline 1 & 1 \\\hline 2 \\\cline{1-1}\end{array}\right), T\left(\begin{array}{|c|c|}\hline 1 & 2 \\\hline 2 \\\cline{1-1}\end{array}\right), T\left(\begin{array}{|c|c|}\hline 1 & 3 \\\hline 2 \\\cline{1-1}\end{array}\right), T\left(\begin{array}{|c|c|}\hline 1 & 1 \\\hline 3 \\\cline{1-1}\end{array}\right),$$

$$T\left(\begin{array}{|c|c|}\hline 1 & 2 \\\hline 3 \\\cline{1-1}\end{array}\right), T\left(\begin{array}{|c|c|}\hline 1 & 3 \\\hline 3 \\\cline{1-1}\end{array}\right), T\left(\begin{array}{|c|c|}\hline 2 & 2 \\\hline 3 \\\cline{1-1}\end{array}\right), T\left(\begin{array}{|c|c|}\hline 2 & 3 \\\hline 3 \\\cline{1-1}\end{array}\right).$$

For arbitrary n, we find

$$N([2,1],n) = \frac{1}{3}n(n+1)(n-1)$$

(see the following section). This gives $N[3] = 8$ as in the above set. Note, however, that the second, sixth and eighth diagrams give a vanishing result acting on any tensor. Some of the Young operators maybe not be useful!

11.7 Kronecker products $[f] \otimes [f']$

We analyze the Kronecker products using the techniques of the Young tableaux. This topic is interesting not only in the case of the symmetric group S_n but also in the study of the classical continuous groups, see Vergados (2017), and their applications, see, for example, Vergados and Moustakidis (2021). The reason is that their representations are labeled by the irreducible representations of S_n.

Suppose we are given two tensor representations $T([f])$ and $T(f')$. The product $T([f]) \otimes T([f'])$ is, in general, a reducible tensor representation, but it can be expressed as a linear combination of irreducible tensors:

$$T([f]) \otimes T([f']) = \sum_{[f'']} \alpha_{[f],[f'],[f'']} T[(f'')].$$

The question of evaluating the coefficients $\alpha_{[f],[f'],[f'']}$ is complicated, and it will not be addressed here[4]. At this point, we would like to see how one can find the possible Young tableaux $[f'']$ corresponding to the "product" $[f] \otimes [f']$. We assume that the first tableau corresponds to the symmetry group S_r and the second to the symmetry group $S_{r'}$ while the product to the group $S_{r+r'}$. This product will be indicated as the "external product". We will not prove the relevant theorems, but we will only state the rules

---

[4]In the simple cases $[f'] = [2]$ and $[1^2]$, which are of practical interest, such coefficients have been obtained see, for example, Vergados (2005), Chen *et al.* (2002).

and apply them in some simple cases of practical interest. The standard rules are as follows:

(1) Select the most complicated symmetry $[f]$, usually containing the largest number of boxes, and write the corresponding Young tableaux.
(2) Put some numbers in the Young tableau corresponding to the symmetry $[f']$, e.g. the number 1 in the first row, the number 2 in the second row, etc. We add the boxes of $[f']$ one by one on $[f]$ in all possible ways to yield an acceptable Young tableaux but with the following restrictions:
(3) The number of boxes in $[f'']$ must be equal to the sum of the number of boxes of $[f]$ and $[f']$.
(4) The boxes with the same number cannot belong to the same column.
(5) In the resulting Young tableaux, when reading the elements of a row from right to left, every number 2 must be preceded by at least one number 1, every number 3 must be preceded by at least one number 2, etc.

Let us begin with the very simple cases:

(i) One of the Young tableaux has only one box.
Let us begin with the two simple cases $\boxed{}\otimes\boxed{}$ and $\boxed{}\;\boxed{}\otimes\boxed{}$. Since the second tableaux in both cases has only one box, there is no need to put any numbers in the second box. Thus,

$$\boxed{}\otimes\boxed{} \rightarrow \boxed{} + \boxed{}.$$

From this reduction, we can find the dimension of [2,1] in the case of n objects. All we need to do is to equate the dimensions of the left and right of the above equation:

$$N([1^2, n]) \times N([1, n]) = N([[1, 1, 1], n]) + N([2, 1], n) \Rightarrow$$

$$\frac{1}{2}n(n-1) \times n = \frac{1}{6}n(n-1)(n-2) + N([2, 1], n) \Rightarrow$$

$$N([2, 1], n) = \frac{1}{3}n(n+1)(n-1).$$

Similarly, we find

$$\boxed{}\otimes\boxed{} \rightarrow \boxed{} + \boxed{}.$$

The reductions

$$[3] \otimes [1] \to [4] + [3,1],$$

$$[1,1,1] \otimes [1] \to [1,1,1] + [2,1,1],$$

$$[2,1] \otimes [1] \to [3,1] + [2,2],$$

can be obtained equally simply. Applying the rule of dimensions, one can find the dimensions of $N([3,1],n)$, $N([2,1,1],n)$ and $N([2,2],n)$ from $N([2,1],n)$ found above and the known dimensions of $N([4],n)$, $N([1,1,1]n)$ and $N([1,1,1,1],n)$.

(ii) Let us consider the case $[2,1] \otimes [2]$.

First, we write it as $[2,1] \otimes 1\,1$. Then,

$$[2,1] \otimes 1\,1 \to [3,2] + [3,1,1] + [2,2,1].$$

The middle diagram is not consistent with the above rules, and it is omitted. Thus,

$$[2,1] \otimes [2] \to [3,2] + [2,2,1]. \tag{11.46}$$

All representations that appear in the product space are single.

(iii) Let us now consider a more complicated case: $[2,1] \otimes [2,1]$. We write it as $[2,1] \otimes \begin{smallmatrix}1&1\\2\end{smallmatrix}$. Then, we proceed successively for each row of the second tableau.

As a first step, we have

$$\begin{array}{c}\square\\\square\square\end{array}\otimes\boxed{1\ 1}\;\rightarrow\;\boxed{1\ 1}\;+\;\boxed{1}\;+\;\boxed{1}\;+\;\boxed{1}\;+\;\boxed{1}\;.$$

The last Young tableau does not satisfy the above rules and is omitted. Thus, the correct expression is

$$\begin{array}{c}\square\\\square\square\end{array}\otimes\boxed{1\ 1}\;\rightarrow\;\boxed{1\ 1}\;+\;\boxed{1}\;+\;\boxed{1}\;+\;\boxed{1}\;. \qquad (11.47)$$

We now continue with the second step:

$$\boxed{1\ 1}\otimes\boxed{2}\;\rightarrow\;\boxed{\begin{smallmatrix}1\ 1\\2\end{smallmatrix}}\;+\;\boxed{1\ 1\ 2}\;,$$

$$\boxed{\begin{smallmatrix}1\\1\end{smallmatrix}}\otimes\boxed{2}\;\rightarrow\;\boxed{\begin{smallmatrix}1\\1\ 2\end{smallmatrix}}\;+\;\boxed{\begin{smallmatrix}1\\2\end{smallmatrix}}\;.$$

The diagram

$$\boxed{\begin{smallmatrix}\ \ 1\ 2\\1\end{smallmatrix}}$$

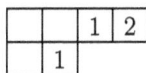

does not satisfy the above rules and has been omitted. Furthermore:

$$\boxed{\begin{smallmatrix}1\\1\end{smallmatrix}}\otimes\boxed{2}\;\rightarrow\;\boxed{\begin{smallmatrix}1\\2\\1\end{smallmatrix}}\;+\;\boxed{\begin{smallmatrix}1\\1\\2\end{smallmatrix}}\;,$$

$$\boxed{\begin{smallmatrix}1\\1\end{smallmatrix}}\otimes\boxed{2}\;\rightarrow\;\boxed{\begin{smallmatrix}1\\1\ 2\end{smallmatrix}}\;+\;\boxed{\begin{smallmatrix}1\\1\\2\end{smallmatrix}}\;.$$

As a result,

$$\begin{array}{ccccc} & & & & & & & & & \\ \end{array} \tag{11.48}$$

Note that the above rules yield the multiplicities, if any. For example, the [3,2,1] obtained in two different ways are

11.8   Problems

11.2.1 In a system of three particles, the available levels (single-particle states) are $\psi_\alpha, \psi_\beta, \psi_\gamma, \psi_\delta$, not necessarily different:

- Construct a completely symmetric function. Is it uniquely defined?
- If not, construct all such functions.
- Construct a completely antisymmetric function.

11.2.2 Consider the single-particle states $\psi_\alpha, \psi_\beta, \psi_\gamma, \psi_\delta$:

- Construct a completely symmetric function of four particles. Is it uniquely defined?
- Do the same if $\psi_\gamma = \psi_\delta$.
- Do the same if $\psi_\beta = \psi_\gamma = \psi_\delta$.
- Construct a completely antisymmetric function of four particles.

11.3.1 Construct a proper function of three Fermion particles involving three orbitals (ordinary space single-particle states) and three spin states:

- when the spin function is completely symmetric.
- when the spin function is completely antisymmetric.
- when the spin function is of mixed symmetry.

11.3.2 Do the same if the particles are Bosons.

11.3.3 Construct a completely antisymmetric function of three single-particle states with spin $3/2$, with total spin $S = 1/2$ and the total spin projection $m_S = 1/2$.

Note: the single-particle states are characterized by m_1, m_2, m_3, respectively, with $m_1 + m_2 + m_3 = m_S$.

11.5.1 Quarks are particles, each with spin $1/2$ and isospin $1/2$. A proton is made up of three such quarks with total spin $S = 1/2, m_S = 1/2$ and isospin $I = 1/2, m_I = 1/2$. Construct the total wave function which is totally symmetric in the combined spin–isospin space with mixed spin symmetry. This is possible in this case, since there exists an additional part called color, in which the function is antisymmetric.

11.6.1 Find the products:

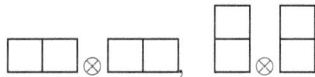

11.6.2 Do the same for the following cases:

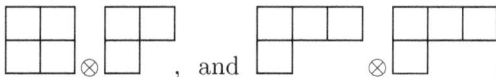

11.6.3 Do the same for the following case:

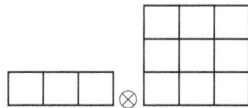

11.6.4 Do the same for the following case:

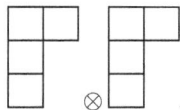

Confirm the obtained result by considering that found in problem 11.6.2 by noting that the latter involves the adjoined diagrams of those encountered here.

11.7.1 Three particles can occupy four different single-particle states $\alpha, \beta, \gamma, \delta$:

- Put these particles in the single-particle states in all possible ways, assuming that the particles are Fermions.
- Write down the wave function for each case.

- Assuming that the Hamiltonian describing the system is invariant under S_n transformations, what are the expected wave functions?

11.7.2 Four particles can occupy four different single-particle states $\alpha, \beta, \gamma, \delta$.

Give an expression yielding the form of the Hamiltonian function describing the system in terms of the elementary single- and two-particle elements of the Hamiltonian, assuming that it remains invariant under S_n.

Chapter 12

Further Applications in Molecular Physics and Crystal Structure — Crystal Harmonics

In this chapter, we consider applications of discrete group theory to molecular physics and crystal structure. We are mainly concern ourselves with the reduction of the representations of the continuous group $SO(3)$, both single-valued and doubled-valued, under the discrete groups of Section 2.4. We also define and explore crystal harmonics.

12.1 Introductory notions

We begin by emphasizing the symmetries

$$C_n, C_{nh}, C_{nv}, S_{2n}, D_n, D_{nh}, D_{nd}, T, T_h, T_d, O, O_h \qquad (12.1)$$

encountered in the study of molecules. In crystal structure, however, we have to deal with translational symmetry as well. Combining the two, one is led to space symmetry, see Chapter 8, which implies that the allowed point symmetries are the following:

$$S_2, C_{2h}, D_{2h} D_{3h}, D_{4h}, D_{6h}, O_h,$$

and their subgroups are known as crystal groups. All these can be viewed as subgroups of the symmetric group S_n studied in Chapter 11.

In Chapter 4, we considered the representations of discrete groups, with emphasis on the irreducible unitary ones. We proved Schur's theorem via the various Schur lemmas, Section 5.4. One characteristic of the irreducible representations is their character, which allows one to define proper relations of orthogonality, Section 5.5. This leads to reduction algorithms,

Section 5.7, and, in particular, the reduction of the direct product of group representations, see Section 5.9. Character tables of the irreducible representations of crystal groups are given in Section 8.5 and Chapter 17.

An irreducible representation $\Gamma^{[f]}$ of a large symmetry G, viewed from a discrete symmetry G_0 of Eq. (12.1), becomes reducible:

$$\Gamma^{[f]}(X) = \sum_i \alpha_{[f],i} \Gamma^i(X), \qquad (12.2)$$

where $\Gamma^i(X)$ is the irreducible representations of G_0. In this chapter, we concern ourselves with G, being the continuous group $SO(3)$, i.e. the group of rotations in 3D.

12.2   Reduction of $SO(3)$ representations under discrete symmetries

The continuous group $SO(3)$ (Vergados, 2017) is somewhat familiar from quantum mechanics. It is about orthogonal transformations in 3D, which includes the set of rotations around any of the axes of the coordinate system. The irreducible representations are defined in the space of the eigenvectors of the angular momentum operators of J^2 J_3:

$$J^2|j,m\rangle = j(j+1)\hbar^2|j,m\rangle, \quad J_3|j,m\rangle = m\hbar|j,m\rangle, \qquad (12.3)$$

$-j \le m \le j$, $(j,m) = $ integers or $(j,m) = $ half-integers.

This basis defines the irreducible representations $D^j_{m',m}(\alpha,\beta,\gamma)$ with dimension $2j+1$ associated with a given rotation $R(\alpha,\beta,\gamma)$ described by the three Euler angles (α,β,γ).

The irreducible representations are characterized by their character. It can be shown that the character of such a representation depends on the parameter j and one angle, which is associated with a rotation around the J_3 axis. It takes the simple form

$$\chi^j(\phi) = \frac{\sin(2j+1)\frac{\phi}{2}}{\sin\frac{\phi}{2}}, \qquad (12.4)$$

see, for example, Vergados (2017) or any book dealing with angular momentum theory.

For discrete values of the angle ϕ, $\phi = (2\pi)/n$, we write $\chi^j(\phi) = \chi^{(j,n)}$. Some characters are given in Tables 12.1 and 12.2. We note that $\chi^{(j+n,n)} = \chi^{(j,n)}$, i.e. the characters for $j < n$ are adequate. To these representations,

Table 12.1. The possible characters $\chi^{(\ell,n)}$, ℓ = integer of the single-valued representations with dimension $2\ell + 1$ of the group $SO(3)$ corresponding to angular momentum ℓ and a rotation angle $\phi = (2\pi)/n$. Note that $\chi^{(\ell+n,n)} = \chi^{(\ell,n)}$.

ℓ	0	1	2	3	4	5	6
$n = \infty$	1	3	5	7	9	11	13
$n = 2$	1	-1	1	-1	1	-1	1
$n = 3$	1	0	-1	1	0	-1	1
$n = 4$	1	1	-1	-1	1	1	-1
$n = 5$	1	$\frac{1}{2}\left(1+\sqrt{5}\right)$	0	$\frac{1}{2}\left(-1-\sqrt{5}\right)$	-1	1	$\frac{1}{2}\left(1+\sqrt{5}\right)$
$n = 6$	1	2	1	-1	-2	-1	1

Table 12.2. The possible characters $\chi^{(j,n)}$, j = half-integral, of the double-valued representations of dimension $2j+1$ of the group $SO(3)$, associated with angular momentum j and rotation angle $\phi = (2\pi)/n$. Note that $\chi^{(j+n,n)} = \chi^{(j,n)}$.

	$\frac{1}{2}$	$\frac{3}{2}$	$\frac{5}{2}$	$\frac{7}{2}$	$\frac{9}{2}$	$\frac{11}{2}$
$n = \infty$	2	4	6	8	10	12
$n = 2$	0	0	0	0	0	0
$n = 3$	1	-1	0	1	-1	0
$n = 4$	$\sqrt{2}$	0	$-\sqrt{2}$	0	$\sqrt{2}$	0
$n = 5$	$\frac{1}{2}\left(1+\sqrt{5}\right)$	1	-1	$\frac{1}{2}\left(-1-\sqrt{5}\right)$	0	$\frac{1}{2}\left(1+\sqrt{5}\right)$
$n = 6$	$\sqrt{3}$	$\sqrt{3}$	0	$-\sqrt{3}$	$-\sqrt{3}$	0

we must add the ones already known to us, i.e. those associated with reflections and inversion as well as their product with rotations.

Finally, these representations can become reducible under the lower discrete symmetries.

12.2.1 *Reduction of single-valued representations of $SO(3)$ under discrete symmetries*

In this case, Eq. (12.2) becomes

$$D^\ell = \sum_i \alpha_{\ell,i}\Gamma^i. \tag{12.5}$$

$$\alpha_{\ell,i} = \frac{1}{g}\sum_{X \in G_0} \chi^\ell(X)\chi^{*i}(X). \tag{12.6}$$

In this case, n is automatically defined by the element $X \in G_0$, and as a result, we write $\chi^\ell(X)$ instead of $\chi^{(\ell,n)}$.

Application 1: We reduce the representation D^ℓ of $SO3$ under the 3D discrete group D_4.

This group has eight elements, one 4-fold axis and four 2-fold axes distributed in five classes: $K_1 = \{E\}, K_2 = \{C_4^2\}, K_3 = \{C_4, C_4^3\}, K_4 = \{C_x, C_y\}, K_5 = \{C_x', C_y'\}$. Consequently, it has five irreducible representations, four 1D $\Gamma_1, \Gamma_2, \Gamma_3, \Gamma_4$ and one 2D Γ_5. Its character table is given in Table 17.6. For the reader's convenience, it is also included here in Table 12.3. Thus,

$$\alpha_{\ell,i} = \frac{1}{8} \left[\chi^\ell(K_1)\chi^i(K_1) + \chi^\ell(K_2)\chi^i(K_2) \right.$$
$$\left. + 2\left(\chi^\ell(K_3)\chi^i(K_3) + \chi^\ell(K_4)\chi^i(K_4) + \chi^\ell(K_5)\chi^i(K_5)\right) \right].$$

Hence:

(i) $\ell = 0$.

$$\alpha_{0,i} = \frac{1}{8} \left[\chi^i(K_1) + \chi^i(K_2) + 2\left(\chi^i(K_3) + \chi^i(K_4) + \chi^i(K_5)\right) \right].$$

As a result,

$$\alpha_{0,1} = \frac{1}{8}[1 + 1 + 2(1 + 1 + 1)] = 1,$$

$$\alpha_{0,2} = \frac{1}{8}[1 + 1 + 2(1 - 1 - 1)] = 0,$$

$$\alpha_{0,3} = \frac{1}{8}[1 + 1 + 2(-1 + 1 - 1)] = 0,$$

$$\alpha_{0,4} = \frac{1}{8}[1 + 1 + 2(-1 - 1 + 1)] = 0,$$

$$\alpha_{0,5} = \frac{1}{8}[2 - 2] = 0,$$

Table 12.3. The character table of the discrete symmetry D_4.

D_4	E or K_1	C_4^2 or K_2	C_4^3, C_4 or K_3	$2C_2'$ or K_4	$2C_2''$ or K_5
$A_1 \Leftrightarrow \chi_1$	1	1	1	1	1
$A_2 \Leftrightarrow \chi_2$	1	1	1	-1	-1
$B_1 \Leftrightarrow \chi_3$	1	1	-1	1	-1
$B_2 \Leftrightarrow \chi_4$	1	1	-1	-1	1
$E \Leftrightarrow \chi_5$	2	-2	0	0	0

i.e.

$$D^0 = \Gamma^1 = A_1.$$

As expected, only one representation is obtained.

(ii) $\ell = 1$. Now, $\chi^1(K_1) = \chi^{1,0} = 3$, $\chi^1(K_2) = \chi^{1,2} = -1$, $\chi^3(K_3) = \chi^{1,4} = 1$, $\chi^3(K_4) = \chi^3(K_5) = \chi^{1,4} = 1$. Hence,

$$\alpha_{1,i} = \frac{1}{8}\left[3\chi^i(K_1) - \chi^i(K_2) + 2\left(\chi^i(K_3) - \chi^i(K_4) - \chi^i(K_5)\right)\right].$$

That is,

$$\alpha_{1,1} = \frac{1}{8}\left[3 - 1 + 2(1 - 1 - 1)\right] = 0,$$

$$\alpha_{1,2} = \frac{1}{8}\left[3 - 1 + 2(1 - 1(-1) - 1(-1))\right] = 1,$$

$$\alpha_{1,3} = \frac{1}{8}\left[3 - 1(1) + 2(1(-1) - 1(1) - 1(-1))\right] = 0,$$

$$\alpha_{1,4} = \frac{1}{8}\left[3 - 1(1) + 2(1(-1) - 1(1) - 1(1))\right] = 0,$$

$$\alpha_{1,5} = \frac{1}{8}\left[3(2) - (-2)\right] = 1.$$

As a result,

$$D^1 = \Gamma^2 + \Gamma^5 = A_2 + E.$$

There exist now two such representations in the reduction. Whenever the Hamiltonian describing the system is invariant under D_4, the eigenfunctions are classified according to Γ^2 or A_2 (i.e. $\underline{1}$) and Γ^5 or E (i.e. $\underline{2}$). Γ^2 and Γ^5 cannot mix, unless they happen to be degenerate. In this case, we say that the state $\ell = 1$ breaks up into two states.

(iii) $\ell = 2$. Now, $\chi^1(K_1) = \chi^{2,0} = 5$, $\chi^1(K_2) = \chi^{2,2} = 1$, $\chi^3(K_3) = \chi^{2,4} = -1$, $\chi^3(K_4) = \chi^3(K_5) = \chi^{2,4} = -1$. Hence,

$$\alpha_{2,i} = \frac{1}{8}\left[\chi^i(K_1) + \chi^i(K_2) + 2\left(-\chi^i(K_3) + \chi^i(K_4) + \chi^i(K_5)\right)\right].$$

That is,

$$\alpha_{2,1} = \frac{1}{8}[5 + 1 + 1 + 1] = 1,$$

$$\alpha_{2,2} = \frac{1}{8}[5 + 1 + 2(-(1) + 1(-1) + 1(-1))] = 0,$$

$$\alpha_{2,3} = \frac{1}{8}[5 + 1(1) + 2(-(-1) + 1(1) + 1(-1))] = 1,$$

$$\alpha_{2,4} = \frac{1}{8}[5 + 1(1) + 2(-(-1) + 1(-1) + 1(1))] = 1,$$

$$\alpha_{2,5} = \frac{1}{8}[5(2) + (-2)] = 1.$$

Thus,

$$D^2 = \Gamma^1 + \Gamma^3 + \Gamma^4 + \Gamma^5 = A_1 + B_1 + B_2 + E.$$

Now, there exist four such sates of D_4, $\Gamma^1, \Gamma^3, \Gamma^4$ (Γ^i, $i = 1, 3, 4$, being (1) and Γ^5 being (2)). Under the same assumption about the Hamiltonian under D_4, the $\ell = 2$ wave functions, originally five-fold degenerate, are classified into three 1 and one 2 dimensional representations of D_4.

These states cannot mix, unless, by chance, they happen to be degenerate. This way, the state $\ell = 2$ breaks up into four states (see Fig. 12.1).

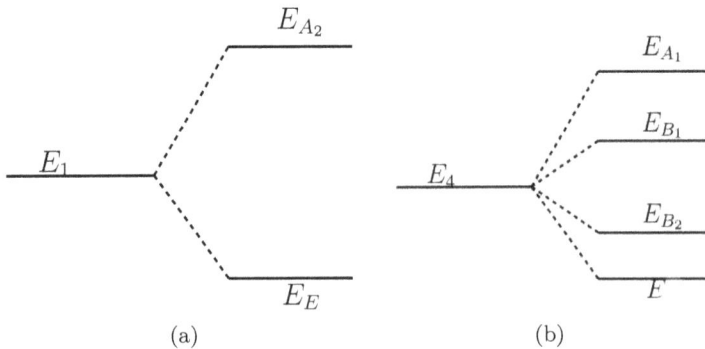

Fig. 12.1. Reduction of $SO(3)$ representations under D_4: $\ell = 1$ (a) and $\ell = 2$ (b).

(iv) $\ell = 3$. In an analogous fashion, we find

$$tr D^3 = \chi^2 + \chi^3 + \chi^4 + 2\chi^5 \Rightarrow D^3 = \Gamma^2 + \Gamma^3 + \Gamma^4 + 2\Gamma^5.$$

$$(12.7)$$

(v) $\ell = 4$.

$$tr D^4 = 2\chi^1 + \chi^2 + \chi^3 + \chi^4 + 2\chi^5 \Rightarrow$$

$$D^4 = 2\Gamma^1 + \Gamma^2 + \Gamma^3 + \Gamma^4 + 2\Gamma^5.$$

$$(12.8)$$

In general, if a representation occurs with multiplicity ≥ 2, i.e. in the reduction it appears more than once, the energies are not defined uniquely; one has to diagonalize the corresponding Hamiltonian in a space with dimension equal to the multiplicity.

Application 2: We reduce the representation D^ℓ of $SO(3)$ with respect to the discrete group O (chiral octahedral or rotational octahedral symmetry), which contains only rotations (no reflections) and is isomorphic to the group S_4.

We remind the reader that the group O has all the rotation elements of a cube, i.e. 3 C_4 axes (through the centers of the faces), $3 \times 3 = 9$ elements, 4 C_3 axes (along the body diagonals) $4 \times 2 = 8$ elements, 6 C_2 axes (through the midpoints of opposite vertices) $6 \times 1 = 6$ elements and the identity, one element, see Section 2.4.2 and Fig. 2.8). They are distributed in five classes: $K_1 = \{E\}, K_2 = \{8C_3\}, K_3 = \{3C_2\}, K_4 = \{6C_4\}, K_5 = \{6C_2'\}$. Thus, it has five irreducible representations, two 1D Γ_1, Γ_2, one 2D Γ_3 and two 3D Γ_4, Γ_5. The character table has already been given in the tables of the appendix, see Chapter 17, but it is also given here in Table 12.4 for the reader's convenience.

Table 12.4. The character table of the discrete group O (octahedral).

O	E	$8C_3$	$3C_2$	$6C_4$	$6C_2'$
	or	or	or	or	or
	K_1	K_2	K_3	K_4	K_5
A_1	1	1	1	1	1
A_2	1	1	1	−1	−1
E	2	−1	2	0	0
F_1	3	0	−1	1	−1
F_2	3	0	−1	−1	1

Thus,

$$\chi^\ell(K_1) = 2\ell + 1, \chi^\ell(K_2) = \chi^\ell\left(\tfrac{2\pi}{3}\right) = \chi^{\ell,3}, \chi^\ell(K_3) = \chi^\ell(K_4) = \chi^\ell(\pi)\chi^{\ell,2},$$
$$\chi^\ell(K_5) = \chi^\ell\left(\tfrac{\pi}{2}\right) = \chi^{\ell,4}.$$

As a result,

$$\alpha_{\ell,i} = \frac{1}{24}\left[\chi^{\ell,0}\chi^i(K_1) + 8\chi^{\ell,3}\chi^i(K_2)\right.$$
$$\left. + 3\chi^{\ell,2}\chi^i(K_3) + 6\chi^{\ell,2}\chi^i(K_4) + 6\chi^{\ell,4}\chi^i(K_5)\right].$$

Let us begin with $\ell = 0$:

$$\alpha_{0,i} = \frac{1}{24}\left[1\chi^i(K_1) + 8\chi^{0,3}(K_3)\chi^i(K_2)\right.$$
$$\left. + 3\chi^{0,2}\chi^i(K_3) + 6\chi^{0,2}\chi^i(K_4) + 6\chi^{0,4}\chi^i(K_5)\right].$$

$$\alpha_{0,1} = \frac{1}{24}\left[1(1) + 8(1)(1) + 3(1)(1) + 6(1)(1) + 0(1)(1)\right] - 1.$$

$$\alpha_{0,2} = \frac{1}{24}\left[1(1) + 8(1)(1) + 3(1)(1) + 6(1)(-1) + 6(1)(-1)\right] = 0.$$

$$\alpha_{0,3} = \frac{1}{24}\left[1(2) + 8(1)(-1) + 3(1)(2) + 6(1)(0) + 6(1)(0)\right] = 0.$$

$$\alpha_{0,4} = \frac{1}{24}\left[1(3) + 8(1)(0) + 3(1)(-1) + 6(1)(1) + 6(1)(-1)\right] = 0.$$

$$\alpha_{0,5} = \frac{1}{24}\left[1(3) + 8(1)(0) + 3(1)(-1) + 6(1)(-1) + 6(1)(1)\right] = 0.$$

$$D^0 = \Gamma^1 = A_1.$$

Proceeding similarly, we find

$$D^1 = \Gamma_5 = E, \quad D^2 = \Gamma_3 + \Gamma_4 = E + F_2,$$
$$D^3 = \Gamma^2 + \Gamma^4 + \Gamma^5 = A_2 + F_1 + F_2,$$
$$D^4 = \Gamma^1 + \Gamma^3 + \Gamma^4 + \Gamma^5 = A_1 + E + F_1 + F_2,$$
$$D^5 = \Gamma^3 + \Gamma^4 + 2\Gamma^5 = E + F_1 + 2F_2.$$

Note that in the case $\ell = 5$, the representation Γ^5 occurs twice i.e. double multiplicity. In this case, the symmetry O is not adequate to determine all the states of the system. For $\ell = 2, 3, 4$ and 5, the states break down into 2,3,4 and 4 states of O, respectively (see Fig. 12.2).

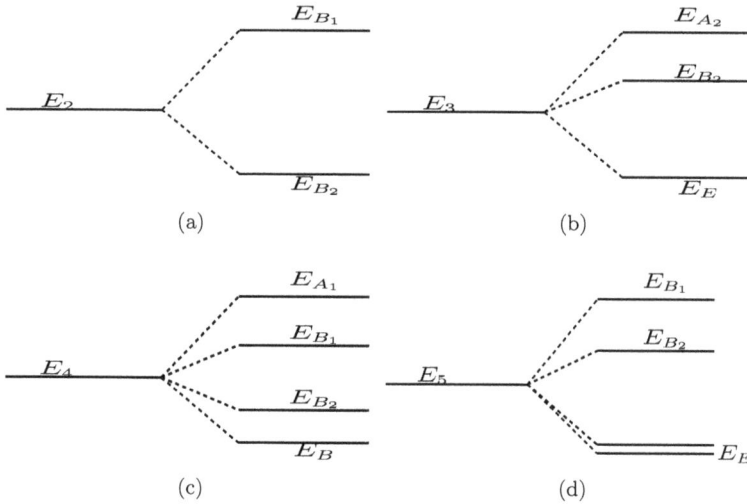

Fig. 12.2. The reduction of the representations of $SO(3)$ with respect to the discrete symmetry O: $\ell = 2$ (a), $\ell = 3$ (b), $\ell = 4$ (c) and $\ell = 5$ (d).

It should be noted that the angular momentum wave functions are characterized by a definite parity, $\pi = (-1)^\ell$, depending on their transformation under inversion with respect to a center, indicated by I, which is an orthogonal transformation with determinant -1. The parity π takes two values, $+1$ and -1, and thus, we often write $D^{\pi,\ell}$. The representations of the discrete group behave similarly, whenever the group contains the inversion generator I. In this case, one uses the symbols $g \Leftrightarrow \pi = 1$ and $u \Leftrightarrow \pi = -1$. Thus, under the discrete group, the symmetry $D^{\pi,\ell}$ breaks up as follows:

$$D^{\pi,\ell} = \sum_i \alpha_{\pi,i} \Gamma^{\pi,i}, \quad \Gamma^{+,i} \to \Gamma^i_g, \quad \Gamma^{-,i} \to \Gamma^i_u.$$

Example 1: We perform the reduction of $SO(3)$ under the discrete group T.

This symmetry is characterized by four axes of order 3 and three axes of order 2. The group therefore contains 12 elements, which are divided into four classes, $K_1 = E$, $K_2 = 4C_3$, $K_2 = 4C_3^2$ $K_4 = 3C_2$. Thus, there exist four irreducible representations with dimensions $\ell_1, \ell_2, \ell_3, \ell_4$ satisfying the condition $\ell_1^2 + \ell_2^2 + \ell_3^2 + \ell_4^2 = 12$. The only solution is $\ell_1 = 1, \ell_2 = 1, \ell_3 = 1, \ell_4 = 3$. The character table of T is given in Table 12.5. In the same table, the characters of $SO(3)$ are also given for $\ell = 0, 1, 2$ and 3.

Table 12.5. The character table of the discrete group T (left) and $SO(3)$ (right).

T	E	$4C_3$	$4C_3^2$	$3C_2$	$SO(3)$	$n=0$	$n=1$	$n=2$	$n=3$
	or K_1	or K_2	or K_3	or K_4					
$A \Leftrightarrow \chi_1$	1	1	1	1	$\ell=0 \Leftrightarrow \chi^0$	1	1	1	1
$E_1 \Leftrightarrow \chi_2$	1	ω	ω^2	1	$\ell=1 \Leftrightarrow \chi^1$	3	0	0	-1
$E_2 \Leftrightarrow \chi_3$	1	ω^2	ω	1	$\ell=2 \Leftrightarrow \chi^2$	5	-1	-1	1
$F \Leftrightarrow \chi_4$	3	0	0	-1	$\ell=3 \Leftrightarrow \chi^3$	7	1	1	-1

From the data of Table 12.5, we find

$\ell = 0$:

$$\alpha_{0,1} = \frac{1}{12}\left[1(1) + 1(4)(1) + 1(4)(1) + 1(3)(1)\right] = 1,$$

$$\alpha_{0,2} = \frac{1}{12}\left[1(1) + 1(4)\omega(1)(1) + 1(4)\omega^2(1)(1) + 1(3)(1)\right] = 0,$$

$$\alpha_{0,3} = \frac{1}{12}\left[1(1) + 1(4)\omega^2(1)(1) + 1(4)\omega(1)(1) + 1(3)(1)\right] = 0,$$

$$\alpha_{0,4} = \frac{1}{12}\left[1(3)1 + 4(1)(0) + 4(1)(0) + (-1)3(1)\right] = 0;$$

$\ell = 1$:

$$\alpha_{1,1} = \frac{1}{12}\left[3(1) + 0(4)(1) + 0(4)(1) - 1(3)(1)\right] = 0,$$

$$\alpha_{1,2} = \frac{1}{12}\left[3(1) + 0(4)\omega(1)(1) + 0(4)\omega^2(1)(1) - 1(3)(1)\right] = 0,$$

$$\alpha_{1,3} = \frac{1}{12}\left[3(1) + 0(4)\omega^2(1)(1) + 0(4)\omega(1)(1) - 1(3)(1)\right] = 0,$$

$$\alpha_{1,4} = \frac{1}{12}\left[3(3)1 + 4(1)(0) + 4(1)(0) + (-1)3(-1)\right] = 1;$$

$\ell = 2$:

$$\alpha_{2,1} = \frac{1}{12}\left[5(1) - 1(4)(1) - 1(4)(1) + 1(3)(1)\right] = 0,$$

$$\alpha_{2,2} = \frac{1}{12}\left[5(1) - 1(4)\omega(1)(1) - 1(4)\omega^2(1)(1) + 1(3)(1)\right] = 1,$$

$$\alpha_{2,3} = \frac{1}{12}\left[5(1) - 1(4)\omega^2(1)(1) - 1(4)\omega(1)(1) + 1(3)(1)\right] = 1,$$

$$\alpha_{2,4} = \frac{1}{12}\left[5(3)1 + 4(1)(0) + 4(1)(0) + (11)3(-1)\right] = 1;$$

$\ell = 3$:

$$\alpha_{3,1} = \frac{1}{12}\left[7(1) + 1(4)(1) + 1(4)(1) - 1(3)(1)\right] = 1,$$

$$\alpha_{3,2} = \frac{1}{12}\left[7(1) + 1(4)\omega(1)(1) + 1(4)\omega^2(1)(1) - 1(3)(1)\right] = 0,$$

$$\alpha_{3,3} = \frac{1}{12}\left[7(1) + 1(4)\omega^2(1)(1) + 1(4)\omega(1)(1) - 1(3)(1)\right] = 0,$$

$$\alpha_{3,4} = \frac{1}{12}\left[7(3)1 + 1(1)(0) + 1(1)(0) - 1(3)(-1)\right] = 2.$$

The above leads to the reductions

$$D^0 = \Gamma^1, \quad D^1 = \Gamma^4, \quad D^2 = \Gamma^2 \oplus \Gamma^3 \oplus \Gamma^4, \quad D^3 = \Gamma^1 \oplus 2\Gamma^4. \qquad (12.9)$$

The above results are presented in Table 12.8. They are also valid in the case of symmetry A_4, which is isomorphic to T and finds applications in particle physics, see Chapter 13.

12.2.2 The reduction of double-valued representations of $SO(3)$ under discrete symmetries

The above procedure cannot be applied in the case of $SO(3)$ representations corresponding to half-integral spin since they are double-valued. The correspondence between $SU(2) \Leftrightarrow U(\alpha, \beta, \gamma)$ and $SO(3) \Leftrightarrow R(\alpha, \beta, \gamma)$ is not one-to-one (Vergados, 2017). It is, therefore, necessary for the discrete group to be extended in connection with rotations. This can be done, see Section 5.10, by extending every rotation as follows:

$$C_n^1, C_n^2, \ldots, C_n^n = E \to u_n^1, u_n^2, \ldots, u_n^n = R,$$

$$u_n^1 R, u_n^2 R, \ldots, u_n^n R = E, \, R = -E.$$

Then, we construct the character table of the extended group and proceed as above.

Application 3: We examine the reduction of the double-valued representations of $SO(3)$ with respect to D_{2h}.

This discrete group has been studied in Section 2.4.2; it is $D_{2h} = \{E, C_2, \sigma_h, \sigma_h C_2\}$, and it has been extended with the inclusion of inversion. Here, we follow a different approach. We consider a representation of this group containing, in addition to the identity, the following elements:

$$R_x = \begin{pmatrix} 1 & 0 & 0 \\ 0 & -1 & 0 \\ 0 & 0 & -1 \end{pmatrix}, \quad R_y = \begin{pmatrix} -1 & 0 & 0 \\ 0 & 1 & 0 \\ 0 & 0 & -1 \end{pmatrix}, \quad R_z = \begin{pmatrix} -1 & 0 & 0 \\ 0 & -1 & 0 \\ 0 & 0 & 1 \end{pmatrix},$$

that is, it is a representation of $SO(3)$ corresponding to rotations around the three coordinate axes by an angle π. To this, we associate the fundamental representation of $SU(2)$, which is the Pauli matrices multiplied by i:

$$u_1 = i\sigma_1 = \begin{pmatrix} 0 & i \\ i & 0 \end{pmatrix}, \quad u_2 = i\sigma_2 = \begin{pmatrix} 0 & 1 \\ -1 & 0 \end{pmatrix}, \quad u_3 = i\sigma_3 = \begin{pmatrix} i & 0 \\ 0 & -i \end{pmatrix}$$

$$(12.10)$$

and

$$u_1 R, \; u_2 R, \; u_3 R, \; R = -E = \begin{pmatrix} -1 & 0 \\ 0 & 1 \end{pmatrix}. \tag{12.11}$$

This group contains eight elements, H $\{E, u_1, u_2, u_3, R, u_1 R, u_2 R, u_3 R\}$, distributed in five classes. This is isomorphic with the group D_4 we encountered above. The character table is given by Table 12.3, while the character table of the group $\{E, R_x, R_y, R_z\}$ has been given by Eq. (5.31).

We note that, with the exclusion of the column that contains the new element R and the last row, the two matrices coincide. Also, the column of the element R coincides with that of the identity element. Thus, the only double-valued representation of the group is Γ_5. As a result, $D^{(j)} = \alpha \Gamma_5$, where α is a constant. The proportionality constant α is determined by $\dim D^{(j)} = (2j + 1)$ and $\dim \Gamma_5 = 2$. Thus,

$$D^{(j)} = \left(j + \frac{1}{2} \right) \Gamma_5.$$

This reduction can, of course, be done directly in the standard way. In this case, Eq. (12.2) becomes

$$D^j = \sum_i \alpha_{j,i} \Gamma^i \Rightarrow, \tag{12.12}$$

where Γ^i are the double-valued representations and

$$\alpha_{j,i} = \frac{1}{8} \left[\chi^j(K_1)\chi^i(K_1) + \chi^j(K_2)\chi^i(K_2) \right.$$
$$\left. + 2 \left(\chi^j(K_3)\chi^i(K_3) + \chi^j(K_4)\chi^i(K_4) + \chi^j(K_5)\chi^i(K_5) \right) \right],$$

where the elements $\chi^j(K_i) \Leftrightarrow \chi^{j,n}$ are given in Table 12.2. Since the only double-valued representation is Γ_5, we obtain

$$\alpha_{j,5} = \frac{1}{8}\left(2\Gamma^j(E) + (-2)\Gamma^j(-E)\right)$$

$$= \frac{1}{8}\left(2(2j+1) + (-2)(-(2j+1))\right) = j + \frac{1}{2}.$$

Application 4: We examine the reduction of the double-valued representations of $SO(3)$ under D_4.

To the elements of D_4, we associate the elements $u_1, u_2, u_3, u_{x_1}, u_{y_1}, u'_{x_1}, u'_{y_1}$, and we extend it by including the additional elements

$$R, u_1 R, u_2 R, u_3 R, u_{x_1} R, u_{y_1} R, u'_{x_1} R, u'_{y_1} R,$$

that is, with the inclusion of the identity, we get the covering group D'_4, see Section 5.10. This has 16 elements distributed in seven classes indicated as follows:

$$K'_1 = \{E\}, \quad K'_2 = \{R\}, \quad K'_3 = \{u_4, u_4^3 R\},$$
$$K'_4 = \{u_4^3, u_4 R\}, \quad K'_5 = \{u_4^2, u_4^2 R\},$$
$$K'_6 = \{u_x, u_y, u_x R, u_y R\}, \quad K'_7 = \{u'_x, u'_y, u'_x R, u'_y R\}. \qquad (12.13)$$

Therefore, this group has seven irreducible representations with dimensions ℓ_i, $i = 1, 2, \ldots, 7$ satisfying the equations

$$\ell_1^2 + \ell_2^2 + \ell_3^2 + \ell_4^2 + \ell_5^2 + \ell_6^2 + \ell_7^2 = 16.$$

The only solution is $\ell_i = 1$, $i = 1, 2, 3, 4$ and $\ell_i = 2$, $i = 5, 6, 7$, i.e. four 1D irreducible representations and three 2D ones. The corresponding character table is given in Table 12.6.

Table 12.6. The character table of the discrete symmetry D'_4, which is the covering group of D_4. The corresponding classes are given in Eq. (12.13).

	K'_1	K'_2	K'_3	K'_4	K'_5	K'_6	K'_7
χ^1	1	1	1	1	1	1	1
χ^2	1	1	1	1	1	-1	-1
χ^3	1	1	-1	-1	1	1	-1
χ^4	1	1	-1	-1	1	-1	1
χ^5	2	2	0	0	-2	0	0
χ^6	2	-2	$\sqrt{2}$	$-\sqrt{2}$	0	0	0
χ^7	2	-2	$-\sqrt{2}$	$\sqrt{2}$	0	0	0

The characters can be obtained relatively easily by comparing the classes of Table 12.6 with those of Table 12.3. If they have a common element, they have the same character. Thus, with the exception of columns K_2', K_3', K_4', the first five representations have the same reduction table as Table 12.3. Consequently, $\Gamma_i, i, 1, \ldots, 5$, are single-valued. The only double-valued representations are Γ_6 and Γ_7.

Furthermore, in the case of single-valued representations, the characters are the same, e.g. $\chi^i(E) = \chi^i(R)$. Thus, $\chi^j(K_3') = \chi^j(K_3)$, $\chi^j(K_4') = \chi^j(K_3)$, $\chi^j(K_5') = \chi^j(K_2)$, $\chi^j(K_6') = \chi^j(K_4)$, $\chi^j(K_7') = \chi^j(K_5)$.

For the double-valued representations, since $R = -E$, $\chi^j(K_3') = -\chi^j(K_4')$, $\chi^j(K_1') = -\chi^j(K_2') = 2$. In addition, since u_4^2 and $u_4^2 R$ belong to the same class, we have $\chi(u_4^2) = \chi(u_4^2 R) = -\chi(u_4^2) = 0$, i.e. $\chi(K_5') = 0$. Similarly, $\chi(K_6') = 0$, $\chi(K_7') = 0$.

So, it only remains to evaluate $\chi^j(K_3')$. The u_4 is a 2×2 matrix corresponding to $\phi = \frac{\pi}{2}$, i.e. $\chi^{1/2}(\frac{\pi}{2}) = \sqrt{2}$. $\chi^6(K_3') = -\chi^7(K_3') = -\chi^6(K_4') = \chi^7(K_4') = \sqrt{2}$.

One can easily verify that $\chi^j(K_1') = -\chi^j(K_2') = 2j+1$, $\chi^j(K_3') = \chi\left(\frac{\pi}{2}\right)$, $\chi^j(K_4') = -\chi\left(\frac{\pi}{2}\right)$. Combining this with Table 12.2, one finds

$$\chi\left(\frac{\pi}{2}\right) = \begin{cases} (-1)^k \sqrt{2}, & \text{if } 2j+1 = 4k+2 \\ 0, & \text{if } 2j+1 = 4k. \end{cases}$$

As a result:

(i) $2j + 1 = 4k$

$\qquad \alpha_{j,6} = \alpha_{j,7} = \frac{1}{16}[2(2j+1) + (-2)(-(2j+1))] = \frac{2j+1}{4}$.

(ii) $2j + 1 = 4k + 2$

$$\alpha_{j,6} = \frac{1}{16}\left[2(2j+1) + (-2)(-(2j+1)) + 2\sqrt{2}\sqrt{2}2(-1)^{(2j+1-2)/4}\right] \Rightarrow$$

$$\alpha_{j,6} = \frac{2j+1}{4} + \frac{1}{2}(-1)^{(2j+1-2)/4}$$

$$\alpha_{j,7} = \frac{2j+1}{4} - \frac{1}{2}(-1)^{(2j+1-2)/4}.$$

Furthermore,

$$D^j = \begin{cases} \dfrac{2j+1}{4}(\Gamma_6 + \Gamma_7), & \text{if } 2j+1 = 4k, k = \text{integer}, \\[2mm] \dfrac{2j+1}{4}(\Gamma_6 + \Gamma_7) & \text{if } 2j+1 = 4k+2, k = \text{integer}, \\[2mm] \qquad + \dfrac{(-1)^{(2j+1-2)/4}}{2}(\Gamma_6 - \Gamma_7), \end{cases}$$

$$(12.14)$$

or, more analytically,

$$D^{1/2} = \Gamma_6, \quad D^{3/2} = \Gamma_6 + \Gamma_7, \quad D^{5/2} = \Gamma_6 + 2\Gamma_7, \quad D^{7/2} = 2\Gamma_6 + 2\Gamma_7,$$

$$D^{9/2} = 3\Gamma_6 + 2\Gamma_7, \quad D^{11/2} = 3\Gamma_6 + 3\Gamma_7, \quad D^{13/2} = 3\Gamma_6 + 4\Gamma_7, \text{ etc.}$$

Note that for $j \geq 5/2$, the symmetry by itself is not sufficient to determine all states, and thus, one needs to diagonalize the Hamiltonian in a space at least equal to the multiplicity of the representation involved.

12.3 Some reduction tables of representations of $SO(3)$ under discrete symmetries

In this section, we provide some reduction tables of $SO(3)$ under the most common discrete symmetries, see Tables 12.7–12.9.

Table 12.7. Reduction tables of representation D^ℓ of $SO(3)$ under the discrete symmetries $C_2, D_2, C_3, D_3, C_4, D_4$.

D^ℓ	C_2	D_2
D^0	Γ^1	Γ^1
D^1	$\Gamma^1 + 2\Gamma^2$	$\Gamma^2 + \Gamma^3 + \Gamma^4$
D^2	$3\Gamma^1 + 2\Gamma^2$	$2\Gamma^1 + \Gamma^2 + \Gamma^3 + \Gamma^4$
D^3	$3\Gamma^1 + 4\Gamma^2$	$\Gamma^1 + 2\Gamma^2 + 2\Gamma^3 + 2\Gamma^4$
D^4	$5\Gamma^1 + 4\Gamma^2$	$3\Gamma^1 + 2\Gamma^2 + 2\Gamma^3 + 2\Gamma^4$
D^5	$5\Gamma^1 + 6\Gamma^2$	$2\Gamma^1 + 3\Gamma^2 + 3\Gamma^3 + 3\Gamma^4$
D^6	$7\Gamma^1 + 6\Gamma^2$	$4\Gamma^1 + 3\Gamma^2 + 3\Gamma^3 + 3\Gamma^4$
$D^{\ell+2}$	$D^\ell + 2(\Gamma^1 + \Gamma^2)$	$D^\ell + \Gamma^1 + \Gamma^2 + \Gamma^3 + \Gamma^4$

D^ℓ	C_3	D_3
D^0	Γ^1	Γ^1
D^1	$\Gamma^1 + \Gamma^2 + \Gamma^3$	$\Gamma^2 + \Gamma^3$
D^2	$3\Gamma^1 + 2\Gamma^2 + 2\Gamma^3$	$2\Gamma^1 + 2\Gamma^3$
D^3	$3\Gamma^1 + 2\Gamma^2 + 2\Gamma^3$	$\Gamma^1 + 2\Gamma^2 + 2\Gamma^3$
D^4	$3\Gamma^1 + 3\Gamma^2 + 3\Gamma^3$	$2\Gamma^1 + 2\Gamma^2 + \Gamma^3 + 2\Gamma^3$
D^5	$3\Gamma^1 + 4\Gamma^2 + 4\Gamma^3$	$\Gamma^1 + 2\Gamma^2 + 4\Gamma^3$
D^6	$5\Gamma^1 + 4\Gamma^2 + 4\Gamma^3$	$3\Gamma^1 + 2\Gamma^2 + 4\Gamma^3$
$D^{\ell+6}$	$D^\ell + 4(\Gamma^1 + \Gamma^2 + \Gamma^3)$	$D^\ell + 2(\Gamma^1 + \Gamma^2 + 2\Gamma^3)$

D^ℓ	C_4	D_4
D^0	Γ^1	Γ^1
D^1	$\Gamma^1 + \Gamma^3 + \Gamma^4$	$\Gamma^2 + \Gamma^5$
D^2	$\Gamma^1 + 2\Gamma^2 + \Gamma^3 + 2\Gamma^4$	$\Gamma^1 + \Gamma^3 + \Gamma^4 + \Gamma^5$
D^3	$\Gamma^1 + 2\Gamma^2 + 2\Gamma^3 + 2\Gamma^4$	$\Gamma^2 + \Gamma^3 + \Gamma^4 + 2\Gamma^3$
D^4	$3\Gamma^1 + 2\Gamma^2 + 2\Gamma^3 + 2\Gamma^3 4$	$2\Gamma^1 + \Gamma^2 + \Gamma^3 + \Gamma^4 + 2\Gamma^5$
D^5	$3\Gamma^1 + 2\Gamma^2 + 3\Gamma^3 + 3\Gamma^4$	$\Gamma^1 + 2\Gamma^2 + \Gamma^3 + \Gamma^4 + 3\Gamma^5$
D^6	$3\Gamma^1 + 4\Gamma^2 + 3\Gamma^3 + 3\Gamma^4$	$2\Gamma^1 + \Gamma^2 + 2\Gamma^3 + 2\Gamma^4 + 3\Gamma^5$
$D^{\ell+4}$	$D^\ell + 2(\Gamma^1 + \Gamma^2 + \Gamma^3 + \Gamma^4)$	$D^\ell + \Gamma^1 + \Gamma^2 + \Gamma^3 + \Gamma^4 + 2\Gamma^5$

Table 12.8. The same as in Table 12.7 for C_6, D_6, T, O.

D^ℓ	C_6	D_6
D^0	Γ^1	Γ^1
D^1	$\Gamma^1 + \Gamma^5 + \Gamma^6$	$\Gamma^2 + \Gamma^5$
D^2	$\Gamma^1 + \Gamma^2 + \Gamma^3 + \Gamma^5 + \Gamma^6$	$\Gamma^1 + \Gamma^5 + \Gamma^6$
D^3	$\Gamma^1 + \Gamma^2 + \Gamma^3 + 2\Gamma^4 + \Gamma^5 + \Gamma^6$	$\Gamma^2 + \Gamma^3 + \Gamma^4 + \Gamma^5 + \Gamma^6$
D^4	$\Gamma^1 + 2\Gamma^2 + 2\Gamma^3 + 2\Gamma^4 + \Gamma^5 + \Gamma^6$	$\Gamma^1 + \Gamma^2 + \Gamma^3 + \Gamma^4 + \Gamma^5 + 2\Gamma^6$
D^5	$\Gamma^1 + 2\Gamma^2 + 2\Gamma^3 + 2\Gamma^4 + 2\Gamma^5 + 2\Gamma^6$	$\Gamma^2 + \Gamma^3 + \Gamma^4 + 2\Gamma^5 + 2\Gamma^6$
D^6	$3\Gamma^1 + 2\Gamma^2 + 2\Gamma^3 + 2\Gamma^4 + 2\Gamma^5 + 2\Gamma^6$	$2\Gamma^1 + \Gamma^2 + \Gamma^3 + \Gamma^4 + 2\Gamma^5 + 2\Gamma^6$
$D^{\ell+6}$	$D^\ell + 2(\Gamma^1 + \Gamma^2 + \Gamma^3 + \Gamma^4 + \Gamma^5 + \Gamma^6)$	$D^\ell + \Gamma^1 + \Gamma^2 + \Gamma^3 + \Gamma^4 + 2\Gamma^5 + 2\Gamma^6$

D^ℓ	T	O
D^0	Γ^1	Γ^1
D^1	Γ^4	Γ^4
D^2	$\Gamma^2 + \Gamma^3 + \Gamma^4$	$\Gamma^3 + \Gamma^5$
D^3	$\Gamma^1 + 2\Gamma^4$	$\Gamma^2 + \Gamma^4 + \Gamma^5$
D^4	$\Gamma^1 + \Gamma^2 + \Gamma^3 + 2\Gamma^4$	$\Gamma^1 + \Gamma^3 + \Gamma^4 + \Gamma^5$
D^5	$\Gamma^2 + \Gamma^3 + 3\Gamma^4$	$\Gamma^3 + 2\Gamma^4 + \Gamma^5$
D^6	$2\Gamma^1 + \Gamma^2 + \Gamma^3 + 3\Gamma^4$	$\Gamma^1 + \Gamma^2 + \Gamma^3 + \Gamma^4 + 2\Gamma^5$
$D^{\ell+6}$	$D^\ell + \Gamma^1 + \Gamma^2 + \Gamma^3 + 3\Gamma^4$	
$D^{\ell+5}$		$-D^{6-\ell} + \Gamma^1 + \Gamma^2 + 2\Gamma^3 + 3\Gamma^4 + 3\Gamma^5$
$D^{\ell+12}$		$D^\ell + \Gamma^1 + \Gamma^2 + 2\Gamma^3 + 3\Gamma^4 + 3\Gamma^5$

At this point, we should mention that the irreducible representations of angular momentum are characterized by a definite parity, which takes two values $\pi = 1, -1$. So, we write $D^{(\pi\ell)}$. In the special case of single-particle states, the parity π is given by $(-1)^\ell$; therefore, it is not explicitly indicated. Sometimes, however, it will be explicitly indicated when the point group under consideration contains the inversion operator I. In this case, we write $\Gamma^{(\pi)(i)}$ for the representations of the discrete group and the reduction as

$$D^{(\pi\ell)} = \sum_i \alpha_{\ell,i} \Gamma^{(\pi)(i)}.$$

Most common, however, is a different notation:

$$\Gamma^{(+)(i)} = \Gamma_g^{(i)}, \quad \Gamma^{(-)(i)} = \Gamma_u^{(i)}.$$

12.4   Crystal harmonics

We begin by considering a basis in the space within which the angular momentum operators act, which is appropriate for applications in the area

Table 12.9. Reduction tables of the representations of D^j of $SO(3)$ under the discrete symmetries $C_2', D_2', C_3', D_3', C_4', D_4', C_6', D_6', T', O'$.

D^j	C_2'	D_2'
D^j	$(j+\frac{1}{2})(\Gamma^3+\Gamma^4)$	$(j+\frac{1}{2})\Gamma^5$
D^j	C_3'	D_3'
$D^{1/2}$	$\Gamma^4+\Gamma^5$	Γ^4
$D^{3/2}$	$3\Gamma^4+\Gamma^5+2\Gamma^6$	$\Gamma^4+\Gamma^5+\Gamma^6$
$D^{5/2}$	$2\Gamma^4+2\Gamma^5+2\Gamma^6$	$2\Gamma^4+\Gamma^5+\Gamma^6$
$D^{7/2}$	$3\Gamma^4+3\Gamma^5+2\Gamma^6$	$3\Gamma^4+\Gamma^5+\Gamma^6$
$D^{9/2}$	$3\Gamma^4+3\Gamma^3+4\Gamma^6$	$4\Gamma^4+2\Gamma^5+2\Gamma^6$
$D^{11/2}$	$4\Gamma^4+4\Gamma^5+4\Gamma^6$	$4\Gamma^4+2\Gamma^5+2\Gamma^6$
$D^{13/2}$	$5\Gamma^4+5\Gamma^5+4\Gamma^6$	$5\Gamma^4+2\Gamma^5+2\Gamma^6$
D^{j+6}	$D^j+4(\Gamma^4+\Gamma^5+\Gamma^6)$	$D^j+2(2\Gamma^4+\Gamma^5+\Gamma^6)$
D^j	C_4'	D_4'
$D^{1/2}$	$\Gamma^5+\Gamma^6$	Γ^6
$D^{3/2}$	$\Gamma^5+\Gamma^6+\Gamma^7+\Gamma^8$	$\Gamma^6+\Gamma^7$
$D^{5/2}$	$\Gamma^5+\Gamma^6+2\Gamma^7+2\Gamma^8$	$\Gamma^6+2\Gamma^7$
$D^{7/2}$	$2\Gamma^1+2\Gamma^6+2\Gamma^7+2\Gamma^8$	$\Gamma^7+\Gamma^8+\Gamma^9$
$D^{9/2}$	$3\Gamma^5+3\Gamma^6+2\Gamma^7+2\Gamma^8$	$3\Gamma^6+2\Gamma^7$
$D^{11/2}$	$3\Gamma^5+3\Gamma^6+3\Gamma^7+3\Gamma^8$	$3\Gamma^6+3\Gamma^7$
$D^{13/2}$	$3\Gamma^5+3\Gamma^6+4\Gamma^7+4\Gamma^8$	$3\Gamma^6+4\Gamma^7$
D^{j+4}	$D^j+2(\Gamma^5+\Gamma^6+\Gamma^7+\Gamma^8)$	$D^j+2(\Gamma^6+\Gamma^7)$
D^j	C_6'	D_j'
$D^{1/2}$	$\Gamma^7+\Gamma^8$	Γ^7
$D^{3/2}$	$\Gamma^7+\Gamma^8+\Gamma^{11}+\Gamma^{12}$	$\Gamma^7+\Gamma^8$
$D^{5/2}$	$\Gamma^7+\Gamma^8+\Gamma^9+\Gamma^{10}+\Gamma^{11}+\Gamma^{12}$	$\Gamma^7+\Gamma^8+\Gamma^8+\Gamma^9$
$D^{7/2}$	$\Gamma^7+\Gamma^8+2\Gamma^9+2\Gamma^{10}+\Gamma^{11}+\Gamma^{12}$	$\Gamma^7+2\Gamma^8+\Gamma^9$
$D^{9/2}$	$\Gamma^7+\Gamma^8+2\Gamma^9+2\Gamma^{10}+2\Gamma^{11}+2\Gamma^{12}$	$\Gamma^7+2\Gamma^8+2\Gamma^9$
$D^{11/2}$	$2\Gamma^7+2\Gamma^8+2\Gamma^9+2\Gamma^{10}+2\Gamma^{11}+2\Gamma^{12}$	$2\Gamma^7+2\Gamma^8+2\Gamma^9$
$D^{13/2}$	$3\Gamma^7+3\Gamma^8+2\Gamma^9+2\Gamma^{10}+2\Gamma^{11}+2\Gamma^{12}$	$3\Gamma^7+2\Gamma^8+2\Gamma^9$
D^{j+6}	$D^j+2(\Gamma^7+\Gamma^8+\Gamma^9+\Gamma^{10}+\Gamma^{11}+\Gamma^{12})$	$D^j+2(\Gamma^7+\Gamma^8+\Gamma^9)$
D^j	T'	O'
$D^{1/2}$	Γ^5	Γ^6
$D^{3/2}$	$\Gamma^6+\Gamma^7$	Γ^8
$D^{5/2}$	$\Gamma^5+\Gamma^6+\Gamma^7$	$\Gamma^7+\Gamma^8$
$D^{7/2}$	$2\Gamma^5+\Gamma^6+\Gamma^7$	$\Gamma^6+\Gamma^7+\Gamma^8$
$D^{9/2}$	$\Gamma^5+2\Gamma^6+2\Gamma^7$	$\Gamma^6+2\Gamma^8$
$D^{11/2}$	$2\Gamma^5+2\Gamma^6+2\Gamma^7$	$\Gamma^6+\Gamma^7+2\Gamma^8$
$D^{13/2}$	$3\Gamma^5+2\Gamma^6+2\Gamma^7$	$\Gamma^6+2\Gamma^7+2\Gamma^8$
D^{j+6}	$D^j+2(\Gamma^5+\Gamma^6+\Gamma^7)$	$D^j+\Gamma^6+\Gamma^7+2\Gamma^8$

of crystal structure. We recall that such a set is formed by the finite and single-valued eigenfunctions of the operators L^2 and L_3:

$$L^2|\ell, m\rangle = \ell(\ell + 1)\hbar^2|\ell, m\rangle, \quad L_3|\ell, m\rangle = m\hbar|\ell, m\rangle, \tag{12.15}$$

where ℓ is a non-negative integer and m takes integer values in the range $-\ell \leq m \leq \ell$. These functions are named spherical harmonics $Y_m^\ell(\theta, \phi)$:

$$Y_m^\ell(\theta, \phi) = \left[\frac{2\ell + 1}{4\pi}\frac{(\ell - m)!}{(\ell + m)!}\right]^{1/2} e^{im\phi} P_m^\ell(\cos\theta), \tag{12.16}$$

where $P_m^\ell(\xi)$ are the adjoined Legendre polynomials. The spherical harmonics are normalized:

$$\int_0^\pi \sin\theta d\theta \int_0^{2\pi} d\phi \left(Y_m^\ell(\theta, \phi)\right)^* Y_{m'}^{\ell'}(\theta, \phi) = \delta_{\ell\ell'}\delta_{m,m'}.$$

In crystal structure, however, one adopts the basis

$$\Phi_{\ell m}^{(1)}(\theta, \phi) = \frac{\sqrt{2}}{\sqrt{1 + \delta_{m0}}}\left[\frac{2\ell + 1}{2}\frac{(\ell - m)!}{(\ell + m)!}\right]^{1/2}\cos(m\phi)P_m^\ell(\cos\theta),$$
$$m = 0, 1, 2, \ldots, \ell, \tag{12.17}$$

$$\Phi_{\ell m}^{(2)}(\theta, \phi) = \sqrt{2}\left[\frac{2\ell + 1}{2}\frac{(\ell - m)!}{(\ell + m)!}\right]^{1/2}\sin(m\phi)P_m^\ell(\cos\theta), \quad m = 1, 2, \ldots, \ell. \tag{12.18}$$

The normalization condition is

$$\int_0^\pi \sin\theta d\theta \int_0^{2\pi} d\phi \Phi_{\ell m}^{(\alpha)}(\theta, \phi)\Phi_{\ell'm'}^{(\alpha')} = 2\pi\delta_{\alpha\alpha'}\delta_{\ell\ell'}\delta_{m,m'}.$$

We note the presence of 2π on the right.

We now study the behavior of these functions under the action of the operators of a discrete group, e.g. C_4, C_x of D_4, we encountered above in Application 1, Section 12.2.1.

The element C_4 causes a rotation by an angle of $\pi/2$ around the z-axis, i.e. $\phi \to \phi + \pi/2$:

$$\Phi_{\ell m}^{(1)}(\theta, \phi) \to \Phi_{\ell m}^{(1)}(\theta, \phi + \pi/2) = \frac{\sqrt{2}}{\sqrt{1 + \delta_{m0}}}\left[\frac{2\ell + 1}{2}\frac{(\ell - m)!}{(\ell + m)!}\right]^{1/2}$$
$$\times \cos m(\phi + \pi/2)P_m^\ell(\cos\theta).$$

This function remains invariant for $m = 0$. For $m > 0$,

$$\cos m(\phi + \pi/2) = \cos(m\phi)\cos m\frac{\pi}{2} - \sin m\phi \sin m\frac{\pi}{2} \Rightarrow$$

$$\cos m(\phi + \pi/2) = \begin{cases} (-1)^{\frac{m}{2}} \cos(m\phi) & m = \text{even} \\ -\sin(m\phi) & m = \text{odd} \end{cases}.$$

Hence,

$$C_4 \Phi_{\ell m}^{(1)}(\theta, \phi) = \begin{cases} (-1)^{\frac{m}{2}} \Phi_{\ell m}^{(1)}(\theta, \phi), & m = \text{even} \\ -\Phi_{\ell m}^{(2)}(\theta, \phi), & m = \text{odd} \end{cases}. \tag{12.19}$$

Similarly,

$$C_4 \Phi_{\ell m}^{(2)}(\theta, \phi) = \begin{cases} (-1)^{\frac{m}{2}} \Phi_{\ell m}^{(2)}(\theta, \phi), & m = \text{even} \\ -\Phi_{\ell m}^{(1)}(\theta, \phi), & m = \text{odd} \end{cases}. \tag{12.20}$$

The action of the operator C_x, which is of second order, is specified by the transformation $\phi \to -\phi$, $\theta \to -\theta + \pi$, and as a result, we get

$$C_x \Phi_{\ell m}^{(1)}(\theta, \phi) = \frac{\sqrt{2}}{\sqrt{1 + \delta_{m0}}} \left[\frac{2\ell + 1}{2} \frac{(\ell - m)!}{(\ell + m)!} \right]^{1/2} \cos(-m\phi) P_m^\ell(\cos\pi - \theta)$$

$$= \frac{\sqrt{2}}{\sqrt{1 + \delta_{m0}}} \left[\frac{2\ell + 1}{2} \frac{(\ell - m)!}{(\ell + m)!} \right]^{1/2} \cos(m\phi) P_m^\ell(-\cos\theta).$$

For the Legendre polynomials, however, we have

$$P_m^\ell(-\cos\theta) = (-1)^{\ell+m} P_m^\ell(\cos\theta),$$
$$C_x \Phi_{\ell m}^{(1)}(\theta, \phi) = (-1)^{\ell+m} \Phi_{\ell m}^{(1)}(\theta, \phi). \tag{12.21}$$

Similarly,

$$C_x \Phi_{\ell m}^{(2)}(\theta, \phi) = (-1)^{\ell+m+1} \Phi_{\ell m}^{(2)}(\theta, \phi). \tag{12.22}$$

We know that, if we limit ourselves to the discrete group G_0, the irreducible representations of $SO(3)$ are reduced under G_0. From the above discussion, one can see that the irreducible representations K_β^{ℓ,Γ^i} of G_0, which depend on ℓ and Γ^i, can be expressed as a superposition of crystal harmonics:

$$K_\beta^{\ell,\Gamma^i}(\theta,\phi) = \sum_{\alpha=1}^{2}\sum_{m=0}^{\ell} C_\beta^{\alpha,m}\Phi_{\ell m}^{(\alpha)}(\theta,\phi),$$

where $\beta = 1, 2, \ldots, \dim\Gamma^i$. The coefficients $C_\beta^{\alpha,m}$ are chosen so that the functions $K_\beta^{\ell,\Gamma^i}(\theta,\phi)$ have the same normalization as that of $\Phi_{\ell m}^{(\alpha)}(\theta,\phi)$.

Let us now explore what happens in the case of D_4. We observe the following:

(i) In the case of Γ^1, the characters of C_4 and C_x are $+1$, see Table 12.3. Thus, Eqs. (12.19) and (12.20) imply $m = 4k$, $k =$ integer. In combination with Eqs. (12.21) and (12.22), we see that the basis contains $\Phi_{\ell m}^{(1)}(\theta,\phi)$ for ℓ even and $\Phi_{\ell m}^{(2)}(\theta,\phi)$ for ℓ odd.

(ii) In the case of Γ^2, from the matrix (12.3), we see that the element C_4 has trace $+1$ and C_x has -1. With arguments as above, we find $m = 4k$ and that the basis is $\Phi_{\ell m}^{(1)}(\theta,\phi)$ for ℓ odd and $\Phi_{\ell m}^{(2)}(\theta,\phi)$ for ℓ even.

(iii) In the case of Γ^3, from Table 12.3, we see that C_4 has trace -1 and C_x has $+1$. Thus, we obtain $m = 4k + 2$ with the basis being $\Phi_{\ell m}^{(1)}(\theta,\phi)$ for ℓ even and $\Phi_{\ell m}^{(2)}(\theta,\phi)$ for ℓ odd.

(iv) In the case of Γ^4, again, $m = 4k + 2$ with the basis being $\Phi_{\ell m}^{(1)}(\theta,\phi)$ for ℓ odd and $\Phi_{\ell m}^{(2)}(\theta,\phi)$ for ℓ even.

(v) In the case of Γ^5, necessarily $m = 4k + 1$, and the two functions are members of a 2D space.

The above results are summarized in Table 12.10. The functions K_β^{ℓ,Γ^i} in the case of the 1D representations of D_4 are proportional to $\Phi_{\ell m}^{(1)}(\theta,\phi)$ or $\Phi_{\ell m}^{(2)}(\theta,\phi)$. This is not true for the 2D representations or other symmetries,

Table 12.10. The functions K_β^{ℓ,Γ^i} in the case of D_4. The 1D ones are proportional to $\Phi_{\ell m}^{(1)}(\theta,\phi)$ or $\Phi_{\ell m}^{(2)}(\theta,\phi)$ for even m. The 2D one lies in the space of these two functions for m, including both ℓ even and odd:

	β	$\ell =$ even	$\ell =$ odd
Γ^1	1	$\Phi_{\ell 4k}^{(1)}(\theta,\phi)$	$\Phi_{\ell 4k}^{(2)}(\theta,\phi)$
Γ^2	1	$\Phi_{\ell 4k}^{(2)}(\theta,\phi)$	$\Phi_{\ell 4k}^{(1)}(\theta,\phi)$
Γ^3	1	$\Phi_{\ell 4k+2}^{(1)}(\theta,\phi)$	$\Phi_{\ell 4k+2}^{(2)}(\theta,\phi)$
Γ^4	1	$\Phi_{\ell 4k+2}^{(2)}(\theta,\phi)$	$\Phi_{\ell 4k+2}^{(1)}(\theta,\phi)$
Γ^5	1	$\Phi_{\ell 4k+1}^{(1)}(\theta,\phi)$	$\Phi_{\ell 4k+1}^{(1)}(\theta,\phi)$
	2	$\Phi_{\ell 4k+1}^{(2)}(\theta,\phi)$	$\Phi_{\ell 4k+1}^{(2)}(\theta,\phi)$

such as the octahedral. In general, the K_β^{ℓ,Γ^i} are linear combinations of $\Phi_{\ell m}^{(1)}(\theta,\phi)$ and $\Phi_{\ell m}^{(2)}(\theta,\phi)$.

We note that, if the Hamiltonian is invariant under D_4, its representation is a matrix with indices m and m', which does not mix different symmetries. Even then, however, the matrix corresponding to the symmetry Γ^5 is not diagonal, but it can be cast in the form

$$H = \begin{pmatrix} \langle \Phi_{\ell 4k'+1}^{(1)}|H|\Phi_{\ell 4k+1}^{(1)}\rangle & 0 \\ 0 & \langle \Phi_{\ell 4k'+1}^{(2)}|H|\Phi_{\ell 4k+1}^{(1)}\rangle \end{pmatrix}.$$

Observe also that in Table 12.10, the allowed values of ℓ are not specified. These can be obtained from the data of Table 12.7. Now, in the special case that $\ell \leq 4$, we get the results of Table 12.11, where

$$P_\ell^{(m)} = \left[\frac{2\ell+1}{2}\frac{(\ell-m)!}{(\ell+m)!}\right]^{1/2} P_m^\ell(\cos\theta). \tag{12.23}$$

This is another notation for the crystal functions frequently used in applications. The examination of more complicated symmetries, such as the octahedral, is very interesting but beyond our goals. The interested reader is referred to the literature, e.g. the text by Hammermesh (1964). Just to give a simple picture of what to expect, we provide some results in Table 12.12.

Table 12.11. The basis K_β^{ℓ,Γ^i} of the irreducible representations D_4 for $\ell \leq 4$. Note that in the case of Γ^1 for $\ell = 4$ and Γ^5 for $\ell = 3, 4$, there exist two bases since these representations appear twice (Eqs. (12.7) and (12.8)).

ℓ	Γ^1	Γ^2	Γ^3	Γ^4	Γ^5
0	1				
1		$P_1^{(0)}$			$\sqrt{2}P_1^{(1)}\cos\phi$ $\sqrt{2}P_1^{(1)}\sin\phi$
2	$P_2^{(0)}$		$\sqrt{2}P_2^{(2)}\sin 2\phi$	$\sqrt{2}P_2^{(2)}\cos 2\phi$	$\sqrt{2}P_2^{(1)}\cos\phi$ $\sqrt{2}P_2^{(1)}\sin\phi$
3	$P_3^{(0)}$		$\sqrt{2}P_3^{(2)}\sin 2\phi$	$\sqrt{2}P_3^{(2)}\cos 2\phi$	$\sqrt{2}P_3^{(1)}\cos\phi$ $\sqrt{2}P_3^{(1)}\sin\phi$ $\sqrt{2}P_3^{(3)}\cos 3\phi$ $\sqrt{2}P_3^{(3)}\sin 3\phi$
4	$P_4^{(0)}$ $\sqrt{2}P_4^{(4)}\sin 4\phi$	$\sqrt{2}P_4^{(4)}\sin 4\phi$	$\sqrt{2}P_4^{(2)}\sin 2\phi$	$\sqrt{2}P_4^{(2)}\cos 2\phi$	$\sqrt{2}P_4^{(1)}\cos\phi$ $\sqrt{2}P_4^{(1)}\sin\phi$ $\sqrt{2}P_4^{(3)}\cos 3\phi$ $\sqrt{2}P_4^{(3)}\sin 3\phi$

Table 12.12. The basis K_β^{ℓ,Γ^i} of the irreducible representations of octahedral symmetry for $\ell \leq 4$. When the basis vectors are linear combinations of crystal harmonics, they are numbered by $|i\rangle$, while for the rest, such a number is understood (the numbering is vertical for a given ℓ).

ℓ	Γ^1	Γ^2	Γ^3	Γ^4	Γ^5
0	1				
1				$P_1^{(0)}$ $\sqrt{2}P_1^{(1)}\cos\phi$ $\sqrt{2}P_1^{(1)}\sin\phi$	
2			$P_2^{(0)}$ $P_2^{(1)}\cos\phi$		$P_2^{(2)}\sin 2\phi$ $P_2^{(1)}\cos\phi$ $P_2^{(1)}\sin\phi$
3		$P_3^{(0)}$ $\lvert 2\rangle = \sqrt{\tfrac{5}{8}}\sqrt{2}P_3^{(3)}\cos 3\phi$ $\qquad -\sqrt{\tfrac{3}{8}}\sqrt{2}P_3^{(1)}\cos\phi$ $\lvert 3\rangle = \sqrt{\tfrac{5}{8}}\sqrt{2}P_3^{(3)}\sin 3\phi$ $\qquad \sqrt{\tfrac{5}{8}}\sqrt{2}P_3^{(1)}\sin\phi$		$P_3^{(0)}$ $\lvert 2\rangle = -\sqrt{\tfrac{5}{8}}\sqrt{2}P_3^{(3)}\cos 3\phi$ $\qquad -\sqrt{\tfrac{3}{8}}\sqrt{2}P_3^{(1)}\cos\phi$ $\lvert 3\rangle = \sqrt{\tfrac{5}{8}}\sqrt{2}P_3^{(3)}\sin 3\phi$ $\qquad \sqrt{\tfrac{3}{8}}\sqrt{2}P_3^{(1)}\sin\phi$	$\sqrt{2}P_3^{(2)}\cos 2\phi$ $\lvert 2\rangle = -\sqrt{\tfrac{3}{8}}\sqrt{2}P_3^{(3)}\cos 3\phi$ $\qquad \sqrt{\tfrac{5}{8}}\sqrt{2}P_3^{(1)}\cos\phi$ $\lvert 3\rangle = -\sqrt{\tfrac{3}{8}}\sqrt{2}P_3^{(3)}\sin 3\phi$ $\qquad -\sqrt{\tfrac{3}{8}}\sqrt{2}P_3^{(1)}\sin\phi$
4	$\sqrt{2}P_4^{(4)}\cos 4\phi$		$\sqrt{2}P_4^{(2)}\cos 2\phi$ $\lvert 2\rangle = -\sqrt{\tfrac{5}{12}}P_4^{(0)}$ $\qquad \sqrt{\tfrac{7}{12}}\sqrt{2}P_4^{(4)}\cos 4\phi$	$\sqrt{2}P_4^{(4)}\sin 4\phi$ $\lvert 2\rangle = \sqrt{\tfrac{7}{8}}\sqrt{2}P_4^{(1)}\cos\phi$ $\qquad -\sqrt{\tfrac{1}{8}}\sqrt{2}P_4^{(3)}\cos 3\phi$ $\lvert 3\rangle = \sqrt{\tfrac{7}{8}}\sqrt{2}P_4^{(1)}\sin\phi$ $\qquad +\sqrt{\tfrac{1}{8}}\sqrt{2}P_4^{(3)}\sin 3\phi$	$\sqrt{2}P_4^{(2)}\sin 2\phi$ $\lvert 2\rangle = -\sqrt{\tfrac{1}{8}}\sqrt{2}P_4^{(1)}\cos\phi$ $\qquad \sqrt{\tfrac{7}{8}}\sqrt{2}P_4^{(3)}\cos 3\phi$ $\lvert 3\rangle = \sqrt{\tfrac{1}{8}}\sqrt{2}P_4^{(1)}\sin\phi$ $\qquad -\sqrt{\tfrac{7}{8}}\sqrt{2}P_4^{(3)}\sin 3\phi$

12.5 Problems

12.2.1.1 Reduce the representations $D^{(+\ell)}$ and $D^{(-\ell)}$ of $O(3)$ with respect to the discrete symmetry $D_{4h} = D_4 \otimes \{E, I\}$. Show in particular that

$$D^{(+2)} = \Gamma^1 + \Gamma^2 + \Gamma^4 + \Gamma^5 \equiv A_{1g} + A_{2g} + B_{2g} + E_g$$

$$D^{(-3)} = A_{2u} + B_{1u} + B_{2u} + 2E_u.$$

12.2.1.2 Do the same with respect to $O_h = O \otimes \{E, I\}$. Show in particular that

$$D^{(+4)} = A_{1g} + E_g + F_{1g} + F_{2g}, \quad D^{-3} = E_u + F_{2u}.$$

12.2.1.3 Reduce the representations of $D^{(6)}$ and $D^{(5)}$ of $SO(3)$ with respect to D_4 and $D^{(+6)}$ and $D^{(-5)}$ of $O(3)$ with respect to D_{4h}.
For the notations see Section 12.3.

12.2.1.4 Do the same in the case that the discrete symmetries are O and O_h, respectively.

12.2.2.1 Find the double-valued representations of $C_3' \Leftrightarrow C_3 = \{E, C_3, C_3^2\}$.

12.2.2.2 Find the classes of D_3', and show that the table of its double-valued characters is

D_3'	E	R	$C_3^2, C_3 R$	$C_3, C_3^2 R$	$C_x(3)$	$C_x R(3)$
χ_1'	1	−1	−1	1	i	$-i$
χ_2'	1	−1	−1	1	$-i$	i
$3pt]\chi_3'$	2	−2	1	−1	0	0.

Note: Whenever a given class contains many elements, the number of elements is indicated in parenthesis next to the class.

12.2.2.3 Find the classes of D_6', and show that the matrix of the double-valued elements is

D_6'	E	R	$C_6^5 R$	$C_6^4 R$	$C_6 R$	$C_6^2 R$	$C_6^3 R$	$C_2(3)$	$C_2' R$
$\chi(E_1')$	2	−2	$\sqrt{3}$	1	$-\sqrt{3}$	−1	0	0	0
$\chi(E_2')$	2	−2	$-\sqrt{3}$	1	$\sqrt{3}$	−1	0	0	0
$\chi(E_3')$	2	−2	0	−2	0	2	0	0	0

(see the above note).

12.2.2.4 Find the classes of O_6', and show that the matrix of the double-valued elements is

O'	E	R	$C_3^2R(4)$	$C_3R(4)$	$C_4^3R(3)$	$C_4R(3)$	$C_4^2R(3)$	C_2R
$\chi(E_1')$	2	-2	1	-1	$\sqrt{2}$	$-\sqrt{2}$	0	0
$\chi(E_2')$	2	-2	1	-1	$-\sqrt{2}$	$\sqrt{2}$	0	0
$\chi(E_3')$	4	-4	-1	1	0	0	0	0

(see the above note).

12.2.2.5 Show that the double-valued representations of $O(3)$ are reduced with respect to D_6' as follows:

$$D^{(1/2)} = E_1', D^{(3/2)} = E_1' + E_3', \quad D^{(5/2)} = E_1' + 2E_2' + E_3',$$
$$D^{(7/2)} = 2E_1' + 2E_2' + E_3', \quad D^{(9/2)} = E_1' + 2E_2' + 2E_3',$$
$$D^{(11/2)} = 2E_1' + 2E_2' + 2E_3'.$$

You may use the table of problem 12.2.2.3.

12.2.2.6 Show that in the case of O', the previous relations become

$$D^{(1/2)} = E_1', D^{(3/2)} = G', \quad D^{(5/2)} = E_2' + G',$$
$$D^{(7/2)} = E_1' + E_2' + G', \quad D^{(9/2)} = E_1' + 2G',$$
$$D^{(11/2)} = 2E_1' + 2E_2' + 2G'.$$

You may use the table of problem 12.2.2.4.

12.4.1 Extend Table 12.11 so that it contains $\ell = 5$ and $\ell = 6$.

12.4.2 Construct the analog of Table 12.11 associated with the symmetry D_4.

12.4.3 Do the same in the case of the discrete symmetry O.

Chapter 13

Applications of Discrete Groups in Particle Physics

A number of point groups are employed in particle physics, in particular some double covering of groups of subgroups of S_n, see Section 5.10. We recall that, as shown in Chapter 11, the set of the even permutations of S_n constitutes a group indicated as A_n. From this, one obtains another group A'_n if, to each element X of A_n, the element XI is adjoined, with I the opposite of the identity. The group A'_n is called the double covering group of A_n. A'_n has the same number of elements as S_n but is not isomorphic to it. Groups of particular interest to particle physics are A_4, A'_4, A_5 and A'_5, see, for example, Wang *et al.* (2021) and Everett and Stuart (2011). In this chapter, we discuss in some detail, from the particle physics point of view, the group A_4, which is isomorphic with the tetrahedral symmetry group.

13.1   The group A_4 as a subgroup of S_4

We have seen that S_4 is the group of permutations of four objects. It contains $4! = 24$ elements, which form four conjugal classes as follows:[1]

(1) The identity element:

$$C_1 \Leftrightarrow E = \begin{pmatrix} 1 & 2 & 3 & 4 \\ 1 & 2 & 3 & 4 \end{pmatrix} = (1\,1),\, (2\,2),\, (3\,3),\, (4\,4).$$

---

[1]For the reader's convenience, this chapter is written in such a way as to be indepen-dent, as much as possible, of the material of the previous chapters.

(2) The six transpositions:

$$C_2 \Leftrightarrow (12), (13), (14), (23), (24).$$

(3) Eight cyclic permutations of length 3:

$$C_3 \Leftrightarrow (123), (321), (124), (421), (134), (431), (234), (432).$$

(4) Six cyclic permutations of length 4:

$$C_4 \Leftrightarrow (1234), (1243), (1324), (1342), (1423), (1432)$$

(5) Three products of commuting transpositions:

$$C_5 \Leftrightarrow (12)(34), (13)(24), (14)(23).$$

One can easily see that the 12 elements of the classes C_1, C_3, C_5, i.e. the set of even permutations, constitute a group, $A_4 = C_1 + C_3 + C_5$, and the elements of the classes C_1, C_5 constitute another Abelian group of four elements known as A_2 or group 4. It can also be seen that the group A_4 is isomorphic to the tetrahedral symmetry (Section 2.4.2).

13.2   The structure of A_4

One simple representation of A_4 can be found with its elements acting in a 4D space x_1, x_2, x_3, x_4, yielding its basic representation. Thus, we obtain

$$(12)(34)x_1 = x_2, (12)(34)x_2 = x_1, (12)(34)x_3 = x_4, (12)(34)x_4 = x_3 \Rightarrow$$

$$T(E) = \begin{pmatrix} 1 & 0 & 0 & 0 \\ 0 & 1 & 0 & 0 \\ 0 & 0 & 1 & 0 \\ 0 & 0 & 0 & 1 \end{pmatrix}, \quad T((12)(34)) = \begin{pmatrix} 0 & 1 & 0 & 0 \\ 1 & 0 & 0 & 0 \\ 0 & 0 & 0 & 1 \\ 0 & 0 & 1 & 0 \end{pmatrix}.$$

Similarly,

$$T((13)(24)) = \begin{pmatrix} 0 & 0 & 1 & 0 \\ 0 & 0 & 0 & 1 \\ 1 & 0 & 0 & 0 \\ 0 & 1 & 0 & 0 \end{pmatrix}, \quad T((14)(23)) = \begin{pmatrix} 0 & 0 & 0 & 1 \\ 0 & 0 & 1 & 0 \\ 0 & 1 & 0 & 0 \\ 1 & 0 & 0 & 0 \end{pmatrix},$$

$$T(123) = \begin{pmatrix} 0 & 0 & 1 & 0 \\ 1 & 0 & 0 & 0 \\ 0 & 1 & 0 & 0 \\ 0 & 0 & 0 & 1 \end{pmatrix}, \quad T(421) = \begin{pmatrix} 0 & 1 & 0 & 0 \\ 0 & 0 & 0 & 1 \\ 0 & 0 & 1 & 0 \\ 1 & 0 & 0 & 0 \end{pmatrix},$$

$$T(134) = \begin{pmatrix} 0 & 0 & 0 & 1 \\ 0 & 1 & 0 & 0 \\ 1 & 0 & 0 & 0 \\ 0 & 0 & 1 & 0 \end{pmatrix}, \quad T(432) = \begin{pmatrix} 1 & 0 & 0 & 0 \\ 0 & 0 & 1 & 0 \\ 0 & 0 & 0 & 1 \\ 0 & 1 & 0 & 0 \end{pmatrix},$$

$$T(321) = \begin{pmatrix} 0 & 1 & 0 & 0 \\ 0 & 0 & 1 & 0 \\ 1 & 0 & 0 & 0 \\ 0 & 0 & 0 & 1 \end{pmatrix}, \quad T(124) = \begin{pmatrix} 0 & 0 & 0 & 1 \\ 1 & 0 & 0 & 0 \\ 0 & 0 & 1 & 0 \\ 0 & 1 & 0 & 0 \end{pmatrix},$$

$$T(431) = \begin{pmatrix} 0 & 0 & 1 & 0 \\ 0 & 1 & 0 & 0 \\ 0 & 0 & 0 & 1 \\ 1 & 0 & 0 & 0 \end{pmatrix}, \quad T(234) = \begin{pmatrix} 1 & 0 & 0 & 0 \\ 0 & 0 & 0 & 1 \\ 0 & 1 & 0 & 0 \\ 0 & 0 & 1 & 0 \end{pmatrix}.$$

From these, one finds the multiplication Table 13.1.

One can consider the basis vectors

$$|1\rangle = \frac{1}{2}(\psi_1 - \psi_2 + \psi_3 - \psi_4), \quad |2\rangle = \frac{1}{2}(\psi_1 - \psi_2 - \psi_3 + \psi_4),$$

$$|3\rangle = \frac{1}{2}(\psi_1 + \psi_2 - \psi_3 - \psi_4). \tag{13.1}$$

Then, one obtains a 3×3 representation given in problem 13.2.1. We note that in this basis, we obtain the elements

$$(123) \Leftrightarrow A = \begin{pmatrix} 0 & -1 & 0 \\ 0 & 0 & -1 \\ 1 & 0 & 0 \end{pmatrix}, \quad (13)(24) \Leftrightarrow B = \begin{pmatrix} 1 & 0 & 0 \\ 0 & -1 & 0 \\ 0 & 0 & -1 \end{pmatrix},$$

$$A^3 = B^2 = E, \tag{13.2}$$

which constitute a set of generators for the group A_4. It is clear that other choices of generators are possible; they are usually employed in groups isomorphic to A_4, e.g. the tetrahedral and C_{4v} groups, see Tables 17.7

Table 13.1. The multiplication table of A_4. The order of the elements p_i, $i = 2, 3, \ldots, 12$, is $t(1), t(2), t(3), a(0), a(1), a(2), a(3), b(0), b(1), b(2), b(3)$. Alternatively, in the base of Eq. (13.1), see problem 13.2.1:

$$(1\,3)(2\,4), (1\,4)(2\,3), (1\,2)(3\,4), (2\,3\,4), (1\,2\,4), (3\,2\,1), (4\,3\,1), (4\,3\,2), (4\,2\,1),$$
$$(1\,2\,3), (1\,3\,4).$$

	E	p_2	p_3	p_4	p_5	p_6	p_7	p_8	p_9	p_{10}	p_{11}	p_{12}
E	E	p_2	p_3	p_4	p_5	p_6	p_7	p_8	p_9	p_{10}	p_{11}	p_{12}
p_2	p_2	E	p_4	p_3	p_8	p_7	p_6	p_5	p_{11}	p_{12}	p_9	p_{10}
p_3	p_3	p_4	E	p_2	p_6	p_5	p_8	p_7	p_{12}	p_{11}	p_{10}	p_9
p_4	p_4	p_3	p_2	E	p_7	p_8	p_5	p_6	p_{10}	p_9	p_{12}	p_{11}
p_5	p_5	p_7	p_8	p_6	p_9	p_{11}	p_{12}	p_{10}	E	p_3	p_4	p_2
p_6	p_6	p_8	p_7	p_5	p_{12}	p_{10}	p_9	p_{11}	p_3	E	p_2	p_4
p_7	p_7	p_5	p_6	p_8	p_{10}	p_{12}	p_{11}	p_9	p_4	p_2	E	p_3
p_8	p_8	p_6	p_5	p_7	p_{11}	p_9	p_{10}	p_{12}	p_2	p_4	p_3	E
p_9	p_9	p_{12}	p_{10}	p_{11}	E	p_4	p_2	p_3	p_5	p_8	p_6	p_7
p_{10}	p_{10}	p_{11}	p_9	p_{12}	p_4	E	p_3	p_2	p_7	p_6	p_8	p_5
p_{11}	p_{11}	p_{10}	p_{12}	p_9	p_2	p_3	E	p_4	p_8	p_5	p_7	p_6
p_{12}	p_{12}	p_9	p_{11}	p_{10}	p_3	p_2	p_4	E	p_6	p_7	p_5	p_8

and 17.9. Another choice, common in particle physics applications, will be considered in the following.

With respect to A_4, the class C_3 breaks up into two:

$$C_{3a} \Leftrightarrow (1\,2\,3), (3\,2\,1), (4\,3\,2), (1\,3\,4), \quad C_{3b} \Leftrightarrow (1\,2\,4), (4\,2\,1), (4\,3\,1), (2\,3\,4).$$

There now exist, therefore, four classes and, as a result, four irreducible representations with ℓ_i positive integers satisfying

$$\sum_{i=1}^{3} \ell_i^2 = 12, \tag{13.3}$$

with the unique solution

$$\ell_1 = 1, \; \ell_2 = 1, \; \ell_3 = 1, \; \ell_4 = 3, \tag{13.4}$$

i.e. we have three 1D and one 3D representations. The last one is the most important. It can be constructed considering the group elements acting on the basis given by Eq. (13.1), see problem 13.2.1. Since, however, this group is isomorphic to the tetrahedral group, one can use Table 17.9.

In particle physics, the following steps are involved:

• The identity element represented by the 3×3 unit matrix.

- The elements of class C_5 commute with each other. Thus, they can be simultaneously diagonalized. Since they are orthogonal with determinant $+1$, they can take the form

$$t_1 = \begin{pmatrix} 1 & 0 & 0 \\ 0 & -1 & 0 \\ 0 & 0 & -1 \end{pmatrix}, \quad t_2 = \begin{pmatrix} -1 & 0 & 0 \\ 0 & 1 & 0 \\ 0 & 0 & -1 \end{pmatrix}, \quad t_3 = \begin{pmatrix} -1 & 0 & 0 \\ 0 & -1 & 0 \\ 0 & 0 & 1 \end{pmatrix}.$$

$$(13.5)$$

In reality, the transformation (13.1) leads directly to this diagonal form with the exception of the sign of t_2. In any case, vectors of the form $(1,0,0)^T$, $(0,1,0)^T$, $(0,0,1)^T$ are eigenvectors of the elements of this class with eigenvalues as indicated by the diagonal elements This representation is known as real. The matrix t_2 is often indicated as S'.

- The operator $(1\,2\,3)$ caused the transformation $(x_1, x_2, x_3) \to (x_2, x_3, x_1)$ and $(1\,3\,2)$ the transformation $(x_1, x_2, x_3) \to (x_3, x_1, x_2)$. Thus, they are represented as

$$(132) \Leftrightarrow a(0) = \begin{pmatrix} 0 & 0 & 1 \\ 1 & 0 & 0 \\ 0 & 1 & 0 \end{pmatrix}, \quad (123) \Leftrightarrow b(0) = \begin{pmatrix} 0 & 1 & 0 \\ 0 & 0 & 1 \\ 1 & 0 & 0 \end{pmatrix}.$$

With their aid, we construct the operators

$$a(i) = t(i)a(0)t(i), \quad b(i) = t(i)b(0)t(i), \quad i = 1, 2, 3.$$

As a result, we find

$$a(1) = \begin{pmatrix} 0 & 0 & -1 \\ -1 & 0 & 0 \\ 0 & 1 & 0 \end{pmatrix}, \quad a(2) = \begin{pmatrix} 0 & 0 & 1 \\ -1 & 0 & 0 \\ 0 & -1 & 0 \end{pmatrix}, \quad a(3) = \begin{pmatrix} 0 & 0 & -1 \\ 1 & 0 & 0 \\ 0 & -1 & 0 \end{pmatrix},$$

$$b(1) = \begin{pmatrix} 0 & -1 & 0 \\ 0 & 0 & 1 \\ -1 & 0 & 0 \end{pmatrix}, \quad b(2) = \begin{pmatrix} 0 & -1 & 0 \\ 0 & 0 & -1 \\ 1 & 0 & 0 \end{pmatrix}, \quad b(3) = \begin{pmatrix} 0 & 1 & 0 \\ 0 & 0 & -1 \\ -1 & 0 & 0 \end{pmatrix}.$$

The matrix $b(2)$ is often called T'.

The multiplication of A_4 can be constructed using either the fundamental representation or the above 3×3 representation. The results are exhibited in Table 13.1.

Note that $p_i^2 = 1$, $i = 2, \ldots, 4$, $p_i^2 = p_{4+i}$, $i = 5, \ldots, 8$, $p_{i+8}^2 = p_{4+i}$, $i = 1, \ldots, 4$, $(p_i p_j)^2 = E$, $ij = 2, \ldots, 4$ and $(p_i p_j)^3 = E$, $i, = 2, \ldots, 4$, $j = 5, \ldots, 12$. Also, all matrices p_i, $i = 2, \ldots, 12$ are unitary with zero trace.

Before proceeding further, we make some remarks illuminating the symbolism employed in particle physics.

(1) $S^2 = T^3 = (ST)^3 = (TS)^3 = 1$
   $ST = (234)$, $TS = (124)$ have as eigenvalues the cubic roots[2] of unity $1, \omega, \omega^2$, where $\omega = e^{(2\pi i)/3}$.
(2) $C_1 = \{E\}$, $g_1 = 1$.
(3) $C_2 = \{S, T^2 ST, TST^2\}$, $g_2 = 3$.
(4) $C_{3a} = \{T, ST, TS, STS\}$, $g_{3a} = 4$.
(5) $C_{3b} = \{T^2, ST^2, T^2 S, ST^2 S\}$, $g_{3b} = 4$.
(6) Possible subgroups:

   - three Z_2:
     $\{1, S\}$, $\{1, T^2 ST\}$, $\{1, TST^2\}$;
   - four Z_3:
     $\{1, T, T^2\}$, $\{1, ST, T^2 S\}$, $\{1, TS, ST^2\}$, $\{1, STS, ST^2 S\}$;
   - The Klein group K_4: $\{1, S, T^2 ST, TST^2\}$.

One can select a basis on which the operator T is diagonal. Indeed,

$$T = U^+ b(0) U = \begin{pmatrix} 1 & 0 & 0 \\ 0 & \omega & 0 \\ 0 & 0 & \omega^2 \end{pmatrix}, \quad A = \sqrt{3} U = \begin{pmatrix} 1 & 1 & 1 \\ \omega & 1 & \omega^2 \\ \omega^2 & 1 & \omega \end{pmatrix}, \quad \omega = e^{\frac{2\pi i}{3}}.$$

$$(13.6)$$

Then,

$$U^+ t_2 U \to S = \frac{1}{3} \begin{pmatrix} -1 & 2 & 2 \\ 2 & -1 & 2 \\ 2 & 2 & -1 \end{pmatrix}, \quad (13.7)$$

$$U^+ t_3 U \to S_3 = -1 + \frac{2}{3} \begin{pmatrix} 1 & \omega & \omega^2 \\ \omega^2 & 1 & \omega \\ \omega & \omega^2 & 1 \end{pmatrix}, \quad (13.8)$$

---

[2]The same is true for all the eight operators of the form $p_i p_j$ $i = 2, 3, 4; j = 5, 6, \ldots, 12$. The rest, except for the identity, have eigenvalues $(1, -1, -1)$.

$$U^+ t_1 U \to S_1 = -1 + \tfrac{2}{3} \begin{pmatrix} 1 & \omega^2 & \omega \\ \omega & 1 & \omega^2 \\ \omega^2 & \omega & 1 \end{pmatrix}. \tag{13.9}$$

This representation is useful in applications (see the following magic matrix). The eigenvalues of S are $1, -1$ and -1, with the diagonalizing matrix being the above matrix A. The eigenvalues of the matrix T are 1, ω and ω^2.

13.3 The character matrix of A_4

The character table of A_4 can be obtained using techniques similar to those in Section 5.7. We number the lines χ_i, $i = 1, 2, \ldots, 4$, and the columns E, C_5, C_{3a}, C_{3b}. The first column is clearly $1, 1, 1, 3$, and the last row is $3, -1, 0, 0$. Let y_1, y_2, y_3, y_4 be the elements of the second column. The orthogonality condition of the first three lines with the last one yields $3 + 3 \times y_i \times (-1) = 0$, giving $y_i = 1$. So, the second column is complete. We can select the first row to be $1, 1, 1, 1$. The missing elements are x_1, x_2 and y_1, y_2 for the second and third rows, respectively. The condition of orthogonality of the second and third rows yields $1 + 3 + 4 \times x_1 + 4 \times x_2 = 0$, or $x_1 + x_2 = -1$. Similarly, the same condition between the second and third columns yields $x_1 y_1 + x_2 y_2 = -1$. From the solutions to these equations, we select $(x_1, x_2) = (-1, 0)$ and $(y_1, y_2) = (1, -2)$. Thus, we obtain the matrix indicated in Table 13.2.

With the aid of Table 13.2, we can construct the 1D representations of A_4.

$$\Gamma_1 = \left(\frac{1}{2\sqrt{3}}, \frac{1}{2\sqrt{3}}, \frac{1}{2\sqrt{3}}, \frac{1}{2\sqrt{3}}, \frac{1}{2\sqrt{3}}, \frac{1}{2\sqrt{3}}, \frac{1}{2\sqrt{3}}, \frac{1}{2\sqrt{3}}, \frac{1}{2\sqrt{3}}, \frac{1}{2\sqrt{3}}, \frac{1}{2\sqrt{3}}, \frac{1}{2\sqrt{3}} \right),$$

Table 13.2. The character table of A_4, not normalized (left) and normalized (right).

	E	C_5	C_{3a}	C_{3b}		E	C_5	C_{3a}	C_{3b}
χ_1	1	1	1	1	χ_1	$\frac{1}{2\sqrt{3}}$	$\frac{\sqrt{3}}{2}$	$\frac{1}{\sqrt{3}}$	$\frac{1}{\sqrt{3}}$
χ_2	1	1	-1	0	χ_2	$\frac{1}{2\sqrt{2}}$	$\frac{\sqrt{3}}{2\sqrt{2}}$	$-\frac{1}{\sqrt{2}}$	0
χ_3	1	1	1	-2	χ_3	$\frac{1}{2\sqrt{6}}$	$\frac{1}{2\sqrt{2}}$	$\frac{1}{\sqrt{6}}$	$-\frac{2}{\sqrt{6}}$
χ_4	3	-1	0	0	χ_4	$\frac{\sqrt{3}}{2}$	$-\frac{1}{2}$	0	0

$$\Gamma_2 = \left(\frac{1}{2\sqrt{2}}, \frac{1}{2\sqrt{2}}, \frac{1}{2\sqrt{2}}, \frac{1}{2\sqrt{2}}, -\frac{1}{2\sqrt{2}}, -\frac{1}{2\sqrt{2}}, -\frac{1}{2\sqrt{2}}, -\frac{1}{2\sqrt{2}}, 0, 0, 0, 0 \right),$$

$$\Gamma_3 = \left(\frac{1}{2\sqrt{6}}, \frac{1}{2\sqrt{6}}, \frac{1}{2\sqrt{6}}, \frac{1}{2\sqrt{6}}, \frac{1}{2\sqrt{6}}, \frac{1}{2\sqrt{6}}, \frac{1}{2\sqrt{6}}, \frac{1}{2\sqrt{6}}, -\frac{1}{\sqrt{6}}, -\frac{1}{\sqrt{6}}, -\frac{1}{\sqrt{6}}, -\frac{1}{\sqrt{6}} \right).$$

The ordering of the elements of the group are as in the multiplication table, see Table 13.1. The 3D representation has already been given above.

From the character table of A_4, one can obtain the representation reduction of $SO(3)$ under A_4. Since the latter is isomorphic to the tetrahedral group T, this has already been accomplished, as given by Eq. 12.9, i.e.

$$D^0 = \Gamma^1, \quad D^1 = \Gamma^4, \quad D^2 = \Gamma^2 \oplus \Gamma^3 \oplus \Gamma^4, \quad D^3 = \Gamma^1 \oplus 2\Gamma^4. \quad (13.10)$$

13.4   Reduction of the Kronecker product of the representations of A_4

The Kronecker product of two representations is, in general, reducible. In the case of A_4, this is true in the case of the 3D representation indicated as $\underline{3}$. We are, thus, interested in the reduction of

$$\underline{3} \otimes \underline{3}.$$

For this purpose, we find it convenient to consider the larger group of rotations in 3D space, $O(3)$. In this case, we know that

$$(\ell = 1) \otimes (\ell = 1) \rightarrow (\ell = 0) + (\ell = 1) + (\ell = 2).$$

The first two are irreducible under A_4, see Chapter 12, Section 12.2. More specifically,

$$\ell = 0 \Leftrightarrow \underline{1} \leftrightarrow u = \mathbf{x}.\mathbf{y}, \quad \ell = 1 \Leftrightarrow \underline{3}_A \Leftrightarrow \tilde{t}_1 = \frac{i}{\sqrt{2}} \mathbf{x} \times \mathbf{y},$$

$$\underline{3}_A = i \frac{1}{\sqrt{2}} \left(x_2 y_3 - x_3 y_2, \, x_3 y_1 - x_1 y_3, \, x_1 y_2 - x_2 y_1 \right).$$

Limiting ourselves to A_4 as a subgroup of $O(3)$, we find that $\ell = 2$ contains two 1D and one 3D representations, i.e.

$$(\ell = 2) \rightarrow \underline{1}' + \underline{1}'' + \underline{3}_S,$$

where

$$\underline{3}_S \Leftrightarrow \tilde{t}_2 = \frac{1}{\sqrt{2}} (x_2 y_3 + x_3 y_2, \, x_1 y_3 + x_3 y_1, \, x_1 y_2 + x_2 y_1),$$

which is orthogonal[3] to $\underline{3}_A$. Thus,

$$\underline{3} \otimes \underline{3} = \underline{1} + \underline{1}' + \underline{1}'' + \underline{3}_A + \underline{3}_S.$$

Instead of $\tilde{t}_1$ and $\tilde{t}_2$, it may be more convenient to use the basis

$$\frac{1}{\sqrt{2}}(-i\tilde{t}_1 + \tilde{t}_2) = (x_2y_3, x_3y_1, x_1y_2), \quad \frac{1}{\sqrt{2}}(i\tilde{t}_1 - \tilde{t}_2) = (x_3y_2, x_1y_3, x_2y_1).$$

$$(13.11)$$

The two representations $\underline{1}'$ and $\underline{1}''$ are traceless combinations of x and y, e.g. the choice

$$\underline{1}' \Leftrightarrow u' = x_1y_1 + \omega x_2y_2 + \omega^2 x_3y_3,$$

$$\underline{1}'' \Leftrightarrow u'' = x_1y_1 + \omega^2 x_2y_2 + \omega x_3y_3, \quad \omega = e^{(i2\pi)/3}.$$

We find that

$$1 + \omega + \omega^2 = 0$$

and

$$u' = (x_1, x_2, x_3)\, d'\, (y_1, y_2, y_3)^T, \quad u'' = (x_1, x_2, x_3)\, d''\, (y_1, y_2, y_3)^T,$$

$$d' = \begin{pmatrix} 1 & 0 & 0 \\ 0 & \omega & 0 \\ 0 & 0 & \omega^2 \end{pmatrix}, \quad d'' = \begin{pmatrix} 1 & 0 & 0 \\ 0 & \omega^2 & 0 \\ 0 & 0 & \omega \end{pmatrix},$$

$$d'd' = d'', \; d''d'' = d', \; d'd'' = E. \tag{13.12}$$

In other words, the set (E, d', d'') constitutes a group, and thus, we confirm the reductions:

$$\underline{1}' \otimes \underline{1}' = \underline{1}'', \; \underline{1}' \otimes \underline{1}'' = \underline{1}, \; \underline{1}'' \otimes \underline{1}'' = \underline{1}'.$$

Furthermore,

$$(\underline{1}')^* = \underline{1}'',$$

It is clear that u, being the most symmetric, is invariable under A_4. We stress, however, that the two 1D representations u' and u'' are not invariant

---

[3] Another representation is

$$\underline{3}_S = -\frac{1}{\sqrt{6}}(-2x_1x_1 + x_2y_3 + x_3y_2, -2x_2y_2 + x_1y_3 + x_3y_1, -2x_3y_3 + x_1y_2 + x_2y_1).$$

under the transformations induced by A_4. Indeed, under the permutation $a(0) = (132)$ and $b(0) = (123)$, we find that

$$a(0)u' = \omega u' \neq u', \quad a(0)u'' = \omega^2 u'' \neq u'',$$

$$b(0)u' = \omega^2 u' \neq u', \quad b(0)u'' = \omega u'' \neq u''.$$

In the more standard notation,

$$S_1 = 1, \; S_1' = \omega^2 1' \neq 1', \quad S_1'' = \omega 1'' \neq 1''.$$

From the character table, we deduce $X_1 = 1 \forall X \in A_4$, i.e. $T_1 = 1$.

It is now clear that one can choose the basis

$$X_1 = \frac{1}{\sqrt{3}}(x_1 + x_2 + x_3), \quad X_2 = \frac{1}{\sqrt{3}}(x_1 + \omega x_2 + \omega^2 x_3),$$

$$X_3 = \frac{1}{\sqrt{3}}(x_1 + \omega^2 x_2 + \omega x_3)$$

such that the operator S is diagonal with eigenvalues $1, \omega, \omega^2$, not all being real. Proceeding as above, we find

$$\underline{3}(\mathbf{X}) \otimes \underline{3}(\mathbf{Y}) = \underline{3}_S + \underline{3}_A + \underline{1} + \underline{1}' + \underline{1}'',$$

$$\underline{3}_S = (X_2 Y_3 + X_3 Y_2, \; X_3 Y_1 + X_1 Y_3, \; X_1 Y_2 + X_2 Y_1), \tag{13.13}$$

$$\underline{3}_A = (X_2 Y_3 - X_3 Y_2, \; X_3 Y_1 - X_1 Y_3, \; X_1 Y_2 - X_2 Y_1), \tag{13.14}$$

$$\underline{1} = X_1 Y_1 + X_2 Y_2 + X_3 Y_3, \tag{13.15}$$

$$\underline{1}' = X_1 Y' 1 + \omega X_2 Y_2' + \omega^2 X_3 Y_3, \tag{13.16}$$

$$\underline{1}'' = X_1' Y_1 + \omega^2 X_2' Y_2 + \omega X_3' Y_3. \tag{13.17}$$

The above matrices can be given in a real basis. It is not difficult to show that

$$\underline{1} = \frac{1}{3}(x_1 y_1 + x_3 y_2 + x_2 y_3) \text{ or in matrix form}$$

$$= (x_1, x_2, x_3)^T \frac{1}{3} \begin{pmatrix} 1 & 0 & 0 \\ 0 & 0 & 1 \\ 0 & 1 & 0 \end{pmatrix} \begin{pmatrix} y_1 \\ y_2 \\ y_3 \end{pmatrix}, \tag{13.18}$$

$$\underline{1}' = \frac{1}{3}(x_3 y_1 + x_2 y_2 + x_1 y_3) \text{ or in matrix form}$$

$$= (x_1, x_2, x_3)^T \frac{1}{3} \begin{pmatrix} 0 & 0 & 1 \\ 0 & 1 & 0 \\ 1 & 0 & 0 \end{pmatrix} \begin{pmatrix} y_1 \\ y_2 \\ y_3 \end{pmatrix}, \tag{13.19}$$

$$\underline{1}'' = \frac{1}{3}(x_2 y_1 + x_1 y_2 + x_3 y_3) \text{ or in matrix form}$$

$$= (x_1, x_2, x_3)^T \frac{1}{3} \begin{pmatrix} 0 & 1 & 0 \\ 1 & 0 & 0 \\ 0 & 0 & 1 \end{pmatrix} \begin{pmatrix} y_1 \\ y_2 \\ y_3 \end{pmatrix}, \tag{13.20}$$

$$\underline{3}_S = \frac{1}{3}[x_1(2y_1 - y_2 - y_3 + x_2(2y_2 - y_1 - y_3) + x_3(2y_3 - y_1 - y_2),$$

$$x_1(2y_1 + y_3 + (y_3 - y_2)\omega) + x_3(y_1 - y_2 + (y_1 + 2y_3)\omega),$$

$$x_1(y_1 + y_3 + (y_1 + y_3)\omega) + x_2(y_1 + y_3)(1 + \omega) - x_3(y_1\omega + y_2 + y_3)]. \tag{13.21}$$

This can also be written in matrix form:

$$\underline{3}_S|_1 = (x_1, x_2, x_3)^T \frac{1}{3} \begin{pmatrix} 2 & -1 & -1 \\ -1 & 2 & -1 \\ -1 & - & 2 \end{pmatrix} \begin{pmatrix} y_1 \\ y_2 \\ y_3 \end{pmatrix}, \tag{13.22}$$

$$\underline{3}_A = \frac{1}{3}[x_1(y_2 + y_3)(1 + 2\omega) + x_2(y_2 - y_1)(1 + \omega) + x_3(y_2 - y_1)(1 + 2\omega),$$

$$x_1(2y_2 + y_3 + (y_2 - y_3)\omega) - x_2(2y_1 + y_3 + (y_1 + 2y_3)\omega)$$

$$+ x_3(y_2 - y_1 + (y_1 + 2y_2)\omega)$$

$$+ x_1(y_3 - y_2 + (y_2 + 2y_3)\omega) + x_2(y_1 - y_3 - (y_1 + 2y_3)\omega)$$

$$+ x_3(2y_1 + y_2 + (y_1 + 2y_2)\omega]. \tag{13.23}$$

13.5 An application in particle physics

The fact that A_4 has three 1D representations and one 3D, one expects that it will be a good tool for the study of leptons, which occur in three flavors; for a review, see Altarelli and Feruglio (2010a), King and Luhn (2013a) and Ishimori *et al.* (2010a).

The mass operators of charged leptons are of dimension 4, while those of neutral leptons (neutrinos) are of order 5:

$$O_4 = \phi^+ \ell^c \psi, \quad O_5 = \xi \tau_2 \psi C \xi' \tau_2 \psi. \tag{13.24}$$

The Pauli matrix τ_2 and the charge conjugation operator C (Vergados, 2018) have no relation to A_4, and they will not be discussed further. The scalar fields ϕ, ξ and ξ' are isodoublets in the standard model, and they can acquire a vacuum expectation value, depending on A_4. The fields ψ are also Fermion isodoublets, while the field ℓ^c is an isosinglet. They also depend on A_4.

Now, it remains to construct the scalar potential, which is a polynomial of fourth degree with respect to the scalar fields, invariant under the transformations of A_4.

What happens depends on the transformation properties of the fields under A_4.

13.5.1   *The scalar potential in the framework of A_4*

We begin with the construction of the scalar potential in the presence of only one complex scalar field, which under A_4 transforms like $\underline{3}$, that is, it has components ϕ_i, $i = 1, 2, 3$. We employ the above notation with the correspondence $x_i \Leftrightarrow \phi_i^+$, $y_i \Leftrightarrow \phi_i$. The quadratic terms will be of the form $\underline{3} \otimes \underline{3}$. We have already seen that

$$\underline{3} \otimes \underline{3} \to \underline{1} + \underline{1}' + \underline{1}'' + \underline{3} + \underline{3}', \ \underline{3} = \underline{3}_A, \ \underline{3}' = \underline{3}_S. \tag{13.25}$$

From these, only $\underline{1}$ is invariant. Thus, the quadratic term is of the form

$$V_2 = \phi_1^+ \phi_1 + \phi_2^+ \phi_2 + \phi_3^+ \phi_3. \tag{13.26}$$

On the other hand, the quartic terms are of the form

$$(\underline{3} \otimes \underline{3}) \otimes (\underline{3} \otimes \underline{3}) \to (\underline{1} + \underline{1}' + \underline{1}'' + \underline{3} + \underline{3}') \otimes (\underline{1} + \underline{1}' + \underline{1}'' + \underline{3} + \underline{3}'). \tag{13.27}$$

The five invariant quantities are

$$\underline{1} \otimes \underline{1}, \ \underline{1}' \otimes \underline{1}'', \ \underline{3} \otimes \underline{3}, \underline{3} \otimes \underline{3}', \ \underline{3}' \otimes \underline{3}'. \tag{13.28}$$

This way, we obtain the invariant quantities:

$$V_{4a} = \left(\phi_1^+ \phi_1 + \phi_2^+ \phi_2 + \phi_3^+ \phi_3 \right)^2, \tag{13.29}$$

$$V'' = \left(\phi_1^+ \phi_1 + \omega \phi_2^+ \phi_2 + \omega^2 \phi_3^+ \phi_3 \right) \left(\phi_1^+ \phi_1 + \omega^2 \phi_2^+ \phi_2 + \omega \phi_3^+ \phi_3 \right). \tag{13.30}$$

The last one is reduced to the previous one, V_{4a}, and the expression becomes

$$V_{4b} = \phi_1^+ \phi_1 \phi_2^+ \phi_2 + \phi_1^+ \phi_1 \phi_3^+ \phi_3 + \phi_2^+ \phi_2 \phi_3^+ \phi_3. \tag{13.31}$$

The remaining are found using the basis (13.11) with the correspondence $x_i \Leftrightarrow \phi_i^+$, $y_i \Leftrightarrow \phi_i$. One convenient choice is

$$q = (V.V^+, V.V, V^+.V^+), \quad \text{where } V = (\phi_1^+ \phi_2, \phi_2^+ \phi_3, \phi_3^+ \phi_1),$$

$$V_{4c} = |\phi_1^+ \phi_2|^2 + |\phi_3^+ \phi_1|^2 + |\phi_2^+ \phi_3|^2, \tag{13.32}$$

$$V_{4d} = \left(\phi_1^+ \phi_2\right)^2 + \left(\phi_3^+ \phi_1\right)^2 + \left(\phi_2^+ \phi_3\right)^2, \tag{13.33}$$

$$V_{4e} = \left(\phi_2^+ \phi_1\right)^2 + \left(\phi_1^+ \phi_3\right)^2 + \left(\phi_3^+ \phi_2\right)^2. \tag{13.34}$$

As a result, the general form of the potential is

$$V = c_2 V_2 + c_{4a} V_{2b} + c_{4b} V_{4b} + c_{4c} V_{4c} + c_{4d} V_{4d} + c_{4e} V_{4e}, \tag{13.35}$$

where $c_2, c_{4a}, c_{4b}, c_{4c}, c_{4d}$ and c_{4e} are constants.

It can be shown, following standard techniques, see, for example, Vergados (2018), that the minimization of this potential occurs when $\langle \phi_1 \rangle = \langle \phi_2 \rangle = \langle \phi_3 \rangle = v$.

13.5.2 *The lepton masses in a simple A_4 model*

Suppose[4] that

$$(\ell_1^c, \ell_2^c, \ell_3^c) \Leftrightarrow (\underline{1}, \underline{1}', \underline{1}''), \phi \Leftrightarrow \underline{3}, \psi \Leftrightarrow \underline{3}. \tag{13.36}$$

The A_4 invariant quantity is

$$\mathcal{L} = h_1 \ell_1^c \left(\phi_1 \psi_1 + \phi_2 \psi_2 + \phi_3 \psi_3\right) + h_2 \ell_2^c \left(\phi_1 \psi_1 + \omega \phi_2 \psi_2 + \omega^2 \phi_3 \psi_3\right)$$
$$+ h_3 \ell_3^c \left(\phi_1 \psi_1 + \omega^2 \phi_2 \psi_2 + \omega \phi_3 \psi_3\right),$$

where $\ell_i^c, \psi_i, i = 1, 2, 3$ are the left- and right-handed lepton Fermion fields and h_1, h_2, h_3 are constants. This can be cast in a convenient form as

$$\mathcal{L} = (\ell_1^c, \ell_2^c, \ell_3^c) \begin{pmatrix} h_1 & 0 & 0 \\ 0 & h_2 & 0 \\ 0 & 0 & h_3 \end{pmatrix} \begin{pmatrix} 1 & 1 & 1 \\ 1 & \omega & \omega^2 \\ 1 & \omega^2 & \omega \end{pmatrix} \begin{pmatrix} \phi_1 & 0 & 0 \\ 0 & \phi_2 & 0 \\ 0 & 0 & \phi_3 \end{pmatrix} \begin{pmatrix} \psi_1 \\ \psi_2 \\ \psi_3 \end{pmatrix}. \tag{13.37}$$

---

[4]Readers not familiar with this formalism common in particle physics is referred to relevant textbooks, e.g. (Vergados, 2018), Chapters 5 and 6.

When the scalar fields acquire a vacuum expectation value, A_4 invariance implies

$$\langle\phi_1\rangle = \langle\phi_2\rangle = \langle\phi_3\rangle = v.$$

Hence,

$$\mathcal{L} = (\ell_1^c, \ell_2^c, \ell_3^c)\, \mathcal{M}_\ell\, (\psi_1, \psi_2, \psi_3)^T,$$ (13.38)

where $\mathcal{M}_\ell$ is the mass matrix of the charged leptons:

$$\mathcal{M}_\ell = \begin{pmatrix} m_e & 0 & 0 \\ 0 & m_\mu & 0 \\ 0 & 0 & m_\tau \end{pmatrix} \begin{pmatrix} 1 & 1 & 1 \\ 1 & \omega & \omega^2 \\ 1 & \omega^2 & \omega \end{pmatrix}$$ (13.39)

and $m_e = h_1 v$, $m_\mu = h_2 v$, $m_\tau = h_3 v$. This matrix is not Hermitian. It can, however, be diagonalized by separate unitary left and right matrices U_R, U_L, see, for example, Vergados (1986):

$$U_L^+ \mathcal{M}_\ell U_R = \text{diag}\,(m_e, m_\mu, m_\tau),$$ (13.40)

where

$$U_L = U_R^*, \quad U_R \equiv U^+ = \frac{1}{\sqrt{3}} A, \quad A = \begin{pmatrix} 1 & 1 & 1 \\ \omega & 1 & \omega^2 \\ \omega^2 & 1 & \omega \end{pmatrix}), \quad \omega = e^{\frac{2\pi i}{3}}$$ (13.41)

(see Eq. (13.6)).

The matrix A is often called "magic matrix" of A_4.

In a similar fashion, we construct the neutrino mass matrix $\mathcal{M}_\nu$ with mixing matrix U_ν.

From the columns of the magic matrix, we construct the two columns

$$\frac{-1}{\sqrt{6}}\left[\begin{pmatrix} 1 \\ \omega \\ \omega^2 \end{pmatrix} + \begin{pmatrix} 1 \\ \omega^2 \\ \omega \end{pmatrix}\right] = \begin{pmatrix} \frac{-2}{\sqrt{6}} \\ \frac{1}{\sqrt{6}} \\ \frac{1}{\sqrt{6}} \end{pmatrix}, \quad \frac{-i}{\sqrt{6}}\left[\begin{pmatrix} 1 \\ \omega \\ \omega^2 \end{pmatrix} - \begin{pmatrix} 1 \\ \omega^2 \\ \omega \end{pmatrix}\right] = \begin{pmatrix} 0 \\ \frac{1}{\sqrt{2}} \\ -\frac{1}{\sqrt{2}} \end{pmatrix}.$$

Combining these with the normalized second column, we obtain the unitary matrix

$$
V = \begin{pmatrix} \frac{-2}{\sqrt{6}} & \frac{1}{\sqrt{3}} & 0 \\ \frac{1}{\sqrt{6}} & \frac{1}{\sqrt{3}} & \frac{1}{\sqrt{2}} \\ \frac{1}{\sqrt{6}} & \frac{1}{\sqrt{3}} & -\frac{1}{\sqrt{2}} \end{pmatrix}.
\tag{13.42}
$$

This is the famous tri-maximal and bi-maximal matrix, which, to a good approximation, i.e. for $\theta_{13} = 0$, describes well the experimental results of neutrino oscillations.

Since, by definition, $V = U_\ell^\dagger U_\nu$, the neutrino mixing matrix U_ν is

$$
U_\nu = U_\ell V = \frac{1}{\sqrt{3}} \begin{pmatrix} 1 & \omega^2 & \omega \\ 1 & 1 & 1 \\ 1 & \omega & \omega^2 \end{pmatrix} \begin{pmatrix} \frac{-2}{\sqrt{6}} & \frac{1}{\sqrt{3}} & 0 \\ \frac{1}{\sqrt{6}} & \frac{1}{\sqrt{3}} & \frac{1}{\sqrt{2}} \\ \frac{1}{\sqrt{6}} & \frac{1}{\sqrt{3}} & -\frac{1}{\sqrt{2}} \end{pmatrix} = \begin{pmatrix} \frac{-1}{\sqrt{2}} & 0 & \frac{-i}{\sqrt{2}} \\ 0 & 1 & 0 \\ \frac{-1}{\sqrt{2}} & 0 & \frac{i}{\sqrt{2}} \end{pmatrix}.
\tag{13.43}
$$

It can easily be verified that

$$
U_\nu^\dagger \, diag(m_1, m_2, m_3) U_\nu = \begin{pmatrix} \frac{m_1 + m_3}{2} & 0 & i\frac{m_1 - m_3}{2} \\ 0 & m_2 & 0 \\ -i\frac{m_1 - m_3}{2} & 0 & \frac{m_1 + m_3}{2} \end{pmatrix}.
$$

In other words, assuming that the eigenvalues are real, the second matrix of Eq. (13.24) can be cast in the form

$$
m_\nu = \begin{pmatrix} \alpha & 0 & i\beta \\ 0 & \gamma & 0 \\ -i\beta & 0 & \alpha \end{pmatrix}, \ \alpha, \beta, \gamma \text{ real} \Rightarrow
$$

$$
(U_\nu)^\dagger \, m_\nu U_\nu = \begin{pmatrix} \alpha + \beta & 0 & 0 \\ 0 & \gamma & 0 \\ 0 & 0 & \alpha - \beta \end{pmatrix}.
\tag{13.44}
$$

Hence, m_ν has the eigenvalues $\alpha + \beta, \gamma$ and $\alpha - \beta$.

The neutrino matrix, however, is not necessarily Hermitian, since it can be a complex symmetric matrix. In this case, it can be diagonalized with different unitary matrices from the left and right.[5] What thus remains is

---

[5]For details, see the review article by Vergados (1986). In this case, the two matrices are related via $U_L = U_R^*$. In other words, the neutrino matrix is of the form

$$m_\nu = U_R^+ \text{diag}(m_1, m_2, m_3) U_L = U_\nu^T \, \text{diag}(m_1, m_2, m_3) U_\nu = \begin{pmatrix} \frac{m_1+m_3}{2} & 0 & i\frac{m_1-m_3}{2} \\ 0 & m_2 & 0 \\ i\frac{m_1-m_3}{2} & 0 & -\frac{m_1+m_3}{2} \end{pmatrix}.$$

$$(13.45)$$

The matrices U_R and U_L are obtained by diagonalizing the Hermitian matrices $U_\nu^+ U_\nu$ and $U_\nu U_\nu^+$. They can be selected as

$$U_L = \begin{pmatrix} -\frac{1}{\sqrt{2}} & 0 & -\frac{i}{\sqrt{2}} \\ 0 & 1 & 0 \\ -\frac{1}{\sqrt{2}} & 0 & \frac{i}{\sqrt{2}} \end{pmatrix}, \quad U_R = U_L^* = \begin{pmatrix} -\frac{1}{\sqrt{2}} & 0 & \frac{i}{\sqrt{2}} \\ 0 & 1 & 0 \\ -\frac{1}{\sqrt{2}} & 0 & -\frac{i}{\sqrt{2}} \end{pmatrix}, \quad (13.46)$$

$$(\nu_e, \nu_\mu, \nu_\tau)^T = U_R (\nu_{1R}, \nu_{2R}, \nu_{3R})^T, \quad (\bar\nu_e, \bar\nu_\mu, \bar\nu_\tau)^T = U_L (\nu_{1L}, \nu_{2L}, \nu_{3L})^T.$$

The resulting masses are not necessarily real. This, however, does not present a problem since they can become real by a suitable rotation of the eigenvectors, e.g.

$$\nu_{kR} \to e^{i\delta_k} \nu_k, \; \nu_{kL} \to \nu_k,$$

hence

$$(\nu_e, \nu_\mu, \nu_\tau)^T = U_\nu (\nu_1, \nu_2, \nu_3)^T, \quad (\bar\nu_e, \bar\nu_\mu, \bar\nu_\tau)^T = U_L (\nu_{1L}, \nu_{2L}, \nu_{3L})^T,$$

where

$$m_k = |m_k| e^{i\delta_k}, \; U_\nu = U_R \text{diag}(e^{i\delta_1}, e^{i\delta_2}, e^{i\delta_3}).$$

The second matrix of Eq. (13.24) can be cast in the form

$$m_\nu = \begin{pmatrix} \alpha & 0 & i\beta \\ 0 & \gamma & 0 \\ -i\beta & 0 & \alpha \end{pmatrix}, \; \alpha, \beta, \gamma \text{ real} \quad (13.47)$$

i.e.

$$(U_\nu)^+ m_\nu U_\nu = \begin{pmatrix} \alpha+\beta & 0 & 0 \\ 0 & \gamma & 0 \\ 0 & 0 & \alpha-\beta \end{pmatrix}; \quad (13.48)$$

as a result, m_ν has eigenvalues $\alpha + \beta, \gamma$ and $\alpha - \beta$.

the problem of the construction of the neutrino mass

$$m_\nu = \frac{1}{2} \begin{pmatrix} \alpha + \beta & 0 & i(\alpha - \beta) \\ 0 & \gamma & 0 \\ i(\alpha - \beta) & 0 & -(\alpha + \beta) \end{pmatrix}, \tag{13.49}$$

see Eq. (13.45), in the framework of A_4 symmetry. This problem can be solved, but it is technically a bit complicated, and we are not going to discuss it here, see Zee (2005). Here, we examine the simple case of ν_e and ν_τ transforming under A_4 as $1' h 1''$, while ν_μ transforms as 1. We also consider the scalar field ϕ transforming as 1. The matrix of the Yukawa couplings becomes

$$\mathcal{L}_Y = \left(\bar{\nu}_e^c, \bar{\nu}_\mu^c, \bar{\nu}_\tau^c \right) \begin{pmatrix} h_{11}1' \times 1' & 0 & h_{13}1' \times 1' \\ 0 & h_{22}1 \times 1 & 0 \\ h_{31}1' \times 1' & 0 & h_{33}1' \times 1' \end{pmatrix} \begin{pmatrix} \nu_e \\ \nu_\mu \\ \nu_\tau \end{pmatrix} \phi + 1' \leftrightarrow 1''.$$

When the field ϕ attains a non-zero vacuum expectation value $\phi = v$, the matrix (13.49) is obtained by choosing

$$h_{11}v = \frac{\alpha + \beta}{2}, \quad h_{22}v = \frac{\gamma}{2}, \quad h_{13}v = h_{31}v = i\frac{\alpha - \beta}{2}, \quad h_{33}v = -\frac{\alpha + \beta}{2}.$$

Quite often, the neutrino mass matrix is constructed in a basis in which the charged lepton mass matrix is diagonal. In this case, the neutrino mass is diagonalized by the matrix V. Then,

$$V \mathrm{diag}(m_1, m_2, m_3) V^+$$

$$= \begin{pmatrix} \frac{1}{6}(4m_1 + m_2 + m_3) & \frac{-2m_1 + m_2 + m_3}{3\sqrt{2}} & \frac{m_2 - m_3}{2\sqrt{3}} \\ \frac{-2m_1 + m_2 + m_3}{3\sqrt{2}} & \frac{1}{3}(m_1 + m_2 + m_3) & \frac{m_2 - m_3}{\sqrt{6}} \\ \frac{m_2 - m_3}{2\sqrt{3}} & \frac{m_2 - m_3}{\sqrt{6}} & \frac{1}{2}(m_2 + m_3) \end{pmatrix}.$$

In this case, one finds quite convenient the expression

$$M_\nu = \begin{pmatrix} \frac{1}{3}(2a + 1)m_0 & \frac{1}{3}(1 - a)m_0 & \frac{1}{3}(1 - a)m_0 \\ \frac{1}{3}(1 - a)m_0 & \frac{1}{3}\left(\frac{a}{2} + \frac{3b}{2} + 1\right)m_0 & \frac{1}{3}\left(\frac{a}{2} - \frac{3b}{2} + 1\right)m_0 \\ \frac{1}{3}(1 - a)m_0 & \frac{1}{3}\left(\frac{a}{2} - \frac{3b}{2} + 1\right)m_0 & \frac{1}{3}\left(\frac{a}{2} + \frac{3b}{2} + 1\right)m_0 \end{pmatrix}, \tag{13.50}$$

with eigenvalues (m_0, am_0, bm_0).

The aim, then, is to generalize the above matrix so that one gets the accepted value of θ_{13} as well as of the mixing angles. A suitable form is

$$
M_\nu = m_0 \begin{pmatrix}
\frac{1}{3}(2a+1) & \frac{1-a}{3}y_2 & \frac{1-a}{3}y_3 \\
\frac{1-a}{3}y_2 & \frac{1}{3}\left(\frac{a}{2}+\frac{3b}{2}+1\right)y_2^2 & \frac{1}{3}\left(\frac{a}{2}-\frac{3b}{2}+1\right)y_2y_3 \\
\frac{1-a}{3}y_3 & \frac{1}{3}\left(\frac{a}{2}-\frac{3b}{2}+1\right)y_2y_3 & \frac{1}{3}\left(\frac{a}{2}+\frac{3b}{2}+1\right)y_3^2
\end{pmatrix}.
$$

$$(13.51)$$

One can show that, with separate left and right rotations, the quantities y_2 and y_3 can be chosen to be real. In this case, one needs to determine the seven parameters m_0, $|a|$, $Arg(a)$, $|b|$, $Arg(b)$, y_1 and y_2.

Leaving aside the question of the Majorana phases and the phase δ, which violates the CP symmetry, six parameters must be specified, i.e. the angles[6] θ_{23}, θ_{12}, θ_{13} and the mass differences Δm_{sun}^2, Δm_{atm}^2 and the sign of the last quantity. Thus, the parameters given by Eq. (13.51) suffice.

13.5.3 *A semirealistic model for the neutrino masses*

The model is better described in the context of supersymmetry (Ahn, 2014). We are not, however, going to deal with such issues, since we are concerned only with the A_4 symmetry. The relevant particles are:

- Higgs particles: Two scalars Φ_T, Φ_S transforming as triplets, $\underline{3}$, and one indicated as Θ transforming as $\underline{1}$. The A_4 symmetry is broken spontaneously when the particles acquire a vacuum expectation value. In the non-real representation:

$$
\langle \Phi_T \rangle = \frac{v_T}{\sqrt{2}}(1,0,0), \quad \langle \Phi_S \rangle = \frac{v_s}{\sqrt{2}}(1,1,1), \quad \langle \Theta \rangle = v_\theta.
$$

- Neutrinos: The usual neutrinos ν_e, ν_μ, ν_τ transforming as $\underline{1}$, $\underline{1}'$, $\underline{1}''$, respectively, and three heavy neutrinos of the Majorana type, N_1^c, N_2^c, N_3^c, transforming like $\underline{3}$.

The invariant Yukawa interactions are of the form

$$
\mathcal{L}_Y = \hat{y}_{1\nu}\bar{\nu}_e[N^c\Phi_T]_{\underline{1}} + \hat{y}_{2\nu}\bar{\nu}_\mu[N^c\Phi_T]_{\underline{1}'} + \hat{y}_{3\nu}\bar{\nu}_\mu[N^c\Phi_T]_{\underline{1}''} + \frac{1}{2}\hat{y}_\Theta\Theta[N^cN^c]_{\underline{1}}
$$

$$
+ \frac{1}{2}\hat{y}_R[N^cN^c]_{\underline{3}}\Phi_S. \tag{13.52}
$$

---

[6]For the determination of these parameters, see, for example, Vergados (2018), Section 11.7.

After the scalars acquire a non-zero vacuum expectation value, the three first terms through Eqs. (13.18)–(13.20) take the form

$$(\bar{\nu}_e, \bar{\nu}_\mu, \bar{\nu}_\tau)(m_D)(N_1^c, N_2^c, N_3^c)^T, \ (m_D) = m_D \begin{pmatrix} y_{1\nu} & 0 & 0 \\ 0 & 0 & y_{2\nu} \\ 0 & y_{3\nu} & 0 \end{pmatrix},$$

$$m_D = \frac{\upsilon_T}{3\sqrt{2}}.$$

The last two terms of Eq. (13.52), combined with Eq. (13.18) lead to

$$(\bar{N}_1^c, \bar{N}_2^c, \bar{N}_3^c)(m_{N_1})(N_1^c, N_2^c, N_3^c)^T, \ (m_{N_1}) = M \begin{pmatrix} 1 & 0 & 0 \\ 0 & 0 & 1 \\ 0 & 1 & 0 \end{pmatrix},$$

$$M = \left| y_\Theta \frac{\upsilon_\Theta}{\sqrt{2}} \right|,$$

while the third one via Eq. (13.22) becomes

$$(\bar{N}_1^c, \bar{N}_2^c, \bar{N}_3^c)(m_{N_2})(N_1^c, N_2^c, N_3^c)^T, \ (m_{N_1}) = M\Lambda \begin{pmatrix} 2 & -1 & -1 \\ -1 & 2 & -1 \\ -1 & -1 & 2 \end{pmatrix},$$

$$\Lambda = \tfrac{1}{3}\bar{\kappa}e^{i\phi},$$

$\bar{\kappa} = (3/(\sqrt{2}M))|y_R\upsilon_S|$ and $\phi = arg(y_R/y_\Theta)$. We suppose that in the general case, the ratio (y_R/y_Θ) may not be real.

Summarizing the above, the neutrino mass matrix takes the form

$$\mathcal{L}_\nu = \frac{1}{2}(\bar{\nu}_L, \bar{N}_L^c) \begin{pmatrix} 0 & (m_D) \\ (m_D)^T & (M_R) \end{pmatrix} \begin{pmatrix} (\nu_R)^c \\ N_R \end{pmatrix},$$

$$\nu_L \equiv (\nu_e, \nu_\mu, \nu_\tau)_L, \ N_R \equiv (N_1, N_2, N_3)_R, \tag{13.53}$$

with $(N_1, N_2, N_3)_R \Leftrightarrow (N_1^c, N_2^c, N_3^c)_L$ and

$$(M_R) = \begin{pmatrix} 1 + \frac{2}{3}\bar{\kappa}e^{i\phi} & -\frac{1}{3}\bar{\kappa}e^{i\phi} & -\frac{1}{3}\bar{\kappa}e^{i\phi} \\ -\frac{1}{3}\bar{\kappa}e^{i\phi} & \frac{2}{3}\bar{\kappa}e^{i\phi} & 1 - \frac{1}{3}\bar{\kappa}e^{i\phi} \\ -\frac{1}{3}\bar{\kappa}e^{i\phi} & 1 - \frac{1}{3}\bar{\kappa}e^{i\phi} & \frac{2}{3}\bar{\kappa}e^{i\phi} \end{pmatrix}. \tag{13.54}$$

We note that we assumed that there is no Majorana mass term in the case of the usual light neutrinos. Furthermore, the above 6×6 matrix is symmetric, but in general, it is non-Hermitian. As we have seen above, it can be diagonalized by transformations utilizing separate unitary matrices from the left and the right. The light neutrinos acquire a mass through the see saw mechanism. Indeed, supposing that all the eigenvalues of (M_R) are much larger than the scale of the m_D matrix, we find

$$\mathcal{M}_\nu = (\bar\nu_L, \bar\nu_\mu, \bar\nu_\tau)(m_\nu)(\nu^c_{eR}, \nu^c_{\mu R}, \nu^c_{\tau R})^T, \quad (m_\nu) = -(m_D)^T(m_R)^{-1}(m_D). \tag{13.55}$$

In the model we discussed, one can show that

$$(m_\nu) = -m_0 \begin{pmatrix} 1+2F & (1-F)y_2 & (1-F)y_3 \\ (1-F)y_2 & \frac{2+F+3G}{2}y_2^2 & \frac{2+F-3G}{2}y_2y_3 \\ (1-F)y_3 & \frac{2+F-3G}{2}y_2y_3 & \frac{2+F+3G}{2}y_3^2 \end{pmatrix}, \tag{13.56}$$

where $m_0 = (y_{1\nu}m_D)^2/M_R$, $F = 1/(1+\bar\kappa e^{i\phi})$ and $G = -1/(1-\bar\kappa e^{i\phi})$.

This matrix for $y_1 = y_2 = 1$ looks like the matrix of Eq. (13.50), and as a result, it is diagonalized via the di-maximal, tri-maximal matrix V, Eq. (13.42). For $y_1 \neq 1$ and/or $y_2 \neq 1$, we find $\theta_{13} \neq 0$.

13.6 Problems

13.2.1 Consider the group A_4:

- Show that beginning with the generators $(123) \Leftrightarrow A$, $(321) \Leftrightarrow B$ of Eq. (13.2), one can find the elements of the 3D representation as follows:

$$\begin{pmatrix} 1&0&0\\0&1&0\\0&0&1 \end{pmatrix}, \begin{pmatrix} 0&-1&0\\0&0&-1\\1&0&0 \end{pmatrix}, \begin{pmatrix} 0&0&1\\-1&0&0\\0&-1&0 \end{pmatrix},$$

$$\begin{pmatrix} -1&0&0\\0&-1&0\\0&0&1 \end{pmatrix}, \begin{pmatrix} 1&0&0\\0&-1&0\\0&0&-1 \end{pmatrix}, \begin{pmatrix} -1&0&0\\0&1&0\\0&0&-1 \end{pmatrix},$$

$$\begin{pmatrix} 0&1&0\\0&0&-1\\-1&0&0 \end{pmatrix}, \begin{pmatrix} 0&1&0\\0&0&1\\1&0&0 \end{pmatrix}, \begin{pmatrix} 0&-1&0\\0&0&1\\-1&0&0 \end{pmatrix},$$

$$\begin{pmatrix} 0&0&1\\1&0&0\\0&1&0 \end{pmatrix}, \begin{pmatrix} 0&0&-1\\1&0&0\\0&-1&0 \end{pmatrix}, \begin{pmatrix} 0&0&-1\\-1&0&0\\0&1&0 \end{pmatrix}.$$

- Show that in the basis of Eq. (13.1), the previous matrices correspond to the elements

$$E, (123), (321), (12)(34), (13)(24), (14)(23)(134), (432), (421),$$

$$(234), (431), (124).$$

13.2.2 Show that the matrix $A \Leftrightarrow (123)$ satisfies the relation $A^3 = 1$. Consequently, its eigenvalues are the roots of the cubic equation $x^3 = 1$. Furthermore:

- Show that the eigenvalues of A are $\omega = e^{(2\pi i)/3}$, ω^2 and $\omega^3 = 1$.
- Show that $\sum_{n=1}^{3} \omega^n = \omega + \omega^2 + \omega^3 = 0$.
- Show that the matrix diagonalizing A is

$$S = \frac{1}{3} \begin{pmatrix} 1 & 1 & 1 \\ 1 & \omega & \omega^2 \\ 1 & \omega^2 & \omega \end{pmatrix} = \begin{pmatrix} \frac{1}{\sqrt{3}} & \frac{1}{\sqrt{3}} & \frac{1}{\sqrt{3}} \\ \frac{1}{\sqrt{3}} & \frac{e^{\frac{2i\pi}{3}}}{\sqrt{3}} & \frac{e^{-\frac{2i\pi}{3}}}{\sqrt{3}} \\ \frac{1}{\sqrt{3}} & \frac{e^{-\frac{2i\pi}{3}}}{\sqrt{3}} & \frac{e^{\frac{2i\pi}{3}}}{\sqrt{3}} \end{pmatrix} \Rightarrow$$

$$S^+ A S = \mathrm{diag}(\omega^3, \omega^2, \omega).$$

13.2.3 Generalize the previous in the case where the matrix A is unitary satisfying the relation $A^p = 1$, where p is a prime positive integer. More specifically, show that:

- The eigenvalues of A are the roots f of the equation $x^p = 1$.
- $\sum_{n=1}^{p} \omega^n = \omega + \omega^2 + \cdots + \omega^p = 0$.
- The matrix diagonalizing A is

$$S = \frac{1}{\sqrt{p}}$$

$$\times \begin{pmatrix} 1 & 1 & 1 & 1 & \cdots & 1 & 1 \\ 1 & \omega & \omega^2 & \omega^3 & \cdots & \omega^{p-2} & \omega^{p-1} \\ 1 & \omega^2 & \omega^4 & \omega^6 & \cdots & \omega^{2(p-2)} & \omega^{2(p-1)} \\ 1 & \omega^3 & \omega^6 & \omega^9 & \cdots & \omega^{3(p-2)} & \omega^{3(p-1)} \\ \cdots & \cdots & \cdots & \cdots & \cdots & \cdots & \cdots \\ 1 & \omega^{p-2} & \omega^{2(p-2)} & \omega^{3(p-2)} & \cdots & \omega^{(p-2)(p-2)} & \omega^{(p-2)(p-1)} \\ 1 & \omega^{p-1} & \omega^{2(p-1)} & \omega^{3(p-1)} & \cdots & \omega^{(p-1)(p-2)} & \omega^{(p-1)(p-1)} \end{pmatrix},$$

with the understanding that exponents greater than p are equivalent to some which are smaller than p.

We recall that two numbers a and b are equivalent under p (modulo p), and we write $a = b\,\mathrm{m}(p)$ or $a = b\,\mathrm{mod}(p)$ if there exists an integer ℓ such that $a = b + \ell p$, i.e.

$$a = b\,\mathrm{m}(p) \Leftrightarrow a = b + \ell p, \ell = \text{integer}.$$

In our case, a, b and ℓ are non-negative. Thus, $12 = 1\,\mathrm{m}(11)$, $\ell = 1$, $27 = 5\,\mathrm{m}(11)$, $\ell = 2$. For $p = 11$, we have

$$S = \frac{1}{\sqrt{11}}
\begin{pmatrix}
1 & 1 & 1 & 1 & 1 & 1 & 1 & 1 & 1 & 1 & 1 \\
1 & \omega & \omega^2 & \omega^3 & \omega^4 & \omega^5 & \omega^6 & \omega^7 & \omega^8 & \omega^9 & \omega^{10} \\
1 & \omega^2 & \omega^4 & \omega^6 & \omega^8 & \omega^{10} & \omega & \omega^3 & \omega^5 & \omega^7 & \omega^9 \\
1 & \omega^3 & \omega^6 & \omega^9 & \omega & \omega^4 & \omega^7 & \omega^{10} & \omega^2 & \omega^5 & \omega^8 \\
1 & \omega^4 & \omega^8 & \omega & \omega^5 & \omega^9 & \omega^2 & \omega^6 & \omega^{10} & \omega^3 & \omega^7 \\
1 & \omega^5 & \omega^{10} & \omega^4 & \omega^9 & \omega^3 & \omega^8 & \omega^2 & \omega^7 & \omega & \omega^6 \\
1 & \omega^6 & \omega & \omega^7 & \omega^2 & \omega^8 & \omega^3 & \omega^9 & \omega^4 & \omega^{10} & \omega^5 \\
1 & \omega^7 & \omega^3 & \omega^{10} & \omega^6 & \omega^2 & \omega^9 & \omega^5 & \omega & \omega^8 & \omega^4 \\
1 & \omega^8 & \omega^5 & \omega^2 & \omega^{10} & \omega^7 & \omega^4 & \omega & \omega^9 & \omega^6 & \omega^3 \\
1 & \omega^9 & \omega^7 & \omega^5 & \omega^3 & \omega & \omega^{10} & \omega^8 & \omega^6 & \omega^4 & \omega^2 \\
1 & \omega^{10} & \omega^9 & \omega^8 & \omega^7 & \omega^6 & \omega^5 & \omega^4 & \omega^3 & \omega^2 & \omega
\end{pmatrix}.$$

For more applications, see Chapter 14.

13.2.4 Consider the matrix

$$a_4 = \begin{pmatrix} -\frac{1}{3} & \frac{2}{3} & \frac{2}{3} \\ \frac{2}{3} & -\frac{1}{3} & \frac{2}{3} \\ \frac{2}{3} & \frac{2}{3} & -\frac{1}{3} \end{pmatrix},$$

with the diagonalizing matrix and eigenvalue matrix, respectively,

$$S = \begin{pmatrix} \frac{1}{\sqrt{3}} & 0 & \sqrt{\frac{2}{3}} \\ \frac{1}{\sqrt{3}} & \frac{1}{\sqrt{2}} & -\frac{1}{\sqrt{6}} \\ \frac{1}{\sqrt{3}} & -\frac{1}{\sqrt{2}} & -\frac{1}{\sqrt{6}} \end{pmatrix}, \quad S^+ a_4 S = \begin{pmatrix} 1 & 0 & 0 \\ 0 & -1 & 0 \\ 0 & 0 & -1 \end{pmatrix}.$$

In other words, the matrix a_4 coincides with one of the generators of A_4, more specifically the matrix t_2, Eqs. (13.5) and (13.7).

• Show that the most general symmetric matrix commuting with a_4 is of the form

$$ft = \begin{pmatrix} u & v+y & y \\ v+y & u+w & y-w \\ y & y-w & u+v+w \end{pmatrix}.$$

- This matrix can be diagonalized simultaneously with the matrix a_4. It is not clear how this can be done since the matrix a_4 has degenerate eigenvectors, and such eigenvectors can be chosen arbitrarily. Show that the non-degenerate unique eigenvector of a_4 is also an eigenvector of ft with eigenvalue $u + v + 2y$.
- The other two eigenvectors are a linear combination of the degenerate eigenvectors of a_4. The reader is encouraged to confirm that the eigenvalues are $u + v + 2y$, $u + w - \sqrt{v^2 + vw + w^2} - y$, $u + w + \sqrt{v^2 + vw + w^2} - y$, and the diagonalizing matrix is

$$
S = \begin{pmatrix}
\frac{1}{\sqrt{3}} & \frac{v}{\sqrt{2}\sqrt{2v^2+(w+X)v+2wX}} & \frac{v+2X}{\sqrt{6}\sqrt{2v^2+(w+X)v+2wX}} \\
\frac{1}{\sqrt{3}} & \frac{X}{\sqrt{2}\sqrt{2v^2+(w+X)v+2wX}} & -\frac{2v+X}{\sqrt{6}\sqrt{2v^2+(w+X)v+2wX}} \\
\frac{1}{\sqrt{3}} & -\frac{v+X}{\sqrt{2}\sqrt{2v^2+(w+X)v+2wX}} & -\frac{X-v}{\sqrt{6}\sqrt{2v^2+(w+X)v+2wX}}
\end{pmatrix},
$$

with $X = \sqrt{v^2 + vw + w^2} + w$. The diagonalization was accomplished in two steps: first, with a similarity transformation via the eigenvectors of a_4. Thus, the matrix ft is reduced to a matrix 1×1, with eigenvalue $u + v + 2y$, and a 2×2 matrix. Afterward, the 2×2 matrix is diagonalized as usual.
- Consider the case $v = 0$, i.e. in the case of the matrix

$$
t = \begin{pmatrix}
u & y & y \\
y & u + w & y - w \\
y & y - w & u + w
\end{pmatrix}.
$$

Show that the above eigenvectors of a_4 are also eigenvectors of t with eigenvalues $u + 2y, u - y, u + 2w - y$. Since the two new eigenvectors of t are non-degenerate, this is rather accidental.

Chapter 14

Exotic Discrete Groups for Quantum Mechanics — Field Theory

Historically, the Heisenberg–Weyl group, which we study in this chapter, led Weyl to the formal proof of the equivalence of the two pictures of quantum mechanics, namely the picture of Schrödinger and that of Heisenberg. He proved this rigorously through the known transformation of Wigner–Weyl or, according to some others, the transformation Weyl–Wigner, which is an invertible mapping between the functions of the quantum phase space of Heisenberg and the Hilbert space on which the operators in the Schrödinger picture act.

The well-known mathematical function θ is a special case of representations of the Heisenberg group. The Jacobi function θ is invariant under the discrete transformations of the Heisenberg–Weyl group.

Beyond their practical use of the above subjects, in this chapter, we discuss some other topics that exhibit mathematical elegance, such as the modular group and, in general, matrix groups with the integer elements $mod(p)$.

14.1 Matrix groups with elements integers $mod(p)$

Often, in applications, we encounter matrices with integer elements. In some cases, these can be generators of groups if the integers can be constrained to be smaller than an integer p, e.g. by setting

$$n \to k = \mathrm{mod}(p) \Leftrightarrow k \text{ the remainder of the division } n/p. \qquad (14.1)$$

The most interesting case appears when p is a prime integer $= 3, 5, 7, 11,$
etc.:

$$R = \begin{pmatrix} 1 & -1 \\ 0 & 1 \end{pmatrix},$$

$$R^2 = \begin{pmatrix} 1 & -2 \\ 0 & 1 \end{pmatrix}, \quad R^3 = \begin{pmatrix} 1 & -3 \\ 0 & 1 \end{pmatrix} \Rightarrow R^3 \big|_{\mathrm{mod}(3)} = \begin{pmatrix} 1 & 0 \\ 0 & 1 \end{pmatrix}.$$

(14.2)

This is an Abelian group of order 3.
    Proceeding further, one gets

$$R^4 = \begin{pmatrix} 1 & -4 \\ 0 & 1 \end{pmatrix}, \quad R^5 = \begin{pmatrix} 1 & -5 \\ 0 & 1 \end{pmatrix} \Rightarrow R^5 \big|_{\mathrm{mod}(5)} = \begin{pmatrix} 1 & 0 \\ 0 & 1 \end{pmatrix}.$$

In this case, we get an Abelian group of order 5.
    Furthermore,

$$R^6 = \begin{pmatrix} 1 & -6 \\ 0 & 1 \end{pmatrix}, \quad R^7 = \begin{pmatrix} 1 & -7 \\ 0 & 1 \end{pmatrix} \Rightarrow R^7 \big|_{\mathrm{mod}(7)} = \begin{pmatrix} 1 & 0 \\ 0 & 1 \end{pmatrix},$$

i.e. an Abelian group of order 7, etc.
    The reader can verify that, proceeding in a similar fashion, one can get
Abelian groups of order p starting with the matrix

$$L = \begin{pmatrix} 1 & 0 \\ 1 & 1 \end{pmatrix}.$$

These form irreducible representations, but they are not unitary. This happens because these matrices have one eigenvalue with degeneracy 2, but
only one eigenvector, which does not cover the space. Therefore, it is impos-
sible via diagonalization to reduce this representation to a unitary one.
Anyway, the theory in Section 5.2 does not apply in this case since the
groups are defined $\mathrm{mod}(p)$ and the matrices $\Gamma(X)$, as given there, do not
have integer entries. One representation of this group, namely the regular
one, is unitary. Since, however, the group is Abelian, this can be obtained
trivially (see Chapter 4). In fact, from the multiplication table of the group,
one can obtain the regular representation, see Section 4.3. This can be trivially reduced, see Table 5.1.

Thus, in the case of the Abelian group of order 3 discussed above, we obtain

$$\Gamma_E = \begin{pmatrix} 1 & 0 & 0 \\ 0 & 1 & 0 \\ 0 & 0 & 1 \end{pmatrix}, \quad \Gamma_A = \begin{pmatrix} 0 & 1 & 0 \\ 0 & 0 & 1 \\ 1 & 0 & 0 \end{pmatrix}, \quad \Gamma_{A^2} = \begin{pmatrix} 0 & 0 & 1 \\ 1 & 0 & 0 \\ 0 & 1 & 0 \end{pmatrix}.$$

One can easily verify that

$$(\Gamma_A)^+ = \Gamma_{A^2} = (\Gamma_A)^{-1}, \quad (\Gamma_{A^2})^+ = (\Gamma_A) = (\Gamma_{A^2})^{-1},$$

i.e. the representation is unitary.

Often, the following matrices are used:

$$a = \begin{pmatrix} 0 & -1 \\ 1 & 0 \end{pmatrix}, \quad b = \begin{pmatrix} 0 & -1 \\ 1 & 1 \end{pmatrix}. \tag{14.3}$$

The usefulness of such groups in cryptography is obvious when employing huge prime integers p.

14.2 The modular group

In the theory of complex variables, we encounter the fractional (Möbius) transformations

$$z \to z' = f(z) = \frac{az + b}{cz + d}, ab - cd = 1.$$

These transformations are very useful, e.g. they transform circles into circles or straight lines. It is quite important to study the joining of two such functions:

$$f_2(f_1(z)) \equiv f_1 * f_2(z) = \frac{a_2 b_1 + a_1 a_2 z + b_2 (c_1 z + d_1)}{c_2 (a_1 z + b_1) + d_2 (c_1 z + d_1)},$$

$$f_i = \frac{a_i + b_i z}{c_i + d_i z}, \quad i = 1, 2.$$

We note that, essentially, the above involves matrix multiplication. Indeed, symbolically, we write

$$T(a, b, c, d, z) = T(a, b, c, d) \begin{pmatrix} z \\ 1 \end{pmatrix}, \quad T(a, b, c, d) = \begin{pmatrix} a & b \\ c & d \end{pmatrix}, \tag{14.4}$$

$$T(a_2, b_2, c_2, d_2) T(a_1, b_1, c_1, d_1) \begin{pmatrix} z \\ 1 \end{pmatrix}$$

$$= T(a_2, b_2, c_2, d_2) \begin{pmatrix} a_1 + b_1 z \\ c_1 + d_1 z \end{pmatrix} = \begin{pmatrix} a_2 (za_1 + b_1) + b_2 (zc_1 + d_1) \\ (za_1 + b_1) c_2 + (zc_1 + d_1) d_2 \end{pmatrix}.$$

$$(14.5)$$

The set of matrices

$$A = \begin{pmatrix} a & b \\ c & d \end{pmatrix} \leftrightarrow T(a, b, c, d) \text{ for } ad - bc = 1 \qquad (14.6)$$

constitutes a group with an inverse given by

$$A^{-1} = \begin{pmatrix} d & -b \\ -c & a \end{pmatrix}.$$

$(T(a, b, c, d))^{-1} = T(d, -b, -c, a).$

This group can be discrete if it can be a subgroup of a continuous group of transformations $SL(2, c)$, i.e. of $SL(2, Z)$ with integer elements

$$SL(2, Z) = \{T(a, b, c, d), ad - bc = 1, \ a, b, c, d \in Z\}.$$

The quotient group $SL(2, Z)/\{E, -E\}$ is indicated as $PSL(2, Z)$, a special projection group, whenever the indices matrices $T(a, b, c, d)$ and $-T(a, b, c, d)$ are identical.

Example 1: As a start, let us consider two very simple conformal mappings:

$$T : z \to -1/z, S : z \to -z.$$

These correspond to the elements

$$T = \begin{pmatrix} -1 & 0 \\ 0 & -1 \end{pmatrix}, \quad S = \begin{pmatrix} 0 & -1 \\ 1 & 0 \end{pmatrix}.$$

We observe that

$$S^2 = T, \quad S^3 = -S, \quad S^4 = E,$$

that is, in this case, one deals with an Abelian subgroup of $SL(2, Z)$ with the elements E, S, S^2, S^3, with a generator S. We note that any subgroups of $PSL(2, Z)$, i.e. $SL(2, Z)/\{E, -E\}$, must necessarily contain the element E

excluding the element $-E$. Of the rest, only one of the opposite elements must be included. Hence, $Z_2 = \{E, S\}$ and $Z'_2 = \{E, -S\}$ are the only acceptable options, since S^2 is identified with E.

Example 2: As a second, more realistic case, let us consider the mappings

$$T : z \to z + 1, S : z \to -1/z.$$

This is very important since the combination of these mappings achieves a wealth of transformations. As a result, every element in the modular group can be represented (in a non-unique way) by the composition of the powers of S and T. Geometrically, S represents inversion in the unit circle followed by reflection with respect to the imaginary axis, while T represents a unit translation to the right. Anyway, these transformations lead to the elements

$$T = \begin{pmatrix} 1 & 1 \\ 0 & 1 \end{pmatrix}, \quad S = \begin{pmatrix} 0 & -1 \\ 1 & 0 \end{pmatrix}. \tag{14.7}$$

The matrix S is not unique. One could have chosen the matrix $-S$. In any case,

$$T^n = \begin{pmatrix} 1 & n \\ 0 & 1 \end{pmatrix}, \quad ST = \begin{pmatrix} 0 & -1 \\ 1 & 1 \end{pmatrix}, \quad TS = \begin{pmatrix} 1 & -1 \\ 1 & 0 \end{pmatrix}.$$

One then obtains relations independently of n:

$$S^2 = (ST)^3 = (TS)^3 = \begin{pmatrix} -1 & 0 \\ 0 & -1 \end{pmatrix}, \quad S^4 = (ST)^6 = (TS)^6 = \begin{pmatrix} 1 & 0 \\ 0 & 1 \end{pmatrix}.$$

For $n = 3 \bmod(3)$, we find that these elements generate a group with the additional relation $T^3 = E$. One can consider as group generators the elements S and ST or TS.

The analog of the group $PSL(2, Z)$ is to consider, of the elements of the previous group, the identity element and, of the rest of the elements, excluding $-E$, all possible combinations, excluding those whose sum yields zero.

Some possible subgroups are

$$Z_2 = \{E, (ST)^2, (ST)^4\}, Z_2 = \{E, (TS)^2, (TS)^4\}, Z_2 = \{E, T, T^2\}.$$

All of them are Abelian of order 3.

The set of Möbius transformations $\tau' = \frac{a\tau+b}{c\tau+d}$ with a, b, c, d integers, $ac - bd = 1$, constitutes a **modular** group Γ (Apostol, 1990). This group is represented by the set of matrices

$$A = \begin{pmatrix} a & b \\ c & d \end{pmatrix}, a, b, c, d \text{ integers } ad - bc = 1. \tag{14.8}$$

This group is very important due to an important theorem.

Theorem: *Every element $A \in \Gamma$ can be generated by the set S and T, as given by Eq. (14.7). That is, it can be generated by products of the type $T^{n_1} S T^{n_2} S \cdots T^{n_k} S$.*

Finally, it should be mentioned that the group $PSL(2, Z)$ is used in many applications, see, for example, Stillwell (2001) and Apostol (1990), especially in hyperbolic geometry.

14.3   The Heisenberg–Weyl group

In the Heisenberg formulation of quantum mechanics, a crucial role is played by the operators of position Q_i and momentum P_i, which satisfy the commutation relations

$$[Q_i, Q_j] = 0, \quad [P_i, P_j] = 0, \quad [P_i, Q_j] = -i\hbar\delta_{i,j}, \tag{14.9}$$

where $[A, B] \equiv AB - BA$.

These relations were subsequently generalized by Heisenberg as follows:

$$[Q_i, Q_j] = 0, \quad [P_i, P_j] = 0, \quad [P_i, Q_j] = C\delta_{i,j}, \tag{14.10}$$

where

$$[P_i, C] = 0, \quad [Q_i, C] = 0. \tag{14.11}$$

The matrices C are members of a vector space of matrices of the form

$$C(r, s, t) = \begin{pmatrix} 0 & 0 & 0 \\ r & 0 & 0 \\ t & s & 0 \end{pmatrix}, \tag{14.12}$$

$$[C(r_1, s_1, t_1), C(r_2, s_2, t_2)] = \begin{pmatrix} 0 & 0 & 0 \\ 0 & 0 & 0 \\ s_1 r_2 - r_1 s_2 & 0 & 0 \end{pmatrix}. \tag{14.13}$$

From the theory of angular momentum, we know that from the angular momentum operators L_k, $k = x, y, z$, which constitute an algebra involving the commutation relations, one can go to the rotation operators R_k, which constitute a group:

$$L_k \rightarrow R_k = e^{-i\theta(L_k/\hbar)}.$$

We can do the same here:

$$C(r, s, t) \rightarrow g(r, s, t) = e^{C(r,s,t)}.$$

Note, however, that $(C(r, s, t))^n = 0, n = 2, 3, \ldots \Rightarrow g(r, s, t) = E + C(r, s, t)$. In other words, the latter is characterized by the matrix representation

$$g(r, s, t) = \begin{pmatrix} 1 & 0 & 0 \\ r & 1 & 0 \\ t & s & 1 \end{pmatrix}, \tag{14.14}$$

where r, s, t are any numbers. These matrices constitute Heisenberg's group H_3, since they have a determinant equal to 1, the inverse given by

$$(g(r, s, t))^{-1} = \begin{pmatrix} 1 & 0 & 0 \\ -r & 1 & 0 \\ rs - t & -s & 1 \end{pmatrix}, \tag{14.15}$$

and the multiplication table

$$g(r_1, s_1, t_1)g(r_2, s_2, t_2) = \begin{pmatrix} 1 & 0 & 0 \\ r_1 + r_2 & 1 & 0 \\ r_2 s_1 + t_1 + t_2 & s_1 + s_2 & 1 \end{pmatrix}. \tag{14.16}$$

The finite Heisenberg group $H_3(Z)$ results when the elements of the matrices are integers with operations restricted as in the previous section, $r, s, t = $ integers $\bmod(p)$. Of special interest are the matrices

$$x = \begin{pmatrix} 1 & 0 & 0 \\ 1 & 1 & 0 \\ 0 & 0 & 1 \end{pmatrix}, \quad y = \begin{pmatrix} 1 & 0 & 0 \\ 0 & 1 & 0 \\ 0 & 1 & 1 \end{pmatrix}, \quad z = \begin{pmatrix} 1 & 0 & 0 \\ 0 & 1 & 0 \\ 1 & 0 & 1 \end{pmatrix}, \tag{14.17}$$

where x and y are the group generators, while z results from the generators and constitutes the center of the group, that is, the following relations hold:

$$z = xyx^{-1}y^{-1}, \quad xz = zx, \quad yz = zy, \quad g(r, s, t) = x^r z^t y^s. \tag{14.18}$$

The reader should first show the relations

$$
x^r = \begin{pmatrix} 1 & 0 & 0 \\ r & 1 & 0 \\ 0 & 0 & 1 \end{pmatrix}, \quad y^s = \begin{pmatrix} 1 & 0 & 0 \\ 0 & 1 & 0 \\ 0 & s & 1 \end{pmatrix}, \quad z^t = \begin{pmatrix} 1 & 0 & 0 \\ 0 & 1 & 0 \\ t & 0 & 1 \end{pmatrix}.
$$

$$(14.19)$$

From these, one can easily obtain the last of Eq. (14.18). If the operations are performed with integers mod(p), Eq. (14.19) leads to $x^p = y^p = z^p = 1$.

Whenever p is an odd prime number, the order of the group is p^3. In this special case, we talk about extra special groups with exponent p.

Example 1: We first study the case $p = 2$.

From Eqs. (14.18) and (14.19), we find that the group elements are the following:

$$E, x, y, z, xy, yx, xz, yz.$$

This group is isomorphic to D_4 (see case 4 in Section 2.4.2). The correspondence with the basis of Table 17.6 is not at all obvious (it can not be obtained via diagonalizations). There exists, however, an additional element of order 4, namely $A = yx$. Indeed, $A.A = z$, $A.A.A = xy$, $A.A.A.A = E$, mod(2). We thus make the correspondence

$$yx \Leftrightarrow C_4, \quad z \Leftrightarrow C_4^2, \quad xy \Leftrightarrow C_4^3.$$

Then, considering one more element, e.g. $x \Leftrightarrow C_2$. we get

$$A.x = y, \ A.A.y = z.y, \ y.z = y.A.A = y(y,x).(y.x) = (xy).x = (A.A.A).x.$$

In other words, we can consider as generators the elements A, x, and thus, we make the correspondence with generators of Table 17.1 as follows:

$$
A = \begin{pmatrix} 1 & 1 & 0 \\ 0 & 1 & 1 \\ 0 & 0 & 1 \end{pmatrix} \Leftrightarrow \begin{pmatrix} 1 & 0 & 0 \\ 0 & 0 & -1 \\ 0 & 1 & 0 \end{pmatrix},
$$

$$
x = \begin{pmatrix} 1 & 1 & 0 \\ 0 & 1 & 0 \\ 0 & 0 & 1 \end{pmatrix} \Leftrightarrow \begin{pmatrix} -1 & 0 & 0 \\ 0 & 1 & 0 \\ 0 & 0 & -1 \end{pmatrix}.
$$

After the generator correspondence, we get a one-to-one correspondence of all the elements of these groups, i.e. the groups are isomorphic. From this, we conclude that the representations of the current group are reducible,[1] as given in Table 17.6. The conclusion is that the present method is powerful in creating discrete groups, but often it does not give physical meaning to their elements. Thus, the correspondence of the two isomorphic groups, gives physical meaning to the elements of the current group, e.g. the operator x is nothing but a rotation by π in the plane (x, z) ($\hat{y}$ axis).

Example 2: We study the case $p = 3$.

In this case, the resulting group G contains the additional elements

$$S_1 = \left\{x, y, xy, yx, x^2, y^2, x^2y, yx^2, xy^2, y^2x, (xy)^2, (yx)^2\right\}.$$

Hence, we find

$$S_2 = \left\{z = xyx^2y^2, z^2, z^3 = E\right\}.$$

Now, z is the center of the group $z.g_i = g_i z, g_i \in G$:

$$G = S_1 \cup S_2 \cup g_i z, \ g_i \in S_1.$$

In other words, we have a total of $3 + 2 \times 12 = 27$ elements.

This group is isomorphic with the group

$$G = \left(\{E, x, x^2\} \otimes \{E, y, y^2\}\right) \otimes \{E, z, z^2\}.$$

It is also isomorphic with the discrete subgroup of $SU(3)$ known as $\Delta(27)$, see, for example, Luhn *et al.* (2007b), with unitary generators. That is,

$$A_1 = \begin{pmatrix} 0 & 1 & 0 \\ 0 & 0 & 1 \\ 1 & 0 & 0 \end{pmatrix}, \quad B_2 = \begin{pmatrix} \omega & 0 & 0 \\ 0 & \omega^2 & 0 \\ 0 & 0 & 1 \end{pmatrix},$$

$$\omega = e^{\frac{2\pi i}{3}}, (B_1)^3 = E, \quad (B_2,)^3 = E,$$

$$x \Leftrightarrow B_1, \ y \Leftrightarrow (B_2), \ z \Leftrightarrow \begin{pmatrix} e^{\frac{2i\pi}{3}} & 0 & 0 \\ 0 & e^{\frac{2i\pi}{3}} & 0 \\ 0 & 0 & e^{\frac{2i\pi}{3}} \end{pmatrix}.$$

The above elements are distributed in 11 classes, and as a result, one has 11 irreducible representations with dimensions fulfilling the relation

$$\sum_{i=1}^{9} \ell_i^2 = 27.$$

---

[1]This is not inconsistent with the discussion in Section 14.3.1, since p is even.

Table 14.1. The multiplication table of the nine 1D representations of the discrete group $\Delta(27)$.

	Γ_1	Γ_2	Γ_3	Γ_4	Γ_5	Γ_6	Γ_7	Γ_8	Γ_9
Γ_1	Γ_1	Γ_2	Γ_3	Γ_4	Γ_5	Γ_6	Γ_7	Γ_8	Γ_9
Γ_2	Γ_2	Γ_3	Γ_1	Γ_6	Γ_4	Γ_5	Γ_8	Γ_9	Γ_7
Γ_3	Γ_3	Γ_1	Γ_2	Γ_5	Γ_6	Γ_4	Γ_9	Γ_7	Γ_8
Γ_4	Γ_4	Γ_6	Γ_5	Γ_7	Γ_9	Γ_8	Γ_1	Γ_2	Γ_3
Γ_5	Γ_5	Γ_4	Γ_6	Γ_9	Γ_8	Γ_7	Γ_3	Γ_1	Γ_2
Γ_6	Γ_6	Γ_5	Γ_4	Γ_8	Γ_7	Γ_9	Γ_2	Γ_3	Γ_1
Γ_7	Γ_7	Γ_8	Γ_9	Γ_1	Γ_3	Γ_2	Γ_4	Γ_6	Γ_5
Γ_8	Γ_8	Γ_9	Γ_7	Γ_2	Γ_1	Γ_3	Γ_6	Γ_5	Γ_4
Γ_9	Γ_9	Γ_7	Γ_8	Γ_3	Γ_2	Γ_1	Γ_5	Γ_4	Γ_6

The possible solutions are

$$\ell_i = 1, \quad i = 1,9, \quad \ell_{10} = \ell_{11} = 3,$$

i.e. nine 1D Γ^i, $i = 1, 2, \ldots, 9$ and two 3D Γ_3 and Γ'_3,

$$\Gamma^{\text{reg}}(X) = \sum_{i=1}^{9} \Gamma^i(X) + 2\Gamma_3(X).$$

This reduction has not been included in the tables of the appendix, Chapter 17, since it is not isomorphic to any crystal group. The multiplication table of the 1D representations with the choice of Abbas and Khalil (2015) is given in Table 14.1.

This group has been employed in the study of neutrinos, see, for example, Ma (2008) and de Medeiros-Varzielas *et al.* (2007).

14.3.1 *A faithful representation of the discrete Heisenberg group*

We are now in position to study the discrete Heisenberg group (Floratos and Leontaris, 2016b). A faithful representation in this case for p, an odd prime number, is given as follows:

$$J_{r,s,t} = \omega^t P^r Q^s, \tag{14.20}$$

where $\omega = e^{2\pi i/p}$, that is, p defines the primitive root of unity, and P, Q are defined as follows:

$$P_{k\ell} = \delta_{k+1,\ell}, \quad Q_{k\ell} = \omega^k \delta_{k\ell}, \tag{14.21}$$

$k, \ell = 0, \ldots, p-1$.

More explicitly,[2] $P_{0k} = \delta_{kp}, P_{k,k-1} = 1, k = 2, 3, \ldots, p-1, P_{m,k} = 0, m \neq k+1$, i.e.

$$P = \begin{pmatrix} 0 & 0 & 0 & \cdots & 0 & 1 \\ 1 & 0 & 0 & \cdots & 0 & 0 \\ 0 & 1 & 0 & \cdots & 0 & 0 \\ 0 & 0 & 1 & \cdots & 0 & 0 \\ \cdots & \cdots & \cdots & \cdots & \cdots & \cdots \\ 0 & 0 & 0 & \cdots & 1 & 0 \end{pmatrix}.$$

As a result,

$$P^2 = \begin{pmatrix} 0 & 0 & 0 & \cdots & 0 & 1 & 0 \\ 0 & 0 & 0 & \cdots & 0 & 0 & 1 \\ 1 & 0 & 0 & \cdots & 0 & 0 & 0 \\ 0 & 1 & 0 & \cdots & 0 & 0 & 0 \\ \cdots & \cdots & \cdots & \cdots & \cdots & \cdots \\ 0 & 0 & 0 & \cdots & 1 & 0 & 0 \end{pmatrix}.$$

(Note the change in the place of some of the 1's.) In addition,

$$P^3 = \begin{pmatrix} 0 & 0 & 0 & \cdots & 0 & 1 & 0 & 0 \\ 0 & 0 & 0 & \cdots & 0 & 0 & 1 & 0 \\ 0 & 0 & 0 & \cdots & 0 & 0 & 0 & 1 \\ 1 & 0 & & \cdots & 0 & 0 & 0 & 0 \\ \cdots & \cdots & \cdots & \cdots & \cdots & \cdots \\ 0 & 0 & 0 & \cdots & 1 & 0 & 0 & 0 \end{pmatrix}, \text{ etc.}$$

The matrix O is the diagonal

$$Q = \text{diag} = \{1, \omega, \omega^2, \ldots, \omega^{p-1}\}, \quad \omega^p = 1$$

$$(PQ)_{k,\ell} = P_{k,j}Q_{j,\ell} = \delta_{k-1,j}\omega^j \delta_{j,\ell} = \delta_{k-1,\ell}\omega^\ell,$$

$$(QP)_{k,\ell} = Q_{k,j}P_{j,\ell} = \omega^j \delta_k, {}_j \delta_{j-1,\ell} = \delta_{k,\ell+1}\omega^{\ell+1} = \omega\delta_{k-1,\ell}\omega^\ell \Rightarrow$$

$$QP = \omega PQ, \quad [Q, P] = (\omega - 1)PQ. \tag{14.22}$$

This is Heisenberg's commutation relation in exponential form.

It is worth mentioning that, if ω is replaced by ω^k, for $k = 1, 2, \ldots, p-1$, the above relations do not change. In addition, so long as p is a prime number, the resulting representations are of dimension p and not equivalent.

---

[2] Some authors use the transpose of our P.

The matrices P and Q are connected via the matrix F, which diagonal-izes P:

$$F^{-1}PF = Q. \tag{14.23}$$

The matrix F is also a discrete Fourier transform:

$$F_{k\ell} = \frac{1}{\sqrt{p}}\omega^{(k\ell)}, \quad \text{with } k, \ell = 0, \ldots, p-1. \tag{14.24}$$

Some properties are

$$\sum_{k=0}^{p-1} \omega^k = 1 + \omega + \omega^2 \cdots \omega^{p-1} = \frac{1-\omega^p}{1-\omega} \underset{\text{mod}[p]}{\Rightarrow} 0,$$

$$\sum_{k=0}^{p-1} \omega^{k\ell} = \sum_{k=0}^{p-1} (\omega^\ell)^k = \frac{1-\omega^{p\ell}}{1-\omega^\ell} \underset{\text{mod}[p]}{\Rightarrow} 0.$$

The matrix Γ generates an Abelian group of order p with elements

$$F, \ S = F^2, \ F^3, \ldots, F^p = E. \tag{14.25}$$

Example 3: Let us consider the case $p = 3$:

$$P = \begin{pmatrix} 0 & 0 & 1 \\ 1 & 0 & 0 \\ 0 & 1 & 0 \end{pmatrix},$$

$$\text{eigenvalues: } 1, \omega, \omega^2, \ \omega = e^{\frac{2\pi i}{3}}, \ F = \frac{1}{\sqrt{3}}\begin{pmatrix} 1 & 1 & 1 \\ 1 & \omega & \omega^2 \\ 1 & \omega^2 & \omega \end{pmatrix}.$$

On the other hand,

	$j=0$	$j=1$	$j=2$
$i=0$	$ij=0$	$ij=0$	$ij=0$
$i=1$	$ij=0$	$ij=1$	$ij=2$
$i=2$	$ij=0$	$ij=2$	$ij=4$

$\underset{\text{mod}[3]}{\Rightarrow}$

	$j=0$	$j=1$	$j=2$
$i=0$	$ij=0$	$ij=0$	$ij=0$
$i=1$	$ij=0$	$ij=1$	$ij=2$
$i=2$	$ij=0$	$ij=2$	$ij=1$

$\Rightarrow$

or

	$j=0$	$j=1$	$j=2$
$i=0$	1	1	1
$i=1$	1	ω	ω^2
$i=2$	1	ω^2	ω

$\Rightarrow \sqrt{3}F.$

The reader is encouraged to confirm that for $p = 5, 7$, $\omega = e^{(2\pi i)/p}$, one gets

$$F = \frac{1}{\sqrt{5}} \begin{pmatrix} 1 & 1 & 1 & 1 & 1 \\ 1 & \omega & \omega^2 & \omega^3 & \omega^4 \\ 1 & \omega^2 & \omega^4 & \omega & \omega^3 \\ 1 & \omega^3 & \omega & \omega^4 & \omega^2 \\ 1 & \omega^4 & \omega^3 & \omega^2 & \omega \end{pmatrix},$$

$$F = \frac{1}{\sqrt{7}} \begin{pmatrix} 1 & 1 & 1 & 1 & 1 & 1 & 1 \\ 1 & \omega & \omega^2 & \omega^3 & \omega^4 & \omega^5 & \omega^6 \\ 1 & \omega^2 & \omega^4 & \omega^6 & \omega & \omega^3 & \omega^5 \\ 1 & \omega^3 & \omega^6 & \omega^2 & \omega^5 & \omega & \omega^4 \\ 1 & \omega^4 & \omega & \omega^5 & \omega^2 & \omega^6 & \omega^3 \\ 1 & \omega^5 & \omega^3 & \omega & \omega^6 & \omega^4 & \omega^2 \\ 1 & \omega^6 & \omega^5 & \omega^4 & \omega^3 & \omega^2 & \omega \end{pmatrix}.$$

respectively, and should verify that for $p = 11$,

$$F = \frac{1}{\sqrt{11}} \begin{pmatrix} 1 & 1 & 1 & 1 & 1 & 1 & 1 & 1 & 1 & 1 & 1 \\ 1 & \omega & \omega^2 & \omega^3 & \omega^4 & \omega^5 & \omega^6 & \omega^7 & \omega^8 & \omega^9 & \omega^{10} \\ 1 & \omega^2 & \omega^4 & \omega^6 & \omega^8 & \omega^{10} & \omega & \omega^3 & \omega^5 & \omega^7 & \omega^9 \\ 1 & \omega^3 & \omega^6 & \omega^9 & \omega & \omega^4 & \omega^7 & \omega^{10} & \omega^2 & \omega^5 & \omega^8 \\ 1 & \omega^4 & \omega^8 & \omega & \omega^5 & \omega^9 & \omega^2 & \omega^6 & \omega^{10} & \omega^3 & \omega^7 \\ 1 & \omega^5 & \omega^{10} & \omega^4 & \omega^9 & \omega^3 & \omega^8 & \omega^2 & \omega^7 & \omega & \omega^6 \\ 1 & \omega^6 & \omega & \omega^7 & \omega^2 & \omega^8 & \omega^3 & \omega^9 & \omega^4 & \omega^{10} & \omega^5 \\ 1 & \omega^7 & \omega^3 & \omega^{10} & \omega^6 & \omega^2 & \omega^9 & \omega^5 & \omega & \omega^8 & \omega^4 \\ 1 & \omega^8 & \omega^5 & \omega^2 & \omega^{10} & \omega^7 & \omega^4 & \omega & \omega^9 & \omega^6 & \omega^3 \\ 1 & \omega^9 & \omega^7 & \omega^5 & \omega^3 & \omega & \omega^{10} & \omega^8 & \omega^6 & \omega^4 & \omega^2 \\ 1 & \omega^{10} & \omega^9 & \omega^8 & \omega^7 & \omega^6 & \omega^5 & \omega^4 & \omega^3 & \omega^2 & \omega \end{pmatrix}.$$

14.3.2 Magnetic translations

One important subsection of the above transformations HW_p is that of the magnetic translations

$$J_{r,s} = \omega^{rs/2} P^r Q^s, \tag{14.26}$$

with $r, s = 0, \ldots, p - 1$. These matrices are unitary, $(J_{r,s}^\dagger = (J_{r,s})^{-1}$, as a product of unitary matrices so long as $(\omega^{rs/2})^* = \omega^{-rs/2}$. They are also traceless, and they constitute a basis in the space of the Lie algebra of the special linear group $SL(p, \mathbb{C})$.

Theorem: *The following relation holds*:

$$J_{r,s} J_{r',s'} = \omega^{(r's - rs')/2} J_{r+r', s+s'}. \tag{14.27}$$

Proof. We observe that with the help of Eq. (14.22), we can transfer to the right all operators Q appearing in the expression $P^{m_1} Q^{m_2} P^{n_1} Q^{n_2}$. Indeed,

$$PQPQ = P\omega PQQ = \omega P^2 Q^2,$$

$$P^2 Q^2 PQ = P^2 Q\omega PQQ = \omega^2 P^2 PQQ^2 = \omega^2 P^3 Q^3.$$

In addition, noting that $Q^2 P^2 = \omega QPQQP = \omega^2 PQQP = \omega^3 PQPQ = \omega^4 P^2 Q^2$, we find that $P^2 Q^2 P^2 = \omega^4 P^4 Q^4$. Generally,

$$P^{m_1} Q^{m_2} P^{n_1} Q^{n_2} = \omega^{n_1 m_2} P^{m_1 + n_1} Q^{m_2 + n_2}.$$

Now,

$$J_{m_1,m_2} J_{n_1,n_2} = \omega^{\frac{m_1 m_2}{2}} \omega^{\frac{n_1 n_2}{2}} \omega^{n_1 m_2} P^{m_1 + m_2} Q^{n_1 + n_2}$$

$$= \omega^{\frac{m_1 m_2}{2}} \omega^{\frac{n_1 n_2}{2}} \omega^{n_1 m_2} \omega^{-\frac{(m_1 + m_2)(n_1 + n_2)}{2}} J_{m_1 + m_2, n_1 + n_2}$$

$$= \omega^{\frac{n_1 m_2 - m_1 n_2}{2}} J_{m_1 + m_2, n_1 + n_2}, \tag{14.28}$$

and the proof is complete.   □

This shows that the magnetic translations form a projective representation of the translation group $\mathbb{Z}_p \times \mathbb{Z}_p$. The factor $(1/2)$ in the exponent of the expression of Eq. (14.27) must be $\mathrm{mod}(p)$. As an example, suppose that $(1/2)\mathrm{mod}(p)$ is x, i.e. $x = (1/2)\,\mathrm{mod}(p) \Rightarrow 2x\,\mathrm{mod}(p) = 1 \Rightarrow 2x = p + 1 \Rightarrow x = \frac{p+1}{2}$. Thus, for example, $(1/2)\,\mathrm{mod}(11) = 6$.

14.3.3   *The metaplectic representation by Weyl*

The linear transformation

$$\begin{pmatrix} m' \\ n' \end{pmatrix} = \begin{pmatrix} a & b \\ c & d \end{pmatrix} \begin{pmatrix} m \\ n \end{pmatrix}, \quad \det \begin{pmatrix} a & b \\ c & d \end{pmatrix} = 1, \mathrm{mod}(p) \tag{14.29}$$

is called special linear transformation in 2D $\mathrm{mod}(p)$ or $SL[2, p]$. A representation of this group is given by matrices R and L in Section 14.1.

The metaplectic transformation is given as follows:

$$A \rightarrow U(A) : U(A) J_{r,s} U^\dagger(A) = J_{r',s'}, \ (r', s') = (r, s) \begin{pmatrix} a & b \\ c & d \end{pmatrix}. \quad (14.30)$$

This relation determines $U(A)$ up to a phase, and in the case of $A \in SL_2(p)$, the phase can be fixed to give an exact (and not projective) unitary representation of $SL_2(p)$.

The detailed formula of $U(A)$ has been given by Balian and Itzykson (1986). Depending on the specific values of the a, b, c, d parameters of the matrix A, we distinguish (Floratos and Leontaris, 2016a) the following cases:

$$\delta \neq 0 : U(A) = \frac{\sigma(1)\sigma(\delta)}{p} \sum_{r,s} \omega^{[br^2 + (d-a)rs - cs^2]/(2\delta)} J_{r,s}, \quad (14.31)$$

$$\delta = 0, \ b \neq 0 : U(A) = \frac{\sigma(-2b)}{\sqrt{p}} \sum_s \omega^{s^2/(2b)} J_{s(a-1)/b,s}, \quad (14.32)$$

$$\delta = b = 0, \ c \neq 0 : U(A) = \frac{\sigma(2c)}{\sqrt{p}} \sum_r \omega^{-r^2/(2c)} P^r, \quad (14.33)$$

$$\delta = b = 0 = c = 0 : \ U(1) = I, \quad (14.34)$$

where $\delta = 2 - a - d$ and $\sigma(a)$ is the quadratic Gauss sum given by

$$\sigma(a) = \frac{1}{\sqrt{p}} \sum_{k=0}^{p-1} \omega^{ak^2} = (a|p) \times \begin{cases} 1 & \text{for } p = 4k + 1 \\ i & \text{for } p = 4k - 1 \end{cases}, \quad (14.35)$$

while the Legendre symbol takes the values $(a|p) = \pm 1$ depending on whether a is or is not a square modulo p.

Definition: a is the square modulo p if $a = a^2 \bmod(p)$.

Thus, for $a \leq 6$, a is the square modulo 7 only if $a = 2$, $7 = 4k - 1 \Leftrightarrow k = 2$, i.e. $(2|7) = 1$ and $\sigma(2) = i$.

Table 14.2. Examples of a square modulo 7.

a :	0	1	2	3	4	5	6
a^2 :	0	1	4	9	16	25	36
$a^2 \bmod(7)$:	0	1	4	2	2	4	1

The above representation gives interesting results whenever the matrix $U(a)$ is generated by the element

$$a = \begin{pmatrix} 0 & -1 \\ 1 & 0 \end{pmatrix}.$$

It is possible to perform explicitly the above Gaussian sums noting that

$$(J_{r,s})_{k,l} = \delta_{r,k-l}\omega^{\frac{k+l}{2}s}, \tag{14.36}$$

where all indices take the values $k, l, r, s = 0, \ldots, p-1$. This has been done by Athanasiu and Floratos (1994), Athanasiu *et al.* (1998) and Floratos and Leontaris (2016a). In the case of $\delta = 2 - a - d \neq 0$ and $c \neq 0$, the result is

$$\delta \neq 0 : U(A)_{k,l} = \frac{(-2c|p)}{\sqrt{p}} \times \begin{Bmatrix} 1 \\ -i \end{Bmatrix} \omega^{-\frac{ak^2 - 2kl + dl^2}{2c}}. \tag{14.37}$$

If $c = 0$, then we transform the matrix A to one with $c \neq 0$. The other case $\delta = 0$ can be worked out easily using the matrix elements of $J_{r,s}$ given in Eq. (14.36).

It is interesting to note that by redefining ω to become ω^k for $k = 1, 2, \ldots, p-1$, the matrix $U(A)$ transforms into the matrix $U(A_k)$, where A_k is the 2×2 matrix $A_k = \begin{pmatrix} a & bk \\ c/k & d \end{pmatrix}$, which belongs to the same conjugacy class with A as long as k is a quadratic residue. If $k = p-1$, we pass from the representation $U(A)$ to the complex conjugate one $U(A)^*$.

The Weyl representation presented above provides the interesting result that the unitary matrix corresponding to the $SL_2(p)$ element $a = \begin{pmatrix} 0 & -1 \\ 1 & 0 \end{pmatrix}$ is — up to a phase — the discrete finite Fourier transform, Eq. (14.24):

$$U(a) = (-1)^{k+1} i^n F,$$

where $n = 0$ for $p = 4k + 1$ and $n = 1$ for $p = 4k - 1$.

The Fourier transform matrix generates a fourth-order Abelian group with the elements

$$F, \quad F^2 = S, \quad F^3 = F^*, \quad F^4 = I. \tag{14.38}$$

The matrix S represents the element $a^2 = \begin{pmatrix} -1 & 0 \\ 0 & -1 \end{pmatrix}$. Its matrix elements are

$$S_{k,l} = \delta_{k,-l}, \quad k, l = 0, \ldots, p-1, \tag{14.39}$$

$$U(a^2)_{k,l} = i^{2n} S_{k,l} = (-)^n \delta_{k,-l}, \quad k, l = 0, \ldots, p-1. \tag{14.40}$$

For $p = 5$, we get

$$S = \begin{pmatrix} 1 & 0 & 0 & 0 & 0 \\ 0 & 0 & 0 & 0 & 1 \\ 0 & 0 & 0 & 1 & 0 \\ 0 & 0 & 1 & 0 & 0 \\ 0 & 1 & 0 & 0 & 0 \end{pmatrix}.$$

With eigenvalues $\{1, 1, 1, -1, -1\}$ and the corresponding eigenvectors,

$$v = \begin{pmatrix} 0 & 1 & 0 & 0 & 0 \\ 0 & 0 & \frac{1}{\sqrt{2}} & 0 & -\frac{1}{\sqrt{2}} \\ \frac{1}{\sqrt{2}} & 0 & 0 & \frac{1}{\sqrt{2}} & 0 \\ \frac{1}{\sqrt{2}} & 0 & 0 & -\frac{1}{\sqrt{2}} & 0 \\ 0 & 0 & \frac{1}{\sqrt{2}} & 0 & \frac{1}{\sqrt{2}} \end{pmatrix}. \tag{14.41}$$

Because the action of S on $J_{r,s}$ changes the signs of r, s, while $\forall A \in SL_2(p)$, the unitary matrix $U(A)$ depends quadratically on r, s in the sum (14.31), it turns out that S commutes with all $U(A)$. Moreover, $S^2 = I$, and we can construct two projectors:

$$P_+ = \frac{1}{2}(I + S), \quad P_- = \frac{1}{2}(I - S),$$

with dimensions of their invariant subspaces $\frac{p+1}{2}$ and $\frac{p-1}{2}$ correspondingly. So, the Weyl p-dimensional representation is the direct sum of two irreducible unitary representations:

$$U_+(A) = U(A)P_+, \quad U_-(A) = U(A)P_-.$$

To obtain the block diagonal form of the above matrices $U_\pm(A)$, we rotate with the orthogonal matrix of the eigenvectors of S. This p-dimensional orthogonal matrix, dubbed here O_p, can be obtained in a maximally symmetric form (along the diagonal as well as along the anti-diagonal) using the eigenvectors of S in the following order: In the first $(p+1)/2$ columns, we put the eigenvectors of S of eigenvalue equal to 1, and in the next $(p-1)/2$ columns, the eigenvectors of eigenvalue equal to

-1 in the specific order given as follows:

$$(e_0)_k = \delta_{k0}, \tag{14.42}$$

$$(e_j^+)_k = \frac{1}{\sqrt{2}}(\delta_{k,j} + \delta_{k,-j}), \quad j = 1, \ldots, \frac{p-1}{2}, \tag{14.43}$$

$$(e_j^-)_k = \frac{1}{\sqrt{2}}(\delta_{k,j} - \delta_{k,-j}), \quad j = \frac{p+1}{2}, \ldots, p, \tag{14.44}$$

where $k = 0, \ldots, p-1$.

Different orderings of eigenvectors may lead to different forms of the matrices $U_\pm(A)$. Thus obtained, the orthogonal matrix O_p has the property

$$O_p^2 = I,$$

due to its symmetric form.

The final block diagonal form of $U_\pm(A)$ is obtained through an O_p rotation:

$$V_\pm(A) = O_p U(A)_\pm O_p. \tag{14.45}$$

14.4 The construction of the $SL_2(p)$ generators

In this section[3] we give explicit expressions for the $SL_2(p)$ generators associated with the matrices a, b, see Eq. (14.3), for any value of p in the $\frac{p\pm 1}{2}$ irreducible representations. We also consider the corresponding matrix expressions of the projective group $PSL_2(p)$.

According to the above construction, the two generators a, b have the following unitary matrix representations:

$$U_\pm(a) = (-1)^{k+1} i^n F \frac{I \pm S}{2} = (-1)^{k+1} i^n \frac{1}{2}(F \pm F^*), \tag{14.46}$$

with matrix elements

$$[U_\pm(a)]_{k,l} = (-1)^{k+1} i^n \frac{1}{2\sqrt{p}} \left(\omega^{kl} \pm \omega^{-kl} \right),$$

where, as noted previously, $n = 0$ for $p = 4k + 1$ and $n = 1$ for $p = 4k - 1$.

---

[3] The authors are indebted to Professor G. K. Leontaris for making available to us his teaching notes on the material of this chapter.

In order to bring this to the block diagonal form, see Eq. (14.45), we need to perform a rotation with O_p:

$$A^{[\frac{p\pm 1}{2}]} = O_p U_\pm(a) O_p. \tag{14.47}$$

For the second generator, b, given in (14.3), we obtain

$$U(b)_{k,l} = \frac{1}{\sqrt{p}}(-1)^{\frac{p^2-1}{8}} \begin{Bmatrix} 1 \\ i \end{Bmatrix} \omega^{-\frac{l^2}{2}+kl}, \tag{14.48}$$

and so,

$$U_\pm(b)_{k,l} = \frac{1}{2\sqrt{p}}(-1)^{\frac{p^2-1}{8}} \begin{Bmatrix} 1 \\ i \end{Bmatrix} \left(\omega^{-\frac{k^2}{2}+kl} \pm \omega^{-\frac{l^2}{2}-kl} \right),$$

where $k, l = 0, \ldots, p - 1$. As noted previously, in order to get the block diagonal form, we have to rotate the matrix thus obtained with O_p:

$$B^{[\frac{p\pm 1}{2}]} = O_p U_\pm(b) O_p. \tag{14.49}$$

Our final goal is to obtain some basic representations of $PSL_2(p)$, which will be used to build higher dimensional ones. We observe that the difference between $SL_2(p)$ and $PSL_2(p)$ in the defining relations of generators a and b is that, for $SL_2(p)$, one has to take $a^2 = b^3 = -I$, while for $PSL_2(p)$, we have the relations $a^2 = b^3 = I$. This last requirement comes from the different actions of $SL_2(p)$ and $PSL_2(p)$, which are linear and Möbius correspondingly.

We can obtain the irreducible representations of $PSL_2(p)$ from the irreducible representations of $SL_2(p)$ in the following way. Taking into consideration the above observation, we must find the representations of $SL_2(p)$ for which $(A^{[\frac{p\pm 1}{2}]})^2 = (B^{[\frac{p\pm 1}{2}]})^3 = I$. We can easily check that this happens for the $\frac{p+1}{2}$-dimensional representation only when $p = 4k + 1$, and for the $\frac{p-1}{2}$ one, only when $p = 4k - 1$. This way, we get the $(2k+1)$- and $(2k-1)$-dimensional irreducible representations of $PSL_2(p)$ correspondingly.

14.5   Examples: The cases $p = 3, 5, 7$

In this section, using the method described above, we present examples, considering the cases $p = 3, 5$ and $p = 7$. It is straightforward to construct similar representations (Floratos and Leontaris, 2016a) of higher values of p.

14.5.1   *The case $p = 3$*

The resulting group is $SL_2(3)$, which has 24 elements, while its projective subgroup $PSL_2(3)$ has 12 elements and is isomorphic to A_4, the symmetry group of the even permutations of four objects,[4] or the symmetry group of the tetrahedron T. The symmetry groups of the cube and octahedron are S_4, which is isomorphic to $PGL_2(3)$, the automorphism group of $SL_2(3)$.

The generators in the doublet representation are the following: The $A^{[2]}$-representation is

$$A^{[2]} = -\frac{i}{\sqrt{3}} \begin{pmatrix} 1 & \sqrt{2} \\ \sqrt{2} & \eta + \eta^2 \end{pmatrix} = -\frac{i}{\sqrt{3}} \begin{pmatrix} 1 & \sqrt{2} \\ \sqrt{2} & -1 \end{pmatrix}, \qquad (14.50)$$

where we have used the fact that $1 + \eta + \eta^2 = 0$. The representation $B^{[2]}$ is given by

$$B^{[2]} = \frac{i}{\sqrt{3}} \begin{pmatrix} 1 & \sqrt{2}\eta \\ \sqrt{2} & 1 + \eta^2 \end{pmatrix} = -\frac{i}{\sqrt{3}} \begin{pmatrix} 1 & \sqrt{2}\eta \\ \sqrt{2} & -\eta \end{pmatrix}. \qquad (14.51)$$

They satisfy the $SL_2(3)$ relations

$$A^{[2]^2} = B^{[2]^3} = (A^{[2]}B^{[2]})^3 = -I. \qquad (14.52)$$

The singlet representations are

$$A^{[1]} = -\frac{i}{\sqrt{3}}(\eta - \eta^2), \qquad (14.53)$$

$$B^{[1]} = -\frac{i}{\sqrt{3}}(\eta^2 - 1). \qquad (14.54)$$

The defining relations are satisfied:

$$A^{[1]^2} = B^{[1]^3} = (A^{[1]}B^{[1]})^3 = 1 \qquad (14.55)$$

and are consistent with $PSL_2(3) \sim A_4$. From Eqs. (14.53) and (14.54), we deduce

$$A^{[1]} \cdot B^{[1]} \equiv \eta, \quad (A^{[1]} \cdot B^{[1]})^2 \equiv \eta^2,$$

---

[4] A_4 is suitable to reproduce the tri-bi-maximal mixing to leading order in the neutrino sector and has been discussed in Chapter 13, see also Ma and Rajasekaran (2001).

so that we can define the singlet representations with the standard multiplication rules:

$$1' : s_{1'} = \eta, \tag{14.56}$$

$$1'' : s_{1''} = \eta^2, \tag{14.57}$$

$$1' \times 1'' = 1 : s_1 = 1. \tag{14.58}$$

14.5.2 *The case* $p = 5$

Next, we elaborate on the case of $PSL_2(5)$, which is isomorphic to the symmetry group I of the dodecahedron and icosahedron, as well as to A_5. The group $SL_2(5)$ is isomorphic to the symmetry group $2I$ of the binary icosahedron. The 60 elements of A_5 are generated by two generators a, b with the properties

$$a^2 = b^3 = (ab)^5 = I.$$

Through the above method, we find two representations of $SL_2(5)$, one of 3D and a second one of 2D.

The first generator is a unitary 3×3 matrix:

$$A^{[3]} = -\frac{1}{\sqrt{5}} \begin{pmatrix} 1 & \sqrt{2} & \sqrt{2} \\ \sqrt{2} & \eta + \eta^4 & \eta^2 + \eta^3 \\ \sqrt{2} & \eta^2 + \eta^3 & \eta + \eta^4 \end{pmatrix}$$

$$= -\frac{1}{\sqrt{5}} \begin{pmatrix} 1 & \sqrt{2} & \sqrt{2} \\ \sqrt{2} & \frac{\sqrt{5}-1}{2} & -\frac{\sqrt{5}+1}{2} \\ \sqrt{2} & -\frac{\sqrt{5}+1}{2} & \frac{\sqrt{5}-1}{2} \end{pmatrix}, \tag{14.59}$$

where in the last form, the matrix elements have been written in terms of the golden ratio since

$$\eta + \eta^4 = \frac{1}{2}\left(\sqrt{5} - 1\right), \quad \eta^2 + \eta^3 = -\frac{1}{2}\left(\sqrt{5} + 1\right).$$

The character of the representation is $\mathrm{Tr}A^{[3]} = -1$, as expected from the character table of $PSL_2(5)$.

The second generator has the following 3D representation:

$$B^{[3]} = -\frac{1}{\sqrt{5}} \begin{pmatrix} 1 & \sqrt{2}\eta^2 & \sqrt{2}\eta^3 \\ \sqrt{2} & \eta^3 + \eta & \eta + 1 \\ \sqrt{2} & \eta^4 + 1 & \eta^4 + \eta^2 \end{pmatrix}, \tag{14.60}$$

while the character is $\mathrm{Tr}B^{[3]} \propto 1 + \eta + \eta^2 + \eta^3 + \eta^4 = 0$. It can be readily checked that $A^{[3]}$ and $B^{[3]}$ satisfy the defining relations of the $PLS_2(5)$ group:

$$A^{[3]\,2} = B^{[3]\,3} = (A^{[3]} \cdot B^{[3]})^5 = I.$$

These generators correspond to the triplet $\underline{3}'$. Indeed, in order to make contact with the form of some generators given in recent literature, we transform the above in the s_5', t_5' basis,[5] setting

$$s_5' \equiv A^{[3]}, \quad t_5' = A^{[3]} \cdot B^{[3]} \rightarrow B^{[3]} = s_5' \cdot t_5'.$$

Hence, the two new generators s_5', t_5' are

$$s_5' = -\frac{1}{\sqrt{5}} \begin{pmatrix} 1 & \sqrt{2} & \sqrt{2} \\ \sqrt{2} & \eta + \eta^4 & \eta^2 + \eta^3 \\ \sqrt{2} & \eta^2 + \eta^3 & \eta + \eta^4 \end{pmatrix}, \quad t_5' = \begin{pmatrix} 1 & 0 & 0 \\ 0 & n^2 & 0 \\ 0 & 0 & n^3 \end{pmatrix}, \tag{14.61}$$

They satisfy the defining relations

$$s_5'^{\,2} = t_5'^{\,5} = (s_5' \cdot t_5')^3 = I,$$

while their characters are

$$\chi_{s_5'} = -1, \quad \chi_{t_5'} = \frac{1 - \sqrt{5}}{2}.$$

It is possible to get the other triplet representation of $SL_2(5)$ (up to equivalence) given by Feruglio and Paris (2011),

$$s_5 = \frac{1}{\sqrt{5}} \begin{pmatrix} 1 & -\sqrt{2} & -\sqrt{2} \\ -\sqrt{2} & \eta^2 + \eta^3 & \eta + \eta^4 \\ -\sqrt{2} & \eta + \eta^4 & \eta^2 + \eta^3 \end{pmatrix}, \quad t_5 = \begin{pmatrix} 1 & 0 & 0 \\ 0 & n^2 & 0 \\ 0 & 0 & n^3 \end{pmatrix} \tag{14.62}$$

by redefining η to η^3 in Eq. (14.61). As discussed in Section 4 after Eq. (14.37), this is equivalent to a rescaling of the appropriate elements of $SL_2(5)$, which gives the generators s_5' and t_5'.

---

[5] See, for example, neutrino models with A_5 family symmetry given by Feruglio and Paris (2011) and Ding *et al.* (2012).

The 2D representation gives

$$A^{[2]} = -\frac{1}{\sqrt{5}} \begin{pmatrix} \eta^4 - \eta & \eta^2 - \eta^3 \\ \eta^2 - \eta^3 & \eta - \eta^4 \end{pmatrix} \quad (14.63)$$

and

$$B^{[2]} = -\frac{1}{\sqrt{5}} \begin{pmatrix} \eta^2 - \eta^4 & \eta^4 - 1 \\ 1 - \eta & \eta^3 - \eta \end{pmatrix} \quad (14.64)$$

satisfying

$$A^{[2]^2} = B^{[2]^3} = (A^{[2]} \cdot B^{[2]})^5 = -I.$$

Obviously, the above 2D matrices are representations of $SL_2(5)$, but not of $PSL_2(5)$.

The 3D and 2D representations of the generators constructed above are unitary matrices, so they generate discrete subgroups of $SU(3)$ and $SU(2)$ Lie groups.

14.5.3 The case $p = 7$

As a final example in this section, we consider the case $p = 7$. The associated groups are $SL_2(7)$ with 336 elements and its projective one $PSL_2(7)$, which has 168 elements and is a discrete simple subgroup (Floratos and Leontaris, 2016a) of $SU(3)$. It is the group preserving the discrete projective geometry of the Fano plane realizing the multiplication structure of the octonionic units.

Using the method described above, we construct the 4D and 3D representations of $SL_2(7)$. The 3D one is also a representation of $PSL_2(7)$:

$$a^2 = b^3 = (ab)^7 = [a, b]^4 = 1,$$

with $[a, b] = a^{-1}b^{-1}ab$.

The $A^{[4]}$ and $B^{[4]}$ generating matrices of the irreducible 4D unitary representation of $SL_2(7)$ are

$$A^{[4]} = \frac{i}{\sqrt{7}} \begin{pmatrix} 1 & \sqrt{2} & \sqrt{2} & \sqrt{2} \\ \sqrt{2} & \eta + \eta^6 & \eta^2 + \eta^5 & \eta^3 + \eta^4 \\ \sqrt{2} & \eta^2 + \eta^5 & \eta^3 + \eta^4 & \eta + \eta^6 \\ \sqrt{2} & \eta^3 + \eta^4 & \eta + \eta^6 & \eta^2 + \eta^5 \end{pmatrix} \quad (14.65)$$

and

$$B^{[4]} = \frac{i}{\sqrt{7}} \begin{pmatrix} 1 & \sqrt{2}\eta^3 & \sqrt{2}\eta^5 & \sqrt{2}\eta^6 \\ \sqrt{2} & \eta^2 + \eta^4 & 1 + \eta^3 & \eta^2 + \eta^3 \\ \sqrt{2} & \eta + \eta^5 & \eta + \eta^2 & 1 + \eta^5 \\ \sqrt{2} & 1 + \eta^6 & \eta^4 + \eta^6 & \eta + \eta^4 \end{pmatrix}. \tag{14.66}$$

These matrices satisfy the relations for $SL_2(7)$, which are

$$A^{[4]^2} = B^{[4]^3} = (A^{[4]}B^{[4]})^7 = [A^{[4]}, B^{[4]}]^4 = -I.$$

The generators in the triplet representation are the following:

$$A^{[3]} = \frac{i}{\sqrt{7}} \begin{pmatrix} \eta^2 - \eta^5 & \eta^6 - \eta & \eta^3 - \eta^4 \\ \eta^6 - \eta & \eta^4 - \eta^3 & \eta^2 - \eta^5 \\ \eta^3 - \eta^4 & \eta^2 - \eta^5 & \eta - \eta^6 \end{pmatrix} \tag{14.67}$$

and

$$B^{[3]} = \frac{i}{\sqrt{7}} \begin{pmatrix} \eta - \eta^4 & \eta^4 - \eta^6 & \eta^6 - 1 \\ \eta^5 - 1 & \eta^2 - \eta & \eta^5 - \eta \\ \eta^2 - \eta^3 & 1 - \eta^3 & \eta^4 - \eta^2 \end{pmatrix}. \tag{14.68}$$

As expected, the $A^{[3]}$ and $B^{[3]}$ satisfy the defining relations

$$A^{[3]^2} = B^{[3]^3} = (A^{[3]}B^{[3]})^7 = [A^{[3]}, B^{[3]}]^4 = I. \tag{14.69}$$

We note that our representations $A^{[3]}, B^{[3]}$ are connected to the conjugate triplet of those of Luhn *et al.* (2007a) and King and Luhn (2009) through the similarity transformation obtained by the diagonal matrix $M_{diag} = (1, -1, -1)$. We also note in passing that the phenomenological implications of $PSL_2(7)$ have been analyzed in several works (see the reviews (Altarelli and Feruglio, 2010b), (Ishimori *et al.*, 2010b), (King and Luhn, 2013b) and references therein, as well as the more recent one by Aliferis *et al.* (2016)).

14.6   Concluding remarks

In this chapter, we have introduced an intriguing relation between the discrete flavor symmetries and the automorphisms of the magnetic translations of the finite and discrete Heisenberg groups. This relation is reminiscent of the discrete symmetries of the quantum Hall effect, where in a toroidal 2D space, the magnetic flux transforms the torus to a phase space and the Hilbert space of a charged particle becomes finite dimensional and the corresponding torus effectively discrete (Zak, 1989). Tori with fluxes in internal extra dimensions appear naturally in the framework of F-theory of elliptic fibrations over Calabi–Yau manifolds, where they generate the GUT gauge groups and other discrete symmetries at particular singularities of the fibration. Phenomenological explorations have shown that such discrete symmetries are particularly successful in predicting the Fermion mass hierarchies and flavor mixing. Inspired by these observations, the discrete Heisenberg group was used to develop a simple and unified method for the derivation of basic non-trivial representations of a large class of non-Abelian finite groups relevant to the flavor symmetries. As an example, a toy model was worked out where the family symmetry is played by the finite group $PSL_2(7)$. It thus appears that it will be important to construct explicit realistic models of elliptic fibrations with fluxes, where the discrete magnetic translations appear naturally and the discrete flavor symmetries are their automorphisms.

Chapter 15

Appendix I: Proofs of Various Theorems

Schur's first lemma: *Let $\Gamma^i(X)$ be an irreducible representation of a group G and a matrix (ρ) such that*

$$\Gamma^i(X)(\rho) = (\rho)\Gamma^i(X) \quad \text{for } X \in G. \tag{15.1}$$

Then, $(\rho) = \lambda(\epsilon)$, where ϵ is the identity matrix. In other words, (ρ) is a multiple of the identity, provided that the representation is irreducible. Otherwise, the representation is necessarily reducible.

This lemma is not directly a consequence of the group aspects. It depends on notions related to finite-dimensional vector spaces, see for example Vergados (2004), Section 6.4.

Proof: Indeed, suppose that $\Gamma^i(X)$ is an $n \times n$ representation of G and Eq. (15.1) holds. The matrix (ρ) has at least one eigenvalue λ with a corresponding eigenvector $|\alpha\rangle$, i.e.

$$(\rho)|\alpha\rangle = \lambda|\alpha\rangle.$$

We define

$$|\alpha_i(X)\rangle = \Gamma^i(X)|\alpha\rangle.$$

Then,

$$(\rho)|\alpha_i(X)\rangle = (\rho)\Gamma^i(X)|\alpha\rangle = \Gamma^i(X)(\rho)|\alpha\rangle = \lambda\Gamma^i(X)|\alpha\rangle \Rightarrow$$

$$(\rho)|\alpha_i(X)\rangle = \lambda|\alpha_i(X)\rangle.$$

In other words, $|\alpha_i(X)\rangle$ is another eigenvector of (ρ) with the same eigenvalue. With the same procedure, we can construct m such eigenvectors with eigenvalue λ. These cover a space R_m, a subspace of R_n on which the representation $\Gamma^i(X)$ has been defined. R_m, however, cannot be a proper subspace of R_n since the representation $\Gamma^i(X)$ was assumed irreducible. Thus, $R_m = R_n$. In this space, the action of (ρ) is as follows:

$$|\alpha\rangle \in R_n, (\rho)|\alpha\rangle = \lambda|\alpha\rangle \Rightarrow$$

$$(\rho) = \lambda(\epsilon).$$

Schur's second lemma: *Given are two representations of a group $\Gamma^i(X)$ and $\Gamma^j(X)$ of dimension $k_i \times k_i$ and $k_j \times k_j$, respectively. Suppose further that there exists a matrix (ρ) with dimension $k_i \times k_j$, which satisfies the relation*

$$\Gamma^i(X)(\rho) = (\rho)\Gamma^j(X). \tag{15.2}$$

Then, the matrix (ρ) is identically zero or it is a square matrix with $\det(\rho) \neq 0$. In other words, if there exists a non-singular matrix (ρ) satisfying Eq. (15.2), the representations $\Gamma^i(X)$ and $\Gamma^j(X)$ are equivalent, i.e. practically the same. Indeed, since $\det(\rho) \neq 0$, then the inverse of (ρ) exists, and Eq. (15.2) can be written as

$$(\rho)^{-1}\Gamma^i(X)(\rho) = \Gamma^j(X).$$

Proof: Equation (15.2) implies:

$$(\rho)^+ \left(\Gamma^i(X)\right)^+ = \left(\Gamma^j(X)\right)^+ (\rho)^+.$$

Since $\Gamma^i(X)$ and $\Gamma^j(X)$ are unitary, these can be written as

$$(\rho)^+ \left(\Gamma^i(X)\right)^{-1} = \left(\Gamma^j(X)\right)^{-1} (\rho)^+ \Rightarrow (\rho)^+\Gamma^i(X^{-1}) = \Gamma^j(X^{-1})(\rho)^+.$$

Now, as X runs the group, so does X^{-1}, and the previous equation can be written as

$$(\rho)^+\Gamma^i(X) = \Gamma^j(X)(\rho)^+,$$

or

$$(\rho)(\rho)^+\Gamma^i(X) = (\rho)\Gamma^j(X)(\rho)^+.$$

Making use of Eq. (15.2), the above equation becomes:

$$(\rho)(\rho)^+\Gamma^i(X) = \Gamma^j(X)(\rho)(\rho)^+.$$

Making use of Schur's first lemma, we obtain:

$$(\rho)(\rho)^+ = \lambda(\epsilon^{k_i}), (\epsilon^{k_i}) \text{ as the identity matrix of dimension } k_i \times k_i.$$

We distinguish now two cases:

(i) $k_i = k_j$. In this, we distinguish the following:

   (a) $(\rho) = 0$, hence $\lambda = 0$, and the theorem is proved.
   (b) $\lambda \neq 0$, then$(\rho)((\rho)^T \lambda^{-1} = (\epsilon^{k_i})$, i.e. the inverse of (ρ) exists, and

$$(\rho)^{-1} = (\rho)^T \lambda^{-1}.$$

(ii) $k_i > k_j$. In this case, we can extend the matrix (ρ) by adding $k_i - k_j$ columns of zeros. Let (ρ') be the resulting matrix and $(\rho')^+$ its adjoined. Then, one finds that

$$(\rho')(\rho')^+ = \lambda(\epsilon^{k_i}).$$

Since, however, at least one column (ρ') contains only zeros, the determinant of (ρ') is zero, and thus, $\lambda = 0$ and $(\rho')(\rho')^+ = 0$. Then, in such a case,

$$\sum_{j=1}^{k_i}(\rho')_{ij}(\rho')_{ji}^+ = \sum_{j=1}^{k_j}|(\rho')_{ij}|^2 = 0 \Rightarrow (\rho)_{ij} = 0 \Rightarrow (\rho) = 0.$$

(iii) $k_i < k_j$. In this case, we can extend the matrix (ρ) by adding $k_j - k_i$ lines of zeros and the matrix (ϵ^{k_i}) so that it becomes (ϵ^{k_j}), i.e. a $k_j \times k_j$ matrix. Then, proceeding as above, we find

$$(\rho')(\rho')^+ = \lambda(\epsilon^{k_j}) \Rightarrow (\rho') = 0 \Rightarrow (\rho) = 0.$$

Remark: Without loss of generality, we can assume that the representations $\Gamma^i(X)$ and $\Gamma^j(X)$ are unitary. Given two representations $T^i(X)$ and $T^j(X)$, we know that there exist unitary matrices S_i and S_j such that $T_i(X) = S_i^{-1}\Gamma^i(X)S_i$ and $T_j(X) = S_j^{-1}\Gamma^j(X)S_j$. Then,

$$T^i(X)(\rho') = (\rho')T^j(X) \Rightarrow S_i^{-1}\Gamma^i(X)S_i(\rho') = (\rho')S_j^{-1}\Gamma^j(X)S_j \Rightarrow$$
$$\Gamma^i(X)S_i(\rho')S_j^{-1} = S_i(\rho')S_j^{-1}\Gamma^j(X).$$

From this, Eq. (15.2) follows for $(\rho) = S_i(\rho')S_j^{-1}$.

We are in position to state a theorem, which is very useful in applications.

Theorem 3: *Let* $\Gamma^i(X), X \in G$ *be one irreducible representation of dimension* ℓ_i *of a group* G *of order* g. *Then, one can show the following:*

$$\sum_{X \in G} \Gamma^i(X)_{\alpha\beta} \left(\Gamma^j(X)_{\gamma\delta}\right)^* = \frac{g}{\ell_i} \delta_{ij} \delta_{\alpha\gamma} \delta_{\beta\delta}. \tag{15.3}$$

This is known as the great orthogonality theorem (GOT).

Indeed, this is an orthogonality in the usual sense for vectors. Let us consider the set of vectors $|i, \alpha\beta\rangle$, defined over a space of g dimensions with projections $|i, \alpha\beta\rangle_k$ corresponding to the elements A_k of the group in such a way that $|i, \alpha\beta\rangle_k \Leftrightarrow \Gamma^i_{\alpha\beta}(A_k)$. Orthogonality means

$$\langle j, \gamma\delta | i, \alpha\beta\rangle = \sum_k (|j, \gamma\delta\rangle_k)^* |i, \alpha\beta\rangle_k = \sum_{A_k \subset G} \Gamma^i(A_k)_{\alpha\beta} \left(\Gamma^j(A_k)_{\gamma\delta}\right)^*.$$

$$\tag{15.4}$$

The proof will be presented in two steps:

(i) With the data of the theorem, we construct the matrix (M):

$$(M) = \frac{1}{g} \sum_{X \in G} \Gamma^i(X)(\Lambda)\Gamma^j(X^{-1}), \quad i \neq j,$$

where Λ is an arbitrary $\ell_i \times \ell_j$ matrix. Multiplying from the left with the matrix $\Gamma(B), B \in G$, we get:

$$\Gamma^i(B)(M) = \frac{1}{g} \sum_{X \in G} \Gamma^i(B)\Gamma^i(X)(\Lambda)\Gamma^j(X^{-1})$$

$$= \frac{1}{g} \sum_{X \in G} \Gamma^i(BX)(\Lambda)\Gamma^j(X^{-1})$$

$$= \frac{1}{g} \sum_{Y \in G} \Gamma^i(Y)(\Lambda)\Gamma^j(Y^{-1}B)$$

$$= \frac{1}{g} \sum_{Y \in G} \Gamma^i(Y)(\Lambda)\Gamma^j(Y^{-1})\Gamma^j(B)$$

$$\Rightarrow \Gamma^i(B)(M) = (M)\Gamma^j(B) \quad \text{for every } B \in G, \ i \neq j.$$

Thus, according to Schur's second lemma $(M) = 0$. Since, however, the representation is unitary, we have

$$\Gamma^j(X^{-1}) = \left(\Gamma^j(X)\right)^{-1} = \left(\Gamma^j(X)\right)^+.$$

Hence,

$$\sum_{X \in G} \Gamma^i(X)(\Lambda)\left(\left(\Gamma^j(X)\right)^+\right) = 0, \quad i \neq j$$

$$\sum_{X \in G} \sum_{pq} \Gamma^i(X)_{\alpha p}(\Lambda)_{pq}\left(\left(\Gamma^j(X)\right)^+\right)_{q\gamma} = 0 \Rightarrow$$

$$\sum_{X \in G} \sum_{pq} \Gamma^i(X)_{\alpha p}(\Lambda)_{pq}\Gamma^{*j}(X)_{\gamma q} = 0. \tag{15.5}$$

Since the matrix (Λ) is arbitrary, it can be selected so that

$$(\Lambda)_{pq} = (e_{\beta\delta})_{pq}, \quad (e_{\beta\delta})_{pq} = \delta_{p\beta}\delta_{q\delta},$$

i.e. having ones in line β column δ and zeros everywhere else. This way Eq. (15.5) becomes:

$$\sum_{X \in G} \Gamma^i(X)_{\alpha\beta}\Gamma^{*j}(X)_{\gamma\delta} = 0, \ \alpha, \beta = 1, 2, \ldots, \ell_i, \ \gamma, \delta, = 1, 2, \ldots, \ell_j, \ i \neq j \tag{15.6}$$

(obviously this is not a matrix multiplication!).
(ii) We now construct the matrix

$$(N) = \frac{1}{g} \sum_{X \in G} \Gamma^i(X)(\Lambda)\Gamma^i(X^{-1}),$$

and proceeding as above, we get

$$\Gamma^i(B)(N) = (N)\Gamma^i(B) \text{ for every } B \in G.$$

In this case, however, Schur's second lemma implies $(N) = \lambda(\epsilon)$. Using now the same matrix (Λ) as above, we find

$$\frac{1}{g} \sum_{X \in G} \Gamma^i(X)(e_{\beta\delta})\Gamma^i(X^{-1}) = \lambda(\epsilon) \Leftrightarrow \frac{1}{g} \sum_{X \in G} \Gamma^i(X)_{\alpha,\beta}\Gamma^{*i}(X)_{\gamma\delta} = \lambda\delta_{\alpha\gamma}.$$

The constant λ can be evaluated taking the trace of the expression on the left:

$$tr\left(\frac{1}{g}\sum_{X\in G}\Gamma^i(X)(e_{\beta\delta})\Gamma^i(X^{-1})\right) = \frac{1}{g}\sum_{X\in G}tr(e_{\beta\delta}\Gamma^i(X)\Gamma^i(X^{-1}))$$

$$= \frac{1}{g}\sum_{X\in G}tr(e_{\beta\delta}\epsilon) = tr(e_{\beta\delta}) = \sum_{\rho}(e_{\beta\delta})_{\rho\rho} = \delta_{\beta\rho}\delta_{\delta\rho} = \delta_{\beta\delta}.$$

On the other hand,

$$tr(\lambda\epsilon) = \lambda\ell_i \Rightarrow \lambda\ell_i = \delta_{\beta\delta} \Rightarrow \lambda = \frac{\delta_{\beta\delta}}{\ell_i}.$$

Consequently,

$$\frac{1}{g}\sum_{X\in G}\Gamma^i(X)_{\alpha,\beta}\Gamma^{*i}(X)_{\gamma\delta} = \frac{1}{\ell_i}\delta_{\alpha\gamma}\delta_{\beta\delta}. \tag{15.7}$$

Equations (15.6) and (15.7) can be combined into one, which is nothing but Eq. (15.3) of the theorem.

Chapter 16

Appendix II: Representation Reduction Via a Chain of Group Operators

Up until now, we have discussed the problem of the reduction of representations of simple groups. The methods we employed, however, are not easy to apply in the case of large representations of complex discrete groups, which often appear in applications. In this chapter, we are going to develop an efficient method by exploring the action of some symmetry operators, a chain of group operators, on the basis of which the representations have been defined. It is, of course, understood that we know the irreducible representations of the group in question.

16.1 Chains of subspaces of basis vectors

Suppose we have a basis e_i, $i = 1, 2, \ldots, n$, which covers a given space R and defines an $n \times n$ representation $\Gamma(X)$ of a group G. We select one of these, e.g. $e_1' = e_1$, and construct the sub space R_1:

$$R_1 \equiv |\Phi_i^{(1)}\rangle \in R_1 \text{ if } |\Phi_i^{(1)}\rangle = \hat{T}(X)e_1', \quad X \in G.$$

The operator $\hat{T}(X)$ acts in R in a way related to the representation $\Gamma(X)$ given by

$$\hat{T}(X)e_i = \sum_j \Gamma(X)_{ji} e_j.$$

If $R_1 = R$, we have one chain, and we are finished. If not, we select a vector e_2' not belonging in the space R_1 and construct the space R_2:

$$R_2 \equiv |\Phi_i^{(2)}\rangle \in R_2 \text{ if } |\Phi_i^{(2)}\rangle = \hat{T}(X)e_2', \quad X \in G.$$

If $R_1 \cup R_2 = R$, we are finished, having two chains. If not, we select a vector not belonging in the space $R_1 \cup R_2$ and continue the above procedure. Clearly, since the basis is complete and defines the representation, this procedure stops after k steps, and we have $R_1 \cup R_2 \cup \cdots \cup R_k = R$.

Each one of the above sets defines a representation of the group with dimension equal to the length of the chain, which, however, is not necessarily irreducible. These can be reduced to irreducible ones via the following theorem.

Theorem: *Consider a reducible representation $\Gamma(X)$ of the group G and one irreducible representation $\Gamma^{(i)}(X)$ of G. Then, the vectors $\Psi_{k\ell}^{(i,j)} = \hat{P}_{k\ell}^{(i)} |\Phi^j\rangle$, with $|\Phi^{(j)}\rangle$ any vector of the space R_j and*

$$\hat{P}_{k\ell}^{(i)} = \mathcal{N} \sum_{X \in G} \tilde{\Gamma}_{k\ell}^{(i)}(X) \hat{T}(X),$$

define a basis which reduces the representation defined in the subspace R_j. $\tilde{\Gamma}_{k\ell}^{(i)}(X)$ is the adjoined defined in Eq. (4.4) and $\mathcal{N}$ is a normalization constant, which plays the role $\sqrt{\ell_i/g}$ whenever $\Gamma(X)$ is the regular representation.

We note that:

- The vectors $\Psi_{k\ell}^{(i,j)}$ are orthogonal, whenever R_j are orthogonal.
- When $\Gamma^{(i)}$ does not exist in $\Gamma(X)$, $\Psi_{k\ell}^{(i,j)}$ is identically zero.
- The indices (k, ℓ) take the value one whenever $\Gamma^{(i)}$ is 1D. Whenever it is of dimension $d \times d$, they take the values $1, 2, \ldots, d$, but some of these values may not yield anything new.

Before proceeding to the proof of the theorem, let us clarify it with some examples.

Example 1: Here, we consider the 4D representation $\Gamma(X)$ of $G = \{E, I\}$, which describes the molecule of acetylenium C_2H_2, defined over the four coordinates of the molecule, assumed to be linear. We find

$$\Gamma(E) = \hat{T}(E) = \begin{pmatrix} 1 & 0 & 0 & 0 \\ 0 & 1 & 0 & 0 \\ 0 & 0 & 1 & 0 \\ 0 & 0 & 0 & 1 \end{pmatrix}, \quad \Gamma(I) = \hat{T}(I) = \begin{pmatrix} 0 & 0 & 0 & -1 \\ 0 & 0 & -1 & 0 \\ 0 & -1 & 0 & 0 \\ -1 & 0 & 0 & 0 \end{pmatrix},$$

$$\hat{T}(E)e_1 = |\Phi_1^{(1)}\rangle, \quad \hat{T}(I)e_1 = |\Phi_2^{(1)}\rangle = -e_4,$$

$$\hat{T}(E)e_2 = |\Phi_1^{(2)}\rangle = e_2, \quad \hat{T}(I)e_2 = |\Phi_2^{(2)}\rangle = -e_3.$$

To proceed further, we need the irreducible representations of the group $G = \{E, I\}$. In fact, the irreducible representations of this Abelian group are 1D given as follows:

$$
\begin{array}{c|cc}
 & E & I \\
\hline
\Gamma_1 = \chi^{(1)} & 1 & 1 \\
\Gamma_2 = \chi^{(2)} & 1 & -1
\end{array}.
$$

Hence:

$$\Psi_1 = P_{11}^1 e_1 = \frac{1}{\sqrt{2}}(e_1 - e_4), \quad \Psi_2 = P_{11}^1 e_2 = \frac{1}{\sqrt{2}}(e_2 - e_3),$$

$$\Psi_3 = P_{11}^2 e_1 = \frac{1}{\sqrt{2}}(e_1 + e_4), \quad \Psi_4 = P_{11}^2 e_2 = \frac{1}{\sqrt{2}}(e_2 + e_3).$$

The positive signs in the last equation arise from the fact that $\tilde{\Gamma}_{11}^{(2)}(I) = -1$, guaranteeing the orthogonality of the corresponding vectors.

The matrix, which reduces the representation $\Gamma(X)$, is the following:

$$
S = \begin{pmatrix}
\frac{1}{\sqrt{2}} & 0 & \frac{1}{\sqrt{2}} & 0 \\
0 & \frac{1}{\sqrt{2}} & 0 & \frac{1}{\sqrt{2}} \\
0 & -\frac{1}{\sqrt{2}} & 0 & \frac{1}{\sqrt{2}} \\
-\frac{1}{\sqrt{2}} & 0 & \frac{1}{\sqrt{2}} & 0
\end{pmatrix}, \quad
S^T \Gamma(I) S = \begin{pmatrix}
1 & 0 & 0 & 0 \\
0 & 1 & 0 & 0 \\
0 & 0 & -1 & 0 \\
0 & 0 & 0 & -1
\end{pmatrix}.
$$

Finally, we get

$$\Gamma(X) = 2\Gamma_1 \oplus 2\Gamma_2.$$

16.2   Some examples

Example 2:   We apply the theorem in the case of the regular representation of C_{3V}, which we have achieved with different methods, but not so general.

Based on Application 1 of Chapter 4, we observe the following:

$$\hat{T}(E)e_1 = e_1, \hat{T}(C_3)e_1 = e_2, \hat{T}(C_3^2)e_1 = e_3,$$

$$\hat{T}(\sigma_1)e_1 = e_4, \hat{T}(\sigma_2)e_1 = e_5, \hat{T}(\sigma_3)e_1 = e_6.$$

In other words, we have only one chain, and there is no reason to change the notation in $|\Phi_i^1\rangle$. Thus, using Eq. (5.17), we find

$$\Psi_1 = P_{11}^1 e_1 = \frac{1}{\sqrt{6}}(e_1 + e_2 + e_3 + e_4 + e_5 + e_6),$$

$$\Psi_2 = P_{11}^2 e_1 = \frac{1}{\sqrt{6}}(e_1 + e_2 + e_3 - e_4 - e_5 - e_6),$$

$$\Psi_3 = P_{11}^3 e_1 = \frac{\sqrt{2}}{\sqrt{6}}\left(e_1 - \frac{1}{2}e_2 - \frac{1}{2}e_3 - e_4 + \frac{1}{2}e_5 + \frac{1}{2}e_6\right),$$

$$= \frac{1}{2\sqrt{3}}(2e_1 - e_2 - e_3 - 2e_4 + e_5 + e_6),$$

$$\Psi_4 = P_{21}^3 e_1 = \frac{\sqrt{2}}{\sqrt{6}}\left(\frac{\sqrt{3}}{2}e_2 - \frac{\sqrt{3}}{2}e_3 + \frac{\sqrt{3}}{2}e_5 - \frac{\sqrt{3}}{2}e_6\right)$$

$$= \frac{1}{2}(e_2 - e_3 + e_5 - e_6)$$

$$\Psi_5 = P_{22}^3 e_1 = \frac{\sqrt{2}}{\sqrt{6}}(e_1 - \frac{1}{2}e_2 - \frac{1}{2}e_3 + e_4 - \frac{1}{2}e_5 - \frac{1}{2}e_6)$$

$$= \frac{1}{2\sqrt{3}}(2e_1 - e_2 - e_3 + 2e_4 - e_5 - e_6),$$

$$\Psi_6 = P_{12}^3 e_1 = \frac{\sqrt{2}}{\sqrt{6}}\left(-\frac{\sqrt{3}}{2}e_2 + \frac{\sqrt{3}}{2}e_3 + \frac{\sqrt{3}}{2}e_5 - \frac{\sqrt{3}}{2}e_6\right)$$

$$= \frac{1}{2}(-e_2 + e_3 + e_5 - e_6).$$

This way, we find again the matrix S we obtained in Example 3 of Chapter 5, which achieves the reduction of the regular representation of C_{3V}.

Example 3: We apply the theorem in the reduction of the 3×3 representation of C_{3V} given in Example 5 of Chapter 4:

$$\hat{T}(E)e_1 = |\Phi_1^{(1)}\rangle, \hat{T}(C_3)e_1 = |\Phi_2^{(1)}\rangle, \hat{T}(C_3^2)e_1 = |\Phi_3^{(1)}\rangle,$$

$$\hat{T}(\sigma_1)e_1 = |\Phi_1^{(1)}\rangle, \hat{T}(\sigma_2)e_1 = |\Phi_3^{(1)}\rangle, \hat{T}(\sigma_3)e_1 = |\Phi_2^{(1)}\rangle$$

$$\Psi_1 = P_{11}^{(1)} e_1 = \frac{1}{2\sqrt{3}} (|\Phi_1^{(1)}\rangle + |\Phi_2^{(1)}\rangle + |\Phi_3^{(1)}\rangle + |\Phi_1^{(1)}\rangle + |\Phi_3^{(1)}\rangle + |\Phi_2^{(1)}\rangle$$

$$= \frac{1}{\sqrt{3}} (|\Phi_2^{(1)}\rangle + |\Phi_2^{(1)}\rangle + |\Phi_3^{(1)}\rangle),$$

$$\Psi_2 = P_{11}^{(2)} e_1 = \frac{1}{\sqrt{6}} (|\Phi_1^{(1)}\rangle + |\Phi_2^{(1)}\rangle + |\Phi_3^{(1)}\rangle$$

$$- |\Phi_1^{(1)}\rangle - |\Phi_3^{(1)}\rangle - |\Phi_2^{(1)}\rangle = 0,$$

$$\Psi_3 = P_{11}^{(3)} e_1 = \frac{1}{\sqrt{6}} (|\Phi_1^{(1)}\rangle - \frac{1}{2}|\Phi_2^{(1)}\rangle - \frac{1}{2}|\Phi_3^{(1)}\rangle - |\Phi_1^{(1)}\rangle$$

$$+ \frac{1}{2}|\Phi_3^{(1)}\rangle + \frac{1}{2}|\Phi_2^{(1)}\rangle = 0,$$

$$\Psi_3 = P_{11}^{(3)} e_1 = \frac{1}{\sqrt{6}} (|\Phi_1^{(1)}\rangle - \frac{1}{2}|\Phi_2^{(1)}\rangle - \frac{1}{2}|\Phi_3^{(1)}\rangle$$

$$- |\Phi_1^{(1)}\rangle + \frac{1}{2}|\Phi_3^{(1)}\rangle + \frac{1}{2}|\Phi_2^{(1)}\rangle) = 0,$$

$$\Psi_4 = P_{12}^{(3)} e_1 = \frac{1}{\sqrt{6}} (-\frac{\sqrt{3}}{2}|\Phi_2^{(1)}\rangle + \frac{\sqrt{3}}{2}|\Phi_3^{(1)}\rangle$$

$$+ \frac{\sqrt{3}}{2}|\Phi_3^{(1)}\rangle - \frac{\sqrt{3}}{2}|\Phi_2^{(1)}\rangle)),$$

$$= \frac{1}{\sqrt{2}} (-|\Phi_2^{(1)}\rangle + |\Phi_3^{(1)}\rangle),$$

$$\Psi_5 = P_{21}^{(3)} e_1 = \frac{1}{\sqrt{6}} \left(\frac{\sqrt{3}}{2}|\Phi_2^{(1)}\rangle - \frac{\sqrt{3}}{2}|\Phi_3^{(1)}\rangle \right.$$

$$\left. + \frac{\sqrt{3}}{2}|\Phi_3^{(1)}\rangle - \frac{\sqrt{3}}{2}|\Phi_2^{(1)}\rangle \right) = 0$$

$$\Psi_6 = P_{22}^{(3)} e_1 = \frac{1}{\sqrt{6}} \left(|\Phi_1^{(1)}\rangle - \frac{1}{2}|\Phi_2^{(1)}\rangle - \frac{1}{2}|\Phi_3^{(1)}\rangle + |\Phi_1^{(1)}\rangle \right.$$

$$\left. - \frac{1}{2}|\Phi_3^{(1)}\rangle - \frac{1}{2}|\Phi_2^{(1)}\rangle \right)$$

$$= \frac{1}{\sqrt{6}} (2|\Phi_1^{(1)}\rangle - |\Phi_2^{(1)}\rangle - |\Phi_3^{(1)}\rangle).$$

The matrix

$$S = \begin{pmatrix} \frac{1}{\sqrt{3}} & 0 & \frac{2}{\sqrt{6}} \\ \frac{1}{\sqrt{3}} & -\frac{1}{\sqrt{2}} & -\frac{1}{\sqrt{6}} \\ \frac{1}{\sqrt{3}} & \frac{1}{\sqrt{2}} & -\frac{1}{\sqrt{6}} \end{pmatrix}$$

reduces the representation of Example 5 of Chapter 4.

More applications will be given in Chapter 7.

16.3   Proof of a theorem

We have

$$\hat{T}(X)\Psi_{k\ell}^{(i,j)} = \mathcal{N} \sum_{X' \in G} \tilde{\Gamma}^{(i)}(X')_{k\ell} \hat{T}(X)\hat{T}(X')|\Phi^{(j)}\rangle$$

$$= \mathcal{N} \sum_{X' \in G} \tilde{\Gamma}^{(i)}(X') \hat{T}(XX')|\Phi^{(j)}\rangle$$

$$= \mathcal{N} \sum_{X'' \in G} \tilde{\Gamma}^{(i)}(X^{-1}X')_{k\ell} \hat{T}(X'')|\Phi^{(j)}\rangle$$

$$= \mathcal{N} \sum_{n} \tilde{\Gamma}^{(i)}(X^{-1})_{kn} \sum_{X'' \in G} \tilde{\Gamma}^{(i)}(X'')_{kn} T(X'')|\Phi^{(j)}\rangle$$

$$= \sum_{n} \tilde{\Gamma}^{(i)}(X^{-1})_{kn} \Psi_{n\ell}^{(i,j)} = \sum_{n} \Gamma^{(i)}(X)_{nk} \Psi_{n\ell}^{(i,j)}.$$

In other words, for given i, j, ℓ, the following holds:

$$\hat{T}\Psi_{k\ell}^{(i,j)} = \sum_{n} \Gamma^{(i)}(X)_{nk} \Psi_{n\ell}^{(i,j)},$$

which means that the vectors $\Psi_{n\ell}^{(i,j)}$ of R_j constitute a basis in the space the operator $\hat{T}(X)$ is acting on, and thus, the proof is complete.

Chapter 17

Appendix III: Generators and Character Tables of Point Groups

In the first part, we supply a set of generators of the crystal groups, selecting a 3×3 basis. The reader is encouraged to find the order of each group, its classes and the number of its irreducible representations, see Section 8.6.

In the second part shown are the irreducible representations of the crystal groups as well as a representative element of each class of the group. In the first column, the representation is indicated as follows: For the 1D ones, the symbol $A(B)$ is used for the symmetric (antisymmetric) basis with respect to C_n, which is the rotation with the maximum order n, with axis taken to be in the z direction. For this purpose, one could as well have chosen the behavior with the respect to the mirror plane σ_h. The indices g (*gerande*) and u (*ungerande*) refer to the even and odd representations of the inversion operator I, respectively (see also Table 8.3). With the indices E and F, we denote the 2D and 3D representations, respectively. In addition to the characters, two more columns have been included. In the next before the last, we have included the translation operators T_x, T_y, T_z along the axes x, y, z and the rotation ones R_x, R_y, R_z around the indicated axes of symmetry, respectively (the generators given are considered in a 3×3 representation). In the last column, we have included the projections α_{ij} of the polarization tensor or a suitable combination of them next to the corresponding operator in the previous column. The conventional notation has been retained as much as possible. For details, we refer to Section 8.5.

Table 17.1. The generators of the groups $C_1, C_2, S_2, S_1, C_3, C_4, S_4, D_2, C_{2h}, C_{2v}$, $C_6, S_6, C_{3h}, D_3, C_{3v}$.

$$C_1 \Leftrightarrow \begin{pmatrix} 1 & 0 & 0 \\ 0 & 1 & 0 \\ 0 & 0 & 1 \end{pmatrix}, \quad C_2 \Leftrightarrow \begin{pmatrix} 1 & 0 & 0 \\ 0 & -1 & 0 \\ 0 & 0 & -1 \end{pmatrix}, \quad S_2 \Leftrightarrow \begin{pmatrix} -1 & 0 & 0 \\ 0 & -1 & 0 \\ 0 & 0 & -1 \end{pmatrix}$$

$$S_1 \Leftrightarrow \begin{pmatrix} -1 & 0 & 0 \\ 0 & 1 & 0 \\ 0 & 0 & 1 \end{pmatrix}, \quad C_3 \Leftrightarrow \begin{pmatrix} 1 & 0 & 0 \\ 0 & -\frac{1}{2} & -\frac{\sqrt{3}}{2} \\ 0 & \frac{\sqrt{3}}{2} & -\frac{1}{2} \end{pmatrix}, \quad C_4 \Leftrightarrow \begin{pmatrix} 1 & 0 & 0 \\ 0 & 0 & -1 \\ 0 & 1 & 0 \end{pmatrix}$$

$$S_4 \Leftrightarrow \begin{pmatrix} -1 & 0 & 0 \\ 0 & 0 & -1 \\ 0 & 1 & 1 \end{pmatrix}, \quad D_2 \Leftrightarrow \begin{pmatrix} 1 & 0 & 0 \\ 0 & -1 & 0 \\ 0 & 0 & -1 \end{pmatrix}, \quad \begin{pmatrix} 1 & 0 & 0 \\ 0 & 1 & 0 \\ 0 & 0 & 1 \end{pmatrix}$$

$$C_{2h} \Leftrightarrow \begin{pmatrix} -1 & 0 & 0 \\ 0 & -1 & 0 \\ 0 & 0 & -1 \end{pmatrix}, \quad \begin{pmatrix} -1 & 0 & 0 \\ 0 & 1 & 0 \\ 0 & 0 & 1 \end{pmatrix}$$

$$C_{2v} \Leftrightarrow \begin{pmatrix} 1 & 0 & 0 \\ 0 & -1 & 0 \\ 0 & 0 & -1 \end{pmatrix}, \quad \begin{pmatrix} 1 & 0 & 0 \\ 0 & -1 & 0 \\ 0 & 0 & 1 \end{pmatrix}$$

$$C_6 \Leftrightarrow \begin{pmatrix} 1 & 0 & 0 \\ 0 & \frac{1}{2} & -\frac{\sqrt{3}}{2} \\ 0 & \frac{\sqrt{3}}{2} & \frac{1}{2} \end{pmatrix}$$

$$S_6 \Leftrightarrow \begin{pmatrix} -1 & 0 & 0 \\ 0 & \frac{1}{2} & -\frac{\sqrt{3}}{2} \\ 0 & \frac{\sqrt{3}}{2} & \frac{1}{2} \end{pmatrix}, \quad C_{3h} \Leftrightarrow \begin{pmatrix} -1 & 0 & 0 \\ 0 & -\frac{1}{2} & -\frac{\sqrt{3}}{2} \\ 0 & \frac{\sqrt{3}}{2} & -\frac{1}{2} \end{pmatrix}$$

$$D_3 \Leftrightarrow \begin{pmatrix} 1 & 0 & 0 \\ 0 & -\frac{1}{2} & -\frac{\sqrt{3}}{2} \\ 0 & \frac{\sqrt{3}}{2} & -\frac{1}{2} \end{pmatrix}, \quad \begin{pmatrix} -1 & 0 & 0 \\ 0 & 1 & 0 \\ 0 & 0 & -1 \end{pmatrix}, \quad C_{3v} \Leftrightarrow \begin{pmatrix} 1 & 0 & 0 \\ 0 & -\frac{1}{2} & -\frac{\sqrt{3}}{2} \\ 0 & \frac{\sqrt{3}}{2} & -\frac{1}{2} \end{pmatrix}, \quad \begin{pmatrix} 1 & 0 & 0 \\ 0 & -1 & 0 \\ 0 & 0 & 1 \end{pmatrix}$$

Table 17.2. A set of generators of the groups $D_4, C_{4v}, D_{2v}, D_{4h}, D_{2h}, D_6, D_{3v},$ C_{6v}, D_{3h}.

$$D_4 \Leftrightarrow \begin{pmatrix} 1 & 0 & 0 \\ 0 & 0 & -1 \\ 0 & 1 & 0 \end{pmatrix}, \quad \begin{pmatrix} -1 & 0 & 0 \\ 0 & 1 & 0 \\ 0 & 0 & -1 \end{pmatrix}, \quad C_{4v} \Leftrightarrow \begin{pmatrix} 1 & 0 & 0 \\ 0 & 0 & -1 \\ 0 & 1 & 0 \end{pmatrix}, \quad \begin{pmatrix} 1 & 0 & 0 \\ 0 & -1 & 0 \\ 0 & 0 & 1 \end{pmatrix}$$

$$D_{2v} \Leftrightarrow \begin{pmatrix} -1 & 0 & 0 \\ 0 & 0 & -1 \\ 0 & 1 & 0 \end{pmatrix}, \quad \begin{pmatrix} 1 & 0 & 0 \\ 0 & -1 & 0 \\ 0 & 0 & 1 \end{pmatrix}, \quad C_{4h} \Leftrightarrow \begin{pmatrix} 1 & 0 & 0 \\ 0 & 0 & -1 \\ 0 & 1 & 0 \end{pmatrix}, \quad \begin{pmatrix} -1 & 0 & 0 \\ 0 & 1 & 0 \\ 0 & 0 & 1 \end{pmatrix}$$

$$C_{2h} \Leftrightarrow \begin{pmatrix} 1 & 0 & 0 \\ 0 & -1 & 0 \\ 0 & 0 & -1 \end{pmatrix}, \quad \begin{pmatrix} -1 & 0 & 0 \\ 0 & 1 & 0 \\ 0 & 0 & -1 \end{pmatrix}, \quad \begin{pmatrix} -1 & 0 & 0 \\ 0 & -1 & 0 \\ 0 & 0 & -1 \end{pmatrix}$$

$$D_6 \Leftrightarrow \begin{pmatrix} 1 & 0 & 0 \\ 0 & \frac{1}{2} & -\frac{\sqrt{3}}{2} \\ 0 & \frac{\sqrt{3}}{2} & \frac{1}{2} \end{pmatrix}, \quad \begin{pmatrix} -1 & 0 & 0 \\ 0 & -1 & 0 \\ 0 & 0 & 1 \end{pmatrix}, \quad D_{3v} \Leftrightarrow \begin{pmatrix} -1 & 0 & 0 \\ 0 & \frac{1}{2} & -\frac{\sqrt{3}}{2} \\ 0 & \frac{\sqrt{3}}{2} & \frac{1}{2} \end{pmatrix}, \quad \begin{pmatrix} 1 & 0 & 0 \\ 0 & -1 & 0 \\ 0 & 0 & 1 \end{pmatrix}$$

$$D_{6v} \Leftrightarrow \begin{pmatrix} 1 & 0 & 0 \\ 0 & \frac{1}{2} & -\frac{\sqrt{3}}{2} \\ 0 & \frac{\sqrt{3}}{2} & \frac{1}{2} \end{pmatrix}, \quad \begin{pmatrix} 1 & 0 & 0 \\ 0 & -1 & 0 \\ 0 & 0 & 1 \end{pmatrix}$$

$$D_{3h} \Leftrightarrow \begin{pmatrix} -1 & 0 & 0 \\ 0 & -\frac{1}{2} & -\frac{\sqrt{3}}{2} \\ 0 & \frac{\sqrt{3}}{2} & -\frac{1}{2} \end{pmatrix}, \quad \begin{pmatrix} 1 & 0 & 0 \\ 0 & 1 & 0 \\ 0 & 0 & -1 \end{pmatrix}$$

Table 17.3. A set of generators of the groups, $T, C_{6h}, D_{4h}, O, T_b, D_{6h}, T_h, O_h$.

$$T \Leftrightarrow \begin{pmatrix} 1 & 0 & 0 \\ 0 & -1 & 0 \\ 0 & 0 & -1 \end{pmatrix}, \begin{pmatrix} 0 & 0 & 1 \\ 1 & 0 & 0 \\ 0 & 1 & 0 \end{pmatrix}, C_{6h} \Leftrightarrow \begin{pmatrix} 1 & 0 & 0 \\ 0 & \frac{1}{2} & -\frac{\sqrt{3}}{2} \\ 0 & \frac{\sqrt{3}}{2} & \frac{1}{2} \end{pmatrix}, \begin{pmatrix} -1 & 0 & 0 \\ 0 & 1 & 0 \\ 0 & 0 & 1 \end{pmatrix}$$

$$D_{4h} \Leftrightarrow \begin{pmatrix} 1 & 0 & 0 \\ 0 & 0 & -1 \\ 0 & 1 & 0 \end{pmatrix}, \begin{pmatrix} -1 & 0 & 0 \\ 0 & -1 & 0 \\ 0 & 0 & -1 \end{pmatrix}, O \Leftrightarrow \begin{pmatrix} 1 & 0 & 0 \\ 0 & 0 & 1 \\ 0 & -1 & 0 \end{pmatrix}, \begin{pmatrix} 0 & 0 & 1 \\ 1 & 0 & 0 \\ 0 & 1 & 0 \end{pmatrix}$$

$$T_b \Leftrightarrow \begin{pmatrix} 1 & 0 & 0 \\ 0 & 0 & -1 \\ 0 & -1 & 0 \end{pmatrix}, \begin{pmatrix} 1 & 0 & 0 \\ 0 & -1 & 0 \\ 0 & 0 & -1 \end{pmatrix}, \begin{pmatrix} 0 & 0 & 1 \\ 1 & 0 & 0 \\ 0 & 1 & 0 \end{pmatrix}$$

$$D_{6h} \Leftrightarrow \begin{pmatrix} 1 & 0 & 0 \\ 0 & \frac{1}{2} & -\frac{\sqrt{3}}{2} \\ 0 & \frac{\sqrt{3}}{2} & \frac{1}{2} \end{pmatrix}, \begin{pmatrix} -1 & 0 & 0 \\ 0 & 1 & 0 \\ 0 & 0 & 1 \end{pmatrix}, \begin{pmatrix} 1 & 0 & 0 \\ 0 & -1 & 0 \\ 0 & 0 & 1 \end{pmatrix}$$

$$T_h \Leftrightarrow \begin{pmatrix} -1 & 0 & 0 \\ 0 & 1 & 0 \\ 0 & 0 & 1 \end{pmatrix}, \begin{pmatrix} 1 & 0 & 0 \\ 0 & -1 & 0 \\ 0 & 0 & -1 \end{pmatrix}, \begin{pmatrix} 0 & 0 & 1 \\ 1 & 0 & 0 \\ 0 & 1 & 0 \end{pmatrix}$$

$$O_h \Leftrightarrow \begin{pmatrix} 1 & 0 & 0 \\ 0 & 0 & 1 \\ 0 & -1 & 0 \end{pmatrix}, \begin{pmatrix} 0 & 0 & 1 \\ 1 & 0 & 0 \\ 0 & 1 & 0 \end{pmatrix}, \begin{pmatrix} -1 & 0 & 0 \\ 0 & -1 & 0 \\ 0 & 0 & 1 \end{pmatrix}$$

Table 17.4. The character tables of the irreducible representations of the groups $S_2, C_i, C_2, C_{2h}, C_{2v}, D_2$.

S_2	E	σ_b			
A'	1	1		$T_x, T_y; R_z$	$\alpha_{xx}, \alpha_{yy}, \alpha_{zz}.\alpha_{xy}$
A''	1	-1		$T_z; R_y, R_z$	α_{xz}, α_{yz}

C_i	E	I			
A_g	1	1		R_x, R_y, R_z	$\alpha_{xx}, \alpha_{yy}\alpha_{zz}\alpha_{xy},$ α_{xz}, α_{yz}
A_u	1	-1		$T_x; , T_y, T_z$	

C_2	E	C_2			
A	1	1		$T_z; R_z$	$\alpha_{xx}, \alpha_{yy}\alpha_{zz}, \alpha_{xy}$
B	1	-1		$T_x, T_y T_z; R_x, R_y$	α_{xz}, α_{yz}

C_{2h}	E	C_2	I	σ_h		
A_g	1	1	1	1	R_z	$\alpha_{xx}, \alpha_{yy}\alpha_{zz}\alpha_{xy}$
B_g	1	-1	1	-1	R_x, R_y	α_{xz}, α_{yz}
A_u	1	1	-1	-1	T_z	
B_u	1	-1	-1	1	T_x, T_y	

C_{2v}	E	C_2	$\sigma_v(zx)$	$\sigma_v(yz)$		
A_1	1	1	1	1	T_z	$\alpha_{xx}, \alpha_{yy}, \alpha_{zz}$
A_2	1	1	-1	-1	T_x, R_z	α_{xy}
B_1	1	-1	1	-1	T_z	α_{xz}
B_2	1	-1	-1	1	T_y, R_z	α_{yz}

$D_2 = V$	E	$C_2(z)$	$C_2(y)$	$C_2(x)$		
A_1	1	1	1	1		$\alpha_{xx}, \alpha_{yy}, \alpha_{zz}$
A_2	1	1	-1	-1	T_z, R_z	α_{xy}
B_1	1	-1	1	-1	T_y, R_y	α_{xz}
B_2	1	-1	-1	1	T_x, R_x	α_{yz}

Table 17.5. The same as in Table 17.4 for the groups $C_3, C_{3v}, D_3, D_{2d}, C_{3h}$.

C_3	E	C_3	C_3^2			$\epsilon = e^{(2\pi i)/3}$
A	1	1	1		T_z, R_z	$\alpha_{xx} + \alpha_{zz}, \alpha_{xy}$
B_1	1	ϵ	ϵ^2		T_x, R_x	$\alpha_{yy} - \alpha_{zz}, \alpha_{xy}$
B_2	1	ϵ^2	ϵ		T_y, R_y	α_{xz}, α_{yz}

C_{3v}	E	$2C_3$	$3\sigma_v$		
A_1	1	1	1	T_z	$\alpha_{xx} + \alpha_{yy}, \alpha_{zz}$
A_2	1	1	-1	R_z	
E	2	-1	0	$(T_x, T_y), (R_x, R_y)$	$(\alpha_{xx} - \alpha_{yy}, \alpha_{xy}),$ $(\alpha_{xz}, \alpha_{yz})$

D_3	E	$2C_3$	$3C_2$		
A_1	1	1	1		$\alpha_{xx} + \alpha_{yy}, \alpha_{zz}$
A_2	1	1	-1	T_z, R_z	
E	2	-1	0	$(T_x, T_y), (R_x, R_y)$	$(\alpha_{xx} - \alpha_{yy}, \alpha_{xy}),$ $(\alpha_{xz}, \alpha_{yz})$

$D_{2d} = V_d$	E	$2S_4$	C_2	$2C_2'$	$2\sigma_d$		
A_1	1	1	1	1	1		$\alpha_{xx} + \alpha_{yy}, \alpha_{zz}$
A_2	1	1	1	-1	-1	R_z	
B_1	1	-1	1	1	-1		$\alpha_{xx} - \alpha_{yy}$
B_2	1	-1	1	-1	1	T_z	α_{xy}
E	2	0	-2	0	0	T_y, R_z	α_{xz}, α_{yz}

$C_{3h} =$	E	C_3	C_3^2	σ_h	$C_3\sigma_h$	$C_3^2\sigma_h$		$\epsilon = e^{(2\pi i)/3}$
A_1'	1	1	1	1	1	1	R_z	$\alpha_{xx} + \alpha_{yy}, \alpha_{zz}$
A_2'	1	ϵ	ϵ^2	1	ϵ	ϵ^2	T_x, T_y	$\alpha_{xx} - \alpha_{yy}, \alpha_{xy}$
A_3'	1	ϵ^2	ϵ	1	ϵ^2	ϵ	T_x, T_y	$\alpha_{xx} - \alpha_{yy}, \alpha_{xy}$
A_1''	1	1	1	-1	-1	-1	T_z	
A_2''	1	ϵ	ϵ^2	-1	$-\epsilon$	$-\epsilon^2$	R_x, R_y	α_{xz}, α_{yz}
A_3''	1	ϵ^2	ϵ	-1	$-\epsilon^2$	$-\epsilon$	R_x, R_y	α_{xz}, α_{yz}

Table 17.6. The same as in Table 17.4 for the groups C_4, D_4, D_{3d}, D_{3h}.

C_4	E	C_4	C_2	C_4^3		
A	1	1	1	1	T_z, R_z	$\alpha_{xx}+\alpha_{yy}, \alpha_{zz}$
B	1	-1	1	-1		
E	$\{1$	i	-1	$-i$	(T_x, T_y)	
	1	$-i$	-1	$i\}$		

D_4	E	C_4^2	C_4, C_4^3	$2C_2'$	$2C_2''$		
A_1	1	1	1	1	1		$\alpha_{xx}+\alpha_{yy}, \alpha_{zz}$
A_2	1	1	1	-1	-1	T_x, R_z	
B_1	1	1	-1	1	-1		$\alpha_{xx}-\alpha_{yy}$
B_2	1	1	-1	-1	1		α_{zy}
E	2	-2	0	0	0	T_z	$(\alpha_{xz}-\alpha_{yz})$

$D_{3d} =$	E	$2C_3$	$3C_2$	I	$2S_6$	$3\sigma_d$		$\epsilon = e^{(2\pi i)/3}$
					$S_6 = C_3 I$	$\sigma_d = C_2 I$		
A_{1g}	1	1	1	1	1	1	R_z	$\alpha_{xx}+\alpha_{yy}, \alpha_{zz}$
A_{2g}	1	1	-1	1	1	1	R_z	
E_g	2	-1	0	2	-1	0	R_x, R_y	$(\alpha_{xx}-\alpha_{yy}, \alpha_{xy}),$
								$(\alpha_{xz}, \alpha_{yz})$
A_{1u}	1	1	1	-1	-1	-1	R_z	
A_{2u}	1	1	-1	-1	-1	1	T_z	
E_u	2	-1	0	-2	1	0	T_x, T_y	

$D_{3h} =$	E	$2C_3$	$3C_2$	σ_h	$2S_3$	$3\sigma_v$		ϵ
A_1'	1	1	1	1	1	1	R_z	$\alpha_{xx}+\alpha_{yy}, \alpha_{zz}$
A_2'	1	1	-1	-1	1	-1	R_z	
E'	2	-1	0	2	-1	0	T_x, T_y	$(\alpha_{xx}-\alpha_{yy}, \alpha_{xy})$
A_1''	1	1	1	-1	-1	-1	R_z	
A_2''	1	1	-1	-1	-1	1	T_z	
E''	2	-1	0	-2	1	0	R_x, R_y	$(\alpha_{xz}, \alpha_{yz})$

Table 17.7. The same as in Table 17.4 for the groups $C_{4v}, C_{4h}, D_{4h}, S_d$.

C_{4v}	E	$2C_4$	C_2	$2\sigma_v$	$2\sigma_d$			
A_1	1	1	1	1	1		T_z	$\alpha_{xx}+\alpha_{yy}, \alpha_{zz}$
A_2	1	1	1	-1	-1		R_z	
B_1	1	-1	1	1	-1			$\alpha_{xx}-\alpha_{yy}$
B_2	1	-1	1	-1	1			α_{xy}
E	2	0	-2	0	0			$(\alpha_{xz}, \alpha_{yz})$

C_{4h}	E	C_4	C_4^2	C_4^3	I	S_4^3	σ_h	S_4		
A_g	1	1	1	1	1	1	1	1	R_z	$\alpha_{xx}+\alpha_{yy}, \alpha_{zz}$
B_g	1	-1	1	-1	1	-1	1	-1		$\alpha_{xx}-\alpha_{yy}, \alpha_{xy}$
E_g	$\{1$	i	-1	$-i$	1	i	-1	$-i$	(R_x, R_y)	$(\alpha_{xz}, \alpha_{yz})$
	1	$-i$	-1	i	1	$-i$	-1	$i\}$		
A_u	1	1	1	1	-1	-1	-1	-1	T_z	
B_u	1	-1	1	-1	-1	1	-1	1		
E_u	$\{1$	i	-1	$-i$	-1	$-i$	1	i	(T_x, T_y)	
	1	$-i$	-1	i	-1	i	1	$-i\}$		

D_{4h}	E	$2C_4$	C_2	$2C_2'$	$2C_2''$	I	$2S_4$	σ_h	$2\sigma_v$	$2\sigma_d$		
A_{1g}	1	1	1	1	1	1	1	1	1	1		$\alpha_{xx}+\alpha_{yy}, \alpha_{zz}$
A_{2g}	1	1	1	-1	-1	1	1	1	-1	-1	R_z	
B_{1g}	1	-1	1	1	-1	1	-1	1	1	-1		$\alpha_{xx}-\alpha_{yy}$
B_{2g}	1	-1	1	-1	1	1	-1	1	-1	1		α_{xy}
E_g	2	0	-2	0	0	2	0	-2	0	0	(R_x, R_y)	$(\alpha_{xz}, \alpha_{yz})$
A_{1u}	1	1	1	1	1	-1	-1	-1	-1	-1		
A_{2u}	1	1	1	-1	-1	-1	-1	-1	1	1	T_z	
B_{1u}	1	-1	1	1	-1	-1	1	-1	-1	1		
B_{2u}	1	-1	1	-1	1	-1	1	-1	1	-1		
E_u	2	0	-2	0	0	-2	0	2	0	0	(T_x, T_y)	

S_4	E	S_4	C_2	S_4^3			
A	1	1	1	1		R_z	$\alpha_{xx}+\alpha_{yy}, \alpha_{zz}$
B	1	-1	1	-1		T_z	$\alpha_{xx}-\alpha_{yy}, \alpha_{xy}$
E	$\{1$	i	-1	$-i$			
	1	$-i$	-1	$i\}$		$(T_x, T_y),$ (R_x, R_y)	$(\alpha_{xz}, \alpha_{yz})$

Table 17.8. The same as in Table 17.4 for the groups C_6, C_{6h}, D_{6h}.

C_6	E	C_6	C_3	C_2	C_3^2	C_6^5			$\epsilon=e^{\frac{2\pi i}{6}}$
A	1	1	1	1	1	1	T_z, R_z	$\alpha_{xx}+\alpha_{yy}, \alpha_{zz}$	
B	1	-1	1	-1	1	-1			
E_1	$\{\,1$	ϵ	$-\epsilon^2$	-1	$-\epsilon$	ϵ^2	$(T_x,T_y);$		
	1	ϵ^2	$-\epsilon$	-1	$-\epsilon^2$	$-\epsilon\,\}$	(R_x,R_y)		
E_2	$\{\,1$	$-\epsilon^2$	$-\epsilon$	1	$-\epsilon^2$	$-\epsilon$	$(T_x,T_y);$	$\alpha_{xx}-\alpha_{yy}, \alpha_{xy}$	
	1	$-\epsilon$	$-\epsilon^2$	1	$-\epsilon$	$-\epsilon^2\,\}$	(R_x,R_y)		

C_{6h}	E	$C_6(z)$	C_3	C_2	C_3^2	C_6^5	I	S_3^5	S_6^5	$\sigma_h(xy)$	S_6	S_3		$\epsilon=e^{\frac{2\pi i}{3}}$
A_g	1	1	1	1	1	1	1	1	1	1	1	1	R_z	$\alpha_{xx}+\alpha_{yy}, \alpha_{zz}$
B_g	1	-1	1	-1	1	-1	1	-1	1	-1	1	-1		
E_{1g}	$\{\,1$	ϵ	$-\epsilon^2$	-1	$-\epsilon$	ϵ^2	1	ϵ	$-\epsilon^2$	-1	$-\epsilon$	ϵ^2	R_x, R_y	$(\alpha_{xz}, \alpha_{yz})$
	1	ϵ^2	$-\epsilon$	-1	$-\epsilon^2$	$-\epsilon$	1	ϵ^2	$-\epsilon$	-1	$-\epsilon^2$	$-\epsilon\,\}$		
E_{2g}	$\{\,1$	$-\epsilon^2$	$-\epsilon$	1	$-\epsilon^2$	$-\epsilon$	1	$-\epsilon^2$	$-\epsilon$	1	$-\epsilon^2$	$-\epsilon$		$(\alpha_{xx}-\alpha_{yy}, \alpha_{yy})$
	1	$-\epsilon$	$-\epsilon^2$	1	$-\epsilon$	$-\epsilon^2$	1	$-\epsilon$	$-\epsilon^2$	1	$-\epsilon$	$-\epsilon^2\,\}$		
A_u	1	1	1	1	1	1	-1	-1	-1	-1	-1	-1	T_z	
B_u	1	-1	1	-1	1	-1	-1	1	-1	1	-1	1		
E_{1u}	$\{\,1$	ϵ	$-\epsilon^2$	-1	$-\epsilon$	ϵ^2	-1	$-\epsilon$	ϵ^2	1	ϵ	$-\epsilon^2$	(T_x, T_y)	
	1	ϵ^2	$-\epsilon$	-1	$-\epsilon^2$	$-\epsilon$	-1	$-\epsilon^2$	ϵ	1	ϵ^2	$-\epsilon\,\}$		
E_{2u}	$\{\,1$	$-\epsilon^2$	$-\epsilon$	1	$-\epsilon^2$	$-\epsilon$	-1	ϵ^2	ϵ	-1	ϵ^2	ϵ		
	1	$-\epsilon$	$-\epsilon^2$	1	$-\epsilon$	$-\epsilon^2$	-1	ϵ	ϵ^2	-1	ϵ	$\epsilon^2\,\}$		

D_{6h}	E	$2C_6$	$2C_3$	C_2	$3C_2'$	$3C_2''$	I	$2S_3$	$2S_6$	σ_h	$3\sigma_d$	$3\sigma_v$		
A_{1g}	1	1	1	1	1	1	1	1	1	1	1	1		$\alpha_{xx}+\alpha_{yy}, \alpha_{zz}$
A_{2g}	1	1	1	1	-1	-1	1	1	1	1	-1	-1	R_z	
B_{1g}	1	-1	1	-1	1	-1	1	-1	1	-1	1	-1		
B_{2g}	1	-1	1	-1	-1	1	1	-1	1	-1	-1	1		
E_{1g}	2	1	-1	-2	0	0	2	1	-1	-2	0	0	(R_x, R_y)	$(\alpha_{xz}, \alpha_{yz})$
E_{2g}	2	-1	-1	2	0	0	2	-1	-1	2	0	0		$(\alpha_{xx}-\alpha_{yy}, \alpha_{xy})$
A_{1u}	1	1	1	1	1	1	-1	-1	-1	-1	-1	-1		
A_{2u}	1	1	1	1	-1	-1	-1	-1	-1	-1	1	1	T_z	
B_{1u}	1	-1	1	-1	1	-1	-1	1	-1	1	-1	1		
B_{2u}	1	-1	1	-1	-1	1	-1	1	-1	1	1	-1		
E_{1u}	2	1	-1	-2	0	0	-2	-1	1	2	0	0	(T_x, T_y)	
E_{2u}	2	-1	-1	2	0	0	-2	1	1	-2	0	0		

Table 17.9. The same as in Table 17.4 for the groups C_{6v}, S_6, T, T_d.

C_{6v}	E	$2C_6$	$2C_3$	C_2	$3\sigma_v$	$3\sigma_d$		
A_1	1	1	1	1	1	1	T_z	$\alpha_{xx}+\alpha_{yy},\alpha_{zz}$
A_2	1	1	1	1	-1	-1	R_z	
B_1	1	-1	1	-1	1	-1		
B_2	1	-1	1	-1	-1	1		
E_1	2	1	-1	-2	0	0	$(T_x,T_y);(R_x,R_y)$	$(\alpha_{xz},\alpha_{yz})$
E_2	2	-1	-1	2	0	0		$(\alpha_{xx}-\alpha_{yy},\alpha_{xy})$

D_6	E	$2C_6$	$2C_3$	C_2	$3C_2'$	$3C_2''$		
A_1	1	1	1	1	1	1		$\alpha_{xx}+\alpha_{yy},\alpha_{zz}$
A_2	1	1	1	1	-1	-1	T_z,R_z	
B_1	1	-1	1	-1	1	-1		
B_2	1	-1	1	-1	-1	1		
E_1	2	1	-1	-2	0	0	$(T_x,T_y);(R_x,R_y)$	$(\alpha_{xz},\alpha_{yz})$
E_2	2	-1	-1	2	0	0		$(\alpha_{xx}-\alpha_{yy},\alpha_{xy})$

S_6	E	C_3	C_3^2	I	S_6^5	S_6		$\epsilon=e^{\frac{2\pi i}{3}}$
A_g	1	1	1	1	1	1	R_z	$\alpha_{xx}+\alpha_{yy},\alpha_{zz}$
E_g	$\Big\{\begin{matrix}1\\1\end{matrix}$	$\begin{matrix}\epsilon\\\epsilon^2\end{matrix}$	$\begin{matrix}\epsilon^2\\\epsilon\end{matrix}$	$\begin{matrix}1\\1\end{matrix}$	$\begin{matrix}\epsilon\\\epsilon^2\end{matrix}$	$\begin{matrix}\epsilon^2\\\epsilon\end{matrix}\Big\}$	R_x,R_y	$(\alpha_{xx}-\alpha_{yy},\alpha_{xy});$ $(\alpha_{xz},\alpha_{yz})$
A_u	1	1	1	-1	-1	-1		
E_u	$\Big\{\begin{matrix}1\\1\end{matrix}$	$\begin{matrix}\epsilon\\\epsilon^2\end{matrix}$	$\begin{matrix}\epsilon^2\\\epsilon\end{matrix}$	$\begin{matrix}-1\\-1\end{matrix}$	$\begin{matrix}-\epsilon\\-\epsilon^2\end{matrix}$	$\begin{matrix}-\epsilon^2\\-\epsilon\end{matrix}\Big\}$	$\begin{matrix}T_z\\T_x,T_y\end{matrix}$	

T	E	$4C_3$	$4C_3^2$	$3C_2$		$\epsilon=e^{\frac{2\pi i}{3}}$
A	1	1	1	1		$\alpha_{xx}+\alpha_{yy}+\alpha_{zz}$
E	$\begin{matrix}1\\1\end{matrix}$	$\begin{matrix}\epsilon\\\epsilon^2\end{matrix}$	$\begin{matrix}\epsilon^2\\\epsilon\end{matrix}$	$\begin{matrix}1\\1\end{matrix}$		$(\alpha_{xx}+\alpha_{yy}-2\alpha_{zz},$ $\alpha_{xx}-\alpha_{yy})$
F	3	0	0	-1	$T_x,T_y,T_z;R_x,R_y,R_z$	$(\alpha_{xy}+\alpha_{xz}+\alpha_{yz})$

T_d	E	$8C_3$	$3C_2$	$6S_4$	$6\sigma_d$		
A_1	1	1	1	1	1		$\alpha_{xx}+\alpha_{yy}+\alpha_{zz}$
A_2	1	1	1	-1	-1		
E	2	-1	2	0	0		$(\alpha_{xx}+\alpha_{yy}-2\alpha_{zz},$ $\alpha_{xx}-\alpha_{yy})$
F_1	3	0	-1	1	-1	R_x,R_y,R_z	
F_2	3	0	-1	-1	1	T_x,T_y,T_z	$(\alpha_{xy},\alpha_{xz},\alpha_{yz})$

Table 17.10. The same as in Table 17.4 for the groups T_h, O, O_h.

T_h	E	$4C_3$	$4C_3^2$	$3C_2$	I	$4S_6^5$	$4S_6$	$3\sigma_h$		$\epsilon = e^{\frac{2\pi i}{3}}$
A_g	1	1	1	1	1	1	1	1	T_z	$\alpha_{xx}+\alpha_{yy}+\alpha_{zz}$
E_g	$\{1$	ϵ	ϵ^2	1	1	ϵ	ϵ^2	1		$(\alpha_{xx}+\alpha_{yy}-2\alpha_{zz},$
	1	ϵ^2	ϵ	1	1	ϵ^2	ϵ	$1\}$		$\alpha_{xx}-\alpha_{yy})$
F_g	3	0	0	-1	3	0	0	-1	R_x, R_y, R_z	$(\alpha_{xy}+\alpha_{xz}, \alpha_{yz})$
A_u	1	1	1	1	-1	-1	-1	-1		
E_u	$\{1$	ϵ	ϵ^2	1	-1	$-\epsilon$	$-\epsilon^2$	-1		
	1	ϵ^2	ϵ	1	-1	$-\epsilon^2$	$-\epsilon$	$-1\}$		
F_u	3	0	0	-1	-3	0	0	1	T_x, T_y, T_z	

O	E	$8C_3$	$3C_2$	$6C_4$	$6C_2'$		
A_1	1	1	1	1	1		$\alpha_{xx}+\alpha_{yy}+\alpha_{zz}$
A_2	1	1	1	-1	-1		
E	2	-1	2	0	0		$(\alpha_{xx}+\alpha_{yy}-2\alpha_{zz}, \alpha_{xx}-\alpha_{yy})$
F_1	3	0	-1	1	-1	R_x, R_y, R_z	
F_2	3	0	-1	-1	1	T_x, T_y, T_z	$(\alpha_{xy}, \alpha_{xz}, \alpha_{yz})$

O_h	E	$8C_3$	$3C_2$	$6C_4$	$6C_2'$	I	$8S_6$	$3\sigma_h$	$6S_4$	$6\sigma_d$		
A_{1g}	1	1	1	1	1	1	1	1	1	1		$\alpha_{xx}+\alpha_{yy}+\alpha_{zz}$
A_{2g}	1	1	1	-1	-1	1	1	1	-1	-1		
E_g	2	-1	2	0	0	2	-1	2	0	0		$(\alpha_{xx}+\alpha_{yy}-2\alpha_{zz}, \alpha_{xx}-\alpha_{yy})$
F_{1g}	3	0	-1	1	-1	3	0	-1	1	-1	R_x, R_y, R_z	
F_{2g}	3	0	-1	-1	1	3	0	-1	-1	1		$(\alpha_{xy}, \alpha_{xz}, \alpha_{yz})$
A_{1u}	1	1	1	1	1	-1	-1	-1	-1	-1		
A_{2u}	1	1	1	-1	-1	-1	-1	-1	1	1		
E_u	2	-1	2	0	0	-2	1	-2	0	0		
F_{1u}	3	0	-1	1	-1	-3	0	1	-1	1	T_x, T_y, T_z	
F_{2u}	3	0	-1	-1	1	-3	0	1	1	-1		

Bibliography

Abbas, M. and Khalil, S. (2015). *Phys. Rev. D* **91**, p. 053003, arXiv:1406.6716 [hep-ph].

Ahn, Y. H. (2014). ArXiv:141.1634 (hep-ph).

Aliferis, G., Leontaris, G., and Vlahos, N. D. (2016). Psl(2,7) representations and their relevance to neutrino physics, *Eur. Phys. Jour. C* **77**, p. 6, [arXiv:1612.06161].

Altarelli, G. and Feruglio, F. (2010a). *Rev. Mod. Phys.* **82**, p. 2701, [arXiv:1002.0211].

Altarelli, G. and Feruglio, F. (2010b). Discrete flavor symmetries and models of neutrino mixing, *Rev. Mod. Phys.* **82**, p. 270, [arXiv:1002.0211].

Apostol, M. (1990). *Modular Functions and Dirichlet Series in Number Theory* (Springer, New York).

Athanasiu, G. G. and Floratos, E. G. (1994). Coherent states in finite quantum mechanics, *Nucl. Phys. B* **425**, p. 343.

Athanasiu, G. G., Floratos, E. G., and Nicolis, S. (1998). Fast quantum maps, *J. Phys. A* **31**, p. L655, [math-ph/9805012].

Balian, B. and Itzykson, C. (1986). *C. R. Acad. Sc. Paris* **303-serie 1-No. 16**, p. 773.

Bishop, D. M. (2017). *Group Theory and Chemistry* (Dover Publications Inc., New York).

Castro-Neto, A. H., Guinea, F., Peres, N. M. R., Novoselov, K. S., and Geim, A. K. (2019). *Rev. Mod. Phys.* **81**, p. 109.

Chen, J.-Q., Ping, J., and Wang, F. (2002). *Group Representation Theory for Physicists* (World Scientific, Singapore).

de Medeiros-Varzielas, I., King, S. F., and Ross, G. G. (2007). *Phys. Lett. B* **648**, p. 201, [hep-ph/0607045].

Ding, G., Everett, L. L., and Stuart, A. J. (2012). *Nucl. Phys.* **857**, p. 219, arXiv:1110.1688.

Dresselhaus, M. S., Dresselhaus, G., and Jorio, A. (2002). *Group Theory: Application to the Physics of Condensed Matter* (Springer).

Everett, L. and Stuart, A. (2011). The double cover of the icosahedral symmetry group and quark mass textures, *Phys. Lett. B* **698**, p. 131, [arXiv:1011.4928].

Feruglio, F. and Paris, A. (2011). *JHEP* **1103**, p. 101, arXiv:1101.0393.

Floratos, E. and Leontaris, G. (2016a). Discrete flavor symmetries from the heisenberg group, *Phys. Lett. B* **755**, p. 155, [arXiv:1511.01875].

Floratos, E. G. and Leontaris, G. (2016b). *Phys. Lett. B* **755**, p. 155, arXiv:1511.01875 [hep-th].

Graef, M. D. (2012). *Structure of Materials: An Introduction to Crystallography, Diffraction and Symmetry* (Cambridge University Press).

Hammermesh, M. (1964). *Group Theory* (Addison-Wesley Publishing Company, Inc., Boston).

Ishimori, H., Kobayashi, T., Ohki, H., Shimizu, Y., Okada, H., and Tanimoto, M. (2010a). *Prog. Theor. Phys. Suppl.* **183**, p. 1, arXiv:1003.3552.

Ishimori, H., Kobayashi, T., Ohki, H., Shimizu, Y., Okada, H., and Tanimoto, M. (2010b). Non-abelian discrete symmetries in particle physics, *Prog. Theor. Phys. Suppl.* **183**, p. 1, [arXiv:1003.3552].

King, S. F. and Luhn, C. (2009). *Nucl. Phys.* **820**, p. 269, arXiv:0905.1686 [hep-ph].

King, S. F. and Luhn, C. (2013a). *Rept. Prog. Phys.* **76**, p. 056201, [arXiv:1301.1340].

King, S. F. and Luhn, C. (2013b). Neutrino mass and mixing with discrete symmetry, *Rep. Prog. Theor. Phys.* **76**, p. 056201, [arXiv:1301.1340].

Luhn, C., Nasri, S., and Ramond, P. (2007a). Tri-bimaximal neutrino mixing and the family symmetry semidirect product of z(7) and z(3), *Phys. Lett. B* **652**, p. 27, [arXiv:0706.2341].

Luhn, C., Nasri, S., and Ramond, P. (2007b). *J. Math. Phys.* **48**, p. 073501, [hep-th/0701188].

Ma, E. (2008). *Phys. Lett. B* **660**, p. 505, [arXiv:0709.0507 [hep-ph]].

Ma, E. and Rajasekaran, G. (2001). Softly broken a(4) symmetry for nearly degenerate neutrino masses, *Phys. Rev. D* **64**, p. 113012, [hep-ph/0106291].

Nowick, A. J. (1995). *Crystal Properties Via Group Theory* (Cambridge University Press).

Stillwell, J. (2001). *Am. Math. Mon.* **180(1)**, p. 70.

Vergados, J. D. (1986). *Phys. Rep.* **133**, p. 1.

Vergados, J. D. (2004). *Mathematical Methods in Physics I* (in Greek) (Crete University Press).

Vergados, J. D. (2005). Clebcsh-Gordan coefficients in the symmric group S_n, *J. Phys. A: Math. Gen.* **14**, p. 85, arXiv:1505.04200.

Vergados, J. (2017). *Group and Representation Theory* (World Scientific, 1st edition, Singapore).

Vergados, J. (2018). *The Standard Model and Beyond: A Lecture Series* (World Scientific, Singapore).

Vergados, J. and Moustakidis, H. (2021). *Subatomic Physics An Introduction to Nuclear and Particle Physics, and Astrophysics* (World Scientific, 1st edition).

Wang, X., Yu, B., and Zhou, S. (2021). Double covering of the modular a_5 group and lepton flavor mixing in the minimal seesaw model, *Phys. Rev. D* **103**, p. 076005, arXiv:2010.10159.

Weinberg, S. (1996). *The Quantum Theory of Fields* (Cambridge University Press).

Wigner, E. P. (1959). *Group Theory and Its Application to the Quantum Mechanics of Atomic Spectra* (Academic Press).

Zak, J. (1989). Weyl-heisenberg group and magnetic translations in finite phase space, *Phys. Rev. B* **39**, p. 694.

Zee, A. (2005). Group Theory: Application to the Physics of Condensed Matter, *Phys. Lett.* **630**, p. 58.

.

Index

www.ingramcontent.com/pod-product-compliance
Lightning Source LLC
Chambersburg PA
CBHW050537190326
41458CB00007B/1819